S0-AKY-172

Reactions on Polymers

NATO ADVANCED STUDY INSTITUTES SERIES

Proceedings of the Advanced Study Institute Programme, which aims at the dissemination of advanced knowledge and the formation of contacts among scientists from different countries

The series is published by an international board of publishers in conjunction with NATO Scientific Affairs Division

A	Life Sciences	Plenum Publishing Corporation
B	Physics	London and New York
C	Mathematical and Physical Sciences	D. Reidel Publishing Company Dordrecht and Boston
D	Behavioral and Social Sciences	Sijthoff International Publishing Company Leiden
E	Applied Sciences	Noordhoff International Publishing Leiden

Series C – Mathematical and Physical Sciences

Volume 4 – Reactions on Polymers

Reactions on Polymers

Proceedings of the NATO Advanced Study Institute held at Rensselaer Polytechnic Institute, Troy, N.Y., U.S.A., July 15–25, 1973

edited by

J. A. MOORE

Rensselaer Polytechnic Institute, Troy, N.Y., U.S.A.

D. Reidel Publishing Company

Dordrecht-Holland / Boston-U.S.A.

Published in cooperation with NATO Scientific Affairs Division

First printing: December 1973

Library of Congress Catalog Card Number 73-91207

ISBN 90 277 0416 3

Published by D. Reidel Publishing Company
P.O. Box 17, Dordrecht, Holland

Sold and distributed in the U.S.A., Canada, and Mexico
by D. Reidel Publishing Company, Inc.
306 Dartmouth Street, Boston, Mass. 02116, U.S.A.

Printed in The Netherlands by D. Reidel, Dordrecht

TABLE OF CONTENTS

PREFACE

The subject matter of this volume has its roots in the early days of polymer chemistry when gun cotton and Parkesine were first developed. Indeed, its roots can ultimately be traced into antiquity, since, in commerce and daily life, man has always carried out reactions on polymers, e.g. in primitive dyeing and tanning operations. In more modern times Prof. Staudinger is commonly acknowledged as the investigator most responsible for the renaissance of interest in "polymer analogous" reactions.

In recent years it has become apparent that the "black art" of conducting chemical reactions on macromolecules is an area which is amenable to basic scientific investigation. Examples of important developments which have come about, in part, as a result of this realization have been the advances in molecular biology (which may be rightly considered as the biochemistry of macromolecules) and by such milestone achievements as solid phase peptide synthesis and affinity chromatography. These few selected examples all have macromolecules (biological and synthetic) in common. The stimulation to wide areas of scientific endeavor resulting from these developments would undoubtedly have occurred sooner, and certainly with much less difficulty, had the various investigators been aware of the problems attending attempts at performing reactions on polymers.

The major purposes of this Advanced Studies Institute were threefold:

1. To treat these processes in depth in light of current chemical knowledge to broaden the backgrounds of specialists in this area, as well as to provide a fundamental understanding of it for new workers.

2. To aid the dissemination of detailed and current ideas across international boundaries.

3. To aid in the dissemination of information about chemical reaction on polymers across interdisciplinary boundries.

The scientific level of development in this field at the present time is at a stage where accomplishment of these goals will, hopefully, cause the germination of new ideas and new achievments, so that the next giant step to be made in those areas utilizing chemical reactions on macromolecules may be made more rapidly and more efficiently.

These goals have, I believe, been accomplished, in part, on a microscale during the time of the Institute. It is our hope that the publication of these proceedings will accomplish them on a macroscale.

The lectures and their associated discussions are not necessarily presented in chronological order but in a sequence which I deem most cohesive. The discussions were recorded on tape, transcribed by the respective discussion leaders and edited by me. Regretfully, because of mechanical and electronic difficulties the bulk of the discussion of Prof. Overberger's paper and all of the discussion from Prof. LeChoux's and Dr. Yaroslavsky's papers could not be included in these proceedings. The pressure of keeping the number of pages within a prescribed, manageable limit necessitated the elimination of much of the spontaneous (often somewhat macabre) good humor which characterized all the sessions, from the final draft. Further, three excellent unscheduled presentations by Prof. Smets, Prof. Bamford and Prof. Higgins on photo-condensation polymerization were eliminated because, despite the fervor of the moment in which they were presented, they are beyond the scope of these proceedings.

The efforts of Mrs. H. Hayes and my wife in typing and retyping portions of the manuscript are thankfully acknowledged. The forbearance of my family during the Institute and the preparation of the final draft of these proceedings was a major component in the ultimate success of the program.

James A. Moore
Troy, New York
September, 1973

ACKNOWLEDGEMENT

In Addition to the financial support of the North Atlantic Treaty Organization, the generosity of the following corporations contributed greatly to the completeness and success of the program.

Allied Chemical Co.

Cabot Corp.

Celanese Co.

Eastman Kodak Co.

Esso Research and Engineering Co.

International Telephone and Telegraph, Inc.

Johnson and Johnson Associated Industries

Lord Corp.

Lubrizol Corp.

P P G Industries

Shell Development Co.

ATTENDEES

Dr. R. Auerbach
Lord Corporation
2000 West Grandview Blvd.
Erie, Pa. 16512

Prof. C. H. Bamford
The University of Liverpool
Dept. of Inorganic, Physical&
Industrial Chemistry
Donnan Laboratories
P.O. Box 147
Liverpool L69 3BX
England

Mr. R. Bopp
Dept of Chemistry
Rensselaer Polytechnic Inst.
Troy, N.Y. 12181

Prof. C. Carraher Jr.
Dept of Chemistry
University of South Dakota
Vermillion, South Dakota 57069

Prof. G. Challa
The University of Groningen
Laboratorium voor
Polymeerchemie
Zernikelaan, Groningen
The Netherlands

Dr. R. Chlanda
Chemical Research Center
Allied Chemical Corp.
Morristown, N.J. 07960

Dr. I. Cho
The Korea Advanced Institut
of Science
P. O. Box 150 Chung Ryang Ni
Seoul, Korea

Dr. R. Corett
Israel Institute for
Biological Research
Tel-Aviv University
Medical School
Ness-Ziona, Israel

Dr. J. Dale
Dynapol
1454 Page Mill Rd.
Palo Alto, Calif. 94304

Prof. W. H. Daly
Dept. of Chemistry
Louisiana State University/
Baton Rouge
Baton Rouge, Louisiana 70803

Dr. H. DeLaMare
Shell Development Co.
Chemical Research Laboratory
Houston, Texas 77079

Dr. D. J. Dunn
Dept. of Chemistry
University of Keele
Staffordshire, England

Prof. R. Epton
Principal Lecturer in Chemistry
The Polytechnic
Wolverhampton, England WV1 1LY

Prof. P. Ferruti
Polytechnic Institute of Milan
Piazza Leonardo DaVinci 32
20133 Milano, Italy

R. B. Fox
Organic Polymers Section, Chem. Div.
Naval Research Laboratory
Washington, D.C. 20390

Dr. T. Furuyama
Asahi Chemical Industry America,Inc.
350 Fifth Ave.
New York, N. Y. 10001

Dr. R. J. Gander
Johnson and Johnson
New Brunswick, N. J. 08903

Mr. C. Gajria
Dept. of Chemistry
University of Connecticut
Storrs, Conn. 06268

Dr. C. Gebelein
Dept. of Chemistry
Youngstown State University
Youngstown, Ohio 44503

Prof. E. J. Goethals
Rijksuniversiteit, Labor.
voor Organische Chemie,
Krijgslaan 271
Ghent, Belgium

Dr. M. Hartman
P.P.G. Industries
Springdale, Pa.

Prof. H. J. Harwood
Institute of Polymer Science
University of Akron
Akron, Ohio 44304

Prof. J. Higgins
Dept. of Chemistry
Illinois State University
Normal, Illinois 61761

Prof. N. Ise
Dept. of Polymer Chemistry
Kyoto University
Kyoto, Japan

Mr. J. E. Kelly
Dept. of Chemistry
Rensselaer Polytechnic Inst.
Troy, N. Y. 12181

Prof. E. Klesper
Institute for
Macromolecular Chem.
University of Freiburg
78 Freiburg i.Br.
Steffan-Meier-Str.31
West Germany

Prof. C. LeChoux
Laboratory of Macromolecular Chem.
Universite des Sciences et
Techniques de Lille
B.P. 36-59 Villeneuve D'Ascq
France

Prof. Gideon Levine
State Univ. Polymer Research Center
College of Forestry at Syracuse U
Syracuse, N.Y. 13210

Prof. M. Lewin
Israel Fiber Institute
Hebrew University
5 Emek Refaim St., P.O.B.8001
Jerusalem, Israel

Dr. A. Lewis
Lord Corporation
2000 West Grandview Blvd
Erie, Pa. 16512

Dr. K. J. Liu
Chemical Research Center
Allied Chemical Corp.
Morristown, N. J. 07960

Prof. E. Mano
Director, Nucleo Macromolecular
Universidade Federal do
Rio de Janeiro
Brazil

Prof. C. G. Overberger
Dept. of Chemistry
University of Michigan
Ann Arbor, Mich. 48104

Dr. I. Parikh
Johns Hopkins University
Dept. of Medicine
Dept. of Pharmacology
725 North Wolfe St.
Baltimore, M.D. 21205

Prof. N.A. Plate'
Dept. of Chemistry
Moscow State University
Moscow 7234
U.S.S.R.

Dr. H.K. Reimschuessel
Chemical Research Center
Allied Chemical Corp.
Morristown, N.J. 07960

Prof. P. Rempp
Centre de Recherches sur les Macromolecules
6, rue Boussingault
67- Strasbourg, France

Dr. D. Rivin
Cabot Corp.
Billerica Research Center
Concord Road
Billerica, Mass. 01821

Prof. J. C. Salamone
Polymer Science Program
Lowell Technological Institute
Lowell, Mass. 01854

Prof. J. Sheats
Dept. of Chemistry
Rider College
Newark, N.J.

Prof. G. Smets
Labor. Macromol.Scheikunde
University of Leuven
Celestynenlaan, 200F,
B-3030 Heverlee, Belgium

Dr. R. Stackman
Celanese Research Co.
Box 1000
Summit, N. J. 07901

Prof. J.M. Stewart
School of Medicine
University of Colorado
4200 East 9th Ave.
Denver, Colorado 80220

Prof. T. St. Pierre
Dept. of Chemistry
University of Alabama/Birmingham
Birmingham, Ala. 35210

Prof. J. Sowa
Dept. of Chemistry
Union College
Schenectady, New York 12308

Dr. R. T. Swiger
General Electric Co.
Research & Development Center
Schenectady, N. Y. 12301

Dr. S. R. Turner
Xerox Research Center
800 Phillips Road
Webster, New York 14580

Dr. C. Yaroslavsky
Department of Biophysics
The Weizmann Institute of Science
Rehovot, Israel

Dr. N. Weinschenker
Dynapol
1454 Page Mill Rd.
Palo Alto, Calif. 94304

Dr. R. E. Williams
Division of Biological Sciences
National Research Council
Ottawa, Ontario

Prof. B. Wunderlich
Dept. of Chemistry
Rensselaer Polytechnic Institute
Troy, N.Y. 12180

CATALYSIS BY POLYMERS

C. G. Overberger and Thomas W. Smith

Department of Chemistry and the Macromolecular Research Center, The University of Michigan, Ann Arbor, Michigan

For more than a decade we have been interested in the reactions of synthetic macromolecules with low molecular weight reagents.[1,2] These systems are of interest because of the analogies which might be drawn with enzymatic processes, and also because of the obvious interest in the unusual reactivities of macromolecules as compared to their low-molecular weight analogs.

The solvolytic reactions of synthetic macromolecules with low molecular weight compounds are in general of two types. One consists in the utilization of a macromolecular surface as the site for adsorption of reactive monomeric species. The other type consists in the catalytically active site being a covalent part of the macromolecular species.

In the first instance, if two low molecular weight ionic species are the reactive reagents, a charged polymer will tend to concentrate and/or repel one or both low molecular weight reagents in its vicinity and, consequently, will either accelerate or inhibit the reaction. This type of macromolecule has not been found to alter the reaction of a neutral substrate. The enhanced or inhibited catalytic action of polyelectrolytes lacking catalytically active functionality on the reaction rates of similarly and/or oppositely charged low molecular weight reagents has been studied by several workers.(3-9)

Morawetz and Shafer (3) found that the hydroxyl ion-catalyzed hydrolysis of positively charged esters was inhibited by partially ionized poly(acrylic acid) and poly(methacrylic acid).

In another study, Morawetz and Vogel (5) observed that the rate of the Hg^{+2}-induced hydration of $Co(NH_3)_5Cl^{+2}$ was accelerated by a factor of 176,000 in the presence of 5 x 10^{-5}<u>N</u> poly(vinylsulfonate).

It is not necessary for the polymer to be charged for catalysis of the first type to occur. Apolar binding can also be sufficient to concentrate low molecular weight species in the domain of a macromolecule. Kabanov <u>et al.</u> (10) have reported that the rate of hydrolysis of <u>p</u>-nitrophenyl acetate catalyzed by N-carbobenzoxyhistidine is increased when poly(oxyethylene) is added. Apparently, the poly(oxyethylene) forms a ternary apolar complex with <u>p</u>-nitrophenyl acetate and N-carbobenzoxyhistidine, thus promoting the catalysis.

Kunitake (11) has reported that the rate of spontaneous hydrolysis of phenyl esters in an alkaline medium is decreased upon addition of cholic acid. The results were explained in terms of complex formation between the phenyl ester and cholic acid, the phenyl ester in the complex being less susceptible to hydrolysis.

Macromolecules which possess the catalytic species as a covalent part of their structure are more closely related to enzymatic systems. Although these synthetic, polymeric catalysts obviously lack the unique tertiary structure of enzymes and usually lack any specific secondary structure, it has been possible to achieve certain similarities between natural (enzymatic) and synthetic macromolecular systems. For example, it has been found that synthetic polymeric systems can be characterized by, (a) higher reactivities than corresponding monomeric systems, (b) specificity of substrate hydrolysis, (c) competitive inhibition by compounds similar to the reactive substrate, (13,14,15) (d) bifunctional catalysis involving the interaction of two pendant functional groups and substrate, (1,2) and (e) saturation phenomena. (4,5,32,13,16)

There are three factors which have been shown to influence the hydrolytic efficiency of synthetic polymers, in general, and polymers containing pendant imidazole groups in particular:

1) Cooperative interactions involving polymers with neutral or neutral and anionic imidazole groups,

2) electrostatic interactions involving a partially charged, catalytically active polymer towards an oppositely charged substrate,

3) and apolar interactions between long-chain substrates and hydrophobic polymers.

Cooperative interactions between catalytically active imidazole groups and other functional groups have been well documented.(17) Poly[4(5)-vinylimidazole] was found to be a better catalyst (2-3 fold) than imidazole towards the neutral ester *p*-nitrophenyl acetate when the reaction was carried out at high pH.(18) This enhanced reactivity was attributed to multiple catalysis by a combination of anionic and neutral imidazole groups along the backbone of the polymer. Also, under somewhat different reaction conditions, a bifunctional interaction of neutral imidazole units was observed.(19) The total rate equation for the polymeric catalysis(20) can be expressed by the following equation,

$$k_{cat} = k_1\alpha_1 + k_2\alpha_1^2 + k_3\alpha_2 + k_4\alpha_1\alpha_2 \qquad (1)$$

where k_1 is a simple nucleophilic catalysis rate constant, k_2 is an imidazole-catalyzed imidazole nucleophilic constant, k_3 is a catalysis rate constant for imidazole anion, and k_4 is a rate constant for anionic imidazole-catalyzed imidazole catalysis; α_1 and α_2 are fractions of neutral and anionic imidazole, respectively. Recently, Overberger and Shen have found that the bifunctional effect is primarily between neighboring imidazole units, such as 1,3, 1,4 and 1,5, rather than between remote groups brought together *via* folding of the macromolecule.(20) They investigated a series of oligo-4(5)-vinylimidazoles and found that the oligomers of *ca.* 5 units had about 71% of the catalytic activity of high molecular weight polymer.

The role of electrostatic interactions in the catalytic activity of synthetic polymers has also been extensively investigated. As early as 1959, Ladenheim and Morawetz had studied the reaction of the α-bromoacetate ions with poly(4-vinylpyridine) (21); the rate of the displacement of the bromide ion increased with increasing ionization of the pyridine groups, while the monomeric analog, 4-methylpyridine had no effect on the rate of bromide displacement. The high reactivity of the polymer was attributed to the electrostatic attraction of the bromoacetate anion by the protonated residues of the polymer, thus accumulating the substrate in the vicinity of a high local concentration of catalytically active neutral pyridine groups.

Letsinger and Savereide (22) studied the hydrolysis of potassium 3-nitro-4-acetoxybenzenesulfonate (NABS) catalyzed by partially protonated poly(4-vinyl pyridine) in a pH-range in which the substrate existed extensively as an anion. A bell-shaped pH-rate profile was obtained with a maximum at pH-4 where approximately 70-80% of the pyridine groups were in the neutral form. At pH values where similar concentrations of neutral pyridine groups were present, NABS was hydrolyzed 9.3 times more efficiently by poly(4-vinylpyridine) than by the model system, picoline.

By designing systems in which electrostatic and bifunctional effects could be utilized, Overberger and his associates have been able to considerably improve the catalytic efficiency of polymeric catalysts based on 4(5)-vinylimidazole. Rate enhancements of up to 50 times the rate exhibited by monomeric imidazole were observed in the poly[4(5)-VIm]-catalyzed hydrolysis of the negatively charged substrates 3-nitro-4-acetoxybenzoic acid (S_2^-) or 3-nitro-4-acetoxybenzenesulfonate (NABS).(23) Bell-shaped pH rate profiles were observed for the hydrolysis of these esters. The maximum occurred at pH 7 where about 25% of the imidazole residues were protonated. In a recently published study of the catalytic activity of oligo[4(5)-vinylimidazole] towards NABS, Overberger and Okamoto (46) report that, generally, the catalytic activity increased with increasing molecular weight. The oligomers with a degree of polymerization, (DP) more than eight showed higher reactivity than imidazole and a bell-shaped pH-rate profile. They concluded that electrostatic interactions between the cationically charged imidazole groups in the oligomer and the anionic substrate were important only for high molecular weight oligomers. They further concluded that short-range interactions between neutral imidazole residues also enhanced the catalytic activity of the oligomer.

Bifunctional and electrostatic effects were also observed in the study of various copolymers of 4(5)-vinylimidazole, including copolymers of 4(5)-VIm with acrylic acid and 4(5)-VIm with p-vinylphenol. Overberger and Maki (24) found that, at high pH, copolymers of 4(5)-VIm with acrylic acid exhibited a marked selectivity of hydrolysis toward the positively charged substrate, 3-acetoxy-N-trimethylanilinium iodide (S_2^+). With the negatively charged substrate, S_2^-, the copolymer acted as an inhibitor of the rate of hydrolysis, presumably due to a strong electrostatic repulsion between the negatively charged substrate and the carboxylate anions along the polymer chain. By subjecting S_2^+ to hydrolysis in the presence of various copolymers of 4(5)-vinylimidazole with acrylic acid encompassing a broad spectrum of copolymer composition, it was found that S_2^+ was hydrolyzed most efficiently by a copolymer containing 50 mole % carboxylic acid groups. This copolymer also contained the highest distribution of isolated imidazole units, suggesting that a neutral imidazole group isolated by acrylic acid groups is the most catalytically active species toward S_2^+. Ancillary investigations with copolymers of 4(5)-VIm and vinyl sulfonic acid (24) were undertaken to substantiate this conclusion. If the carboxylate group served only to electrostatically attract the positively charged substrate, S_2^+, it was anticipated that imidazole copolymers containing sulfonate groups should have similar catalytic reactivity to imidazole copolymers containing carboxylate groups. Since this behavior was not observed, some bifunctional catalysis between neutral imidazole groups and carboxylate groups was suggested.

A copolymer of 4(5)-vinylimidazole and *p*-vinyl phenol showed dramatic indication of bifunctional imidazole-phenolate participation in the hydrolysis of S_2^+.(25) Phenol, poly(*p*-vinylphenol), poly[4(5)-VIm] and a 1:0.48 copolymer of 4(5)-vinylimidazole and *p*-methoxystyrene had no effect on the solvolytic rate in the pH region investigated. A (1:1.95) copolymer of 4(5)-vinylimidazole and *p*-vinylphenol was 63 times as efficient as imidazole for the hydrolysis of S_2^+ at pH 9. Bifunctional attack, involving imidazole and phenolate ions was the only viable explanation for such a rate enhancement. This bifunctional interaction could involve imidazole as a nucleophile attacking the substrate and the anionic phenol acting as a general base on the resulting tetrahedral intermediate or the phenol anion could assist the nucleophilic attack of a neutral imidazole residue. (See Scheme I.)

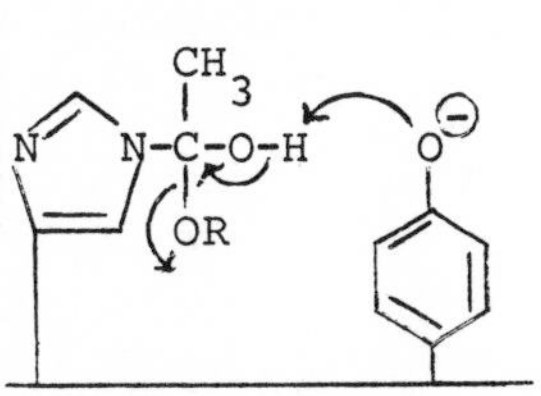

Scheme I

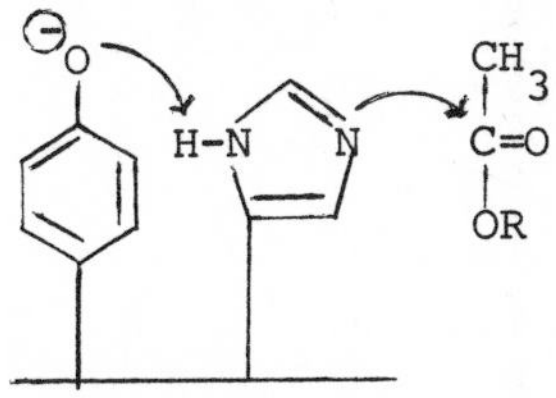

It is only in recent years that the importance of apolar or hydrophobic interactions to enzymatic and synthetic catalysts has been recognized.

Apolar binding is perhaps the most important factor in obtaining rate enhancements with polymeric catalysts. Recent results by Overberger and Morimoto *et al.* (16,26) and by Overberger, Glowaky and Vandewyer (27) strongly support such a possibility.

Overberger and Morimoto have studied the solvolytic reactions of *p*-nitrophenyl acetate and *p*-nitrophenyl heptanoate catalyzed by poly[4(5)-VIm] and imidazole.(26) The kinetics were studied as a function of temperature, pH and the ethanol-water composition of the solvent. It was found that the conformation of the polymer chain of poly[4(5)-VIm] was dramatically affected by the ethanol-water composition and by the degree of neutralization of the pendant imidazole residues. (See Figures 1 and 2).

The enhanced catalytic activity of poly[4(5)-VIm] towards the neutral substrate *p*-nitrophenylacetate in buffer solutions of high and low ethanol composition at pH *ca.* 8 was attributed to increased bifunctional catalysis with the shrinkage of the macromolecule in solution. The enhanced solvolysis of *p*-nitrophenylheptanoate in pH *ca.* 8 buffer solutions of low ethanol composition was attributed to an increased accumulation of substrate in the polymer domain due to hydrophobic interactions between the polymer and the substrate (see Figure 3).

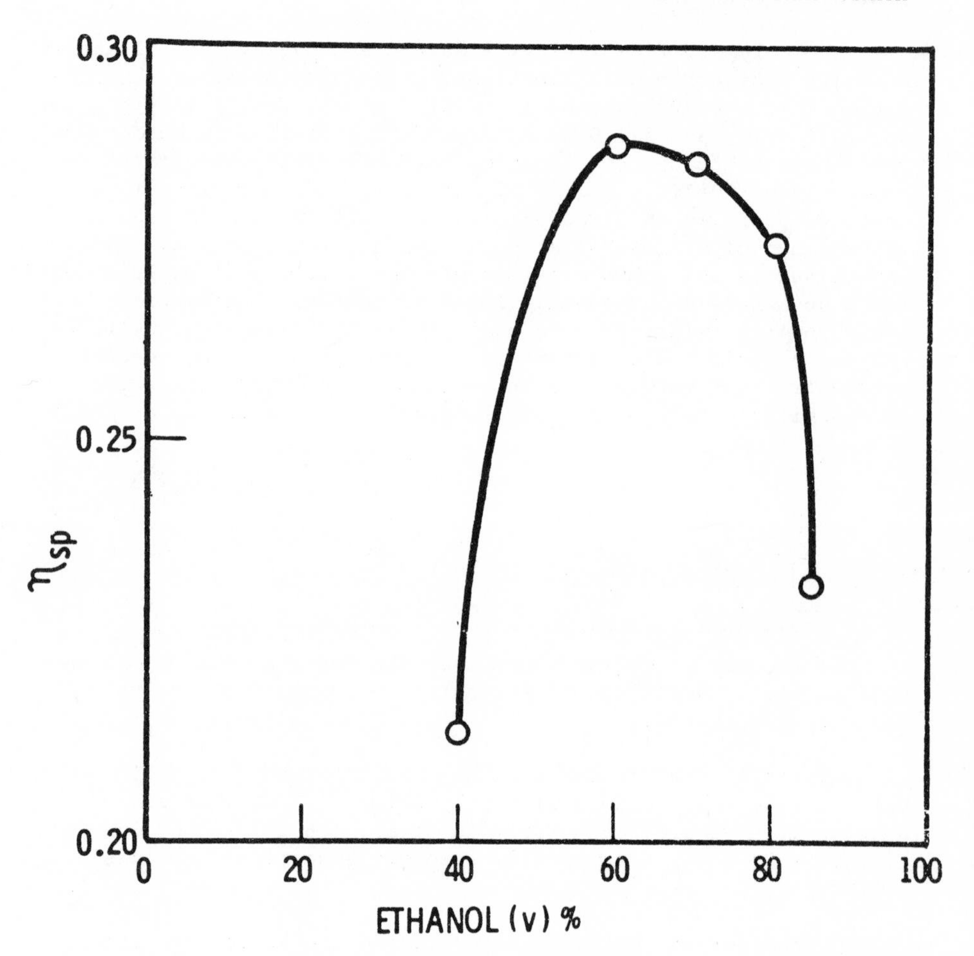

Figure 1. From Reference 26.

Table I[a]
Kinetic Parameters for the Poly[4(5)-VIm]-catalyzed Solvolysis of S_{12}^{-} with Varying Ethanol Content, μ = 0.02, 26°

Ethanol Vol %	pH	Set of Conditions	Fixed, initial Concn. M	Km x 10^4 M	k_2 min^{-1}
43.7	7.99	$[S]_o > [E]_o$	$[E]_o = 1.4 \times 10^{-5}$	4.53±0.64	0.088±0.013
	7.99	$[E]_o > [S]_o$	$[S]_o = 1.1 \times 10^{-4}$	4.77±0.57	0.032±0.004
30	7.90	$[E]_o > [S]$	$[S]_o = 1.78 \times 10^{-4}$	3.11±0.8	20.7 ±4.0
20	7.90	$[E]_o > [S]_o$	$[S]_o = 8.90 \times 10^{-5}$	0.38±0.11	11.4 ±1.8

[a] Taken from Reference 16.

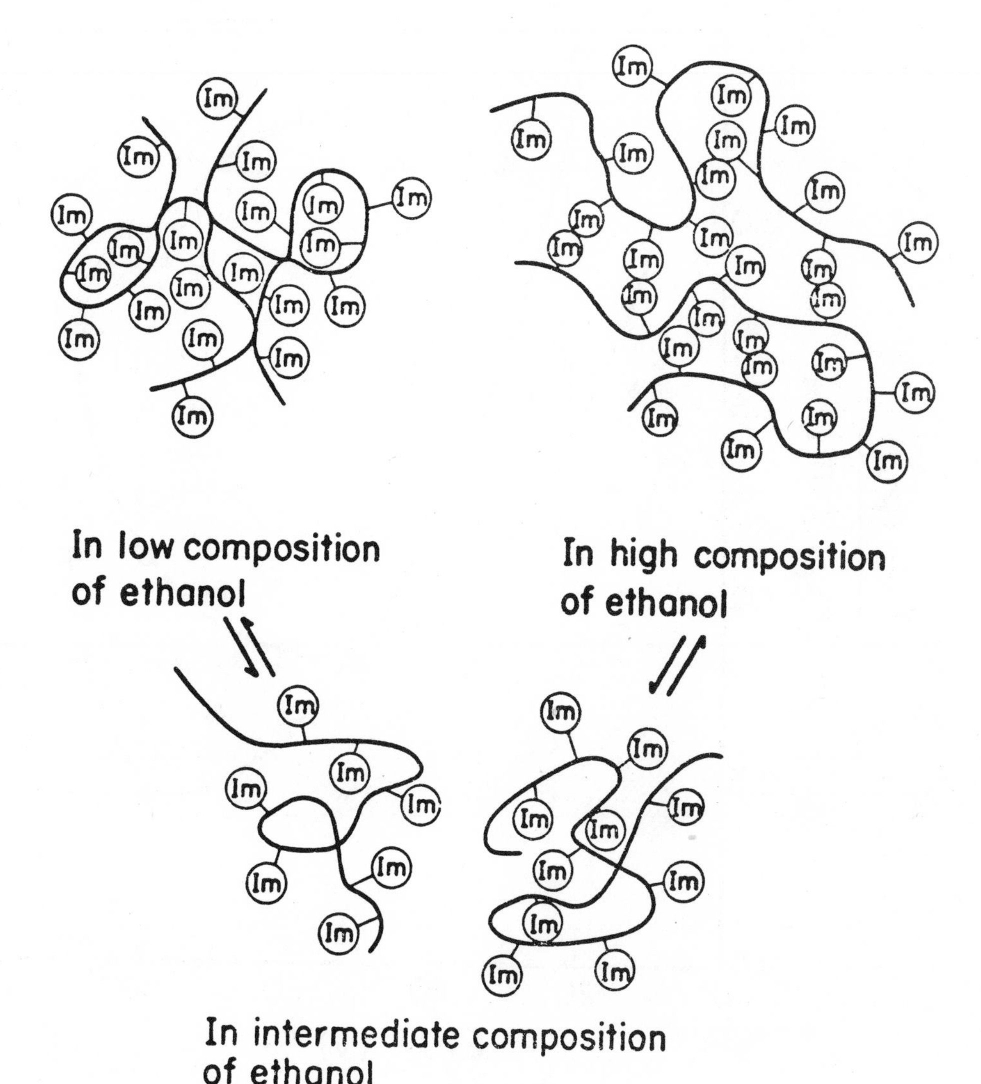

Figure 2. From Reference 26.

Overberger, Morimoto, Cho and Salamone (16) have also reported on the effect of ethanol-water solvent composition on the poly[4(5)-VIm]-catalyzed hydrolysis of 3-nitro-4-dodecanoyl-oxybenzoic acid (S_{12}^-). The poly[4(5)-VIm]-catalyzed hydrolysis of S_{12}^- in solvent systems of low ethanol composition exhibited a saturation phenomenon, with the rate of hydrolysis of S_{12}^- in 30% ethanol-water being 1730 times faster than the rate of

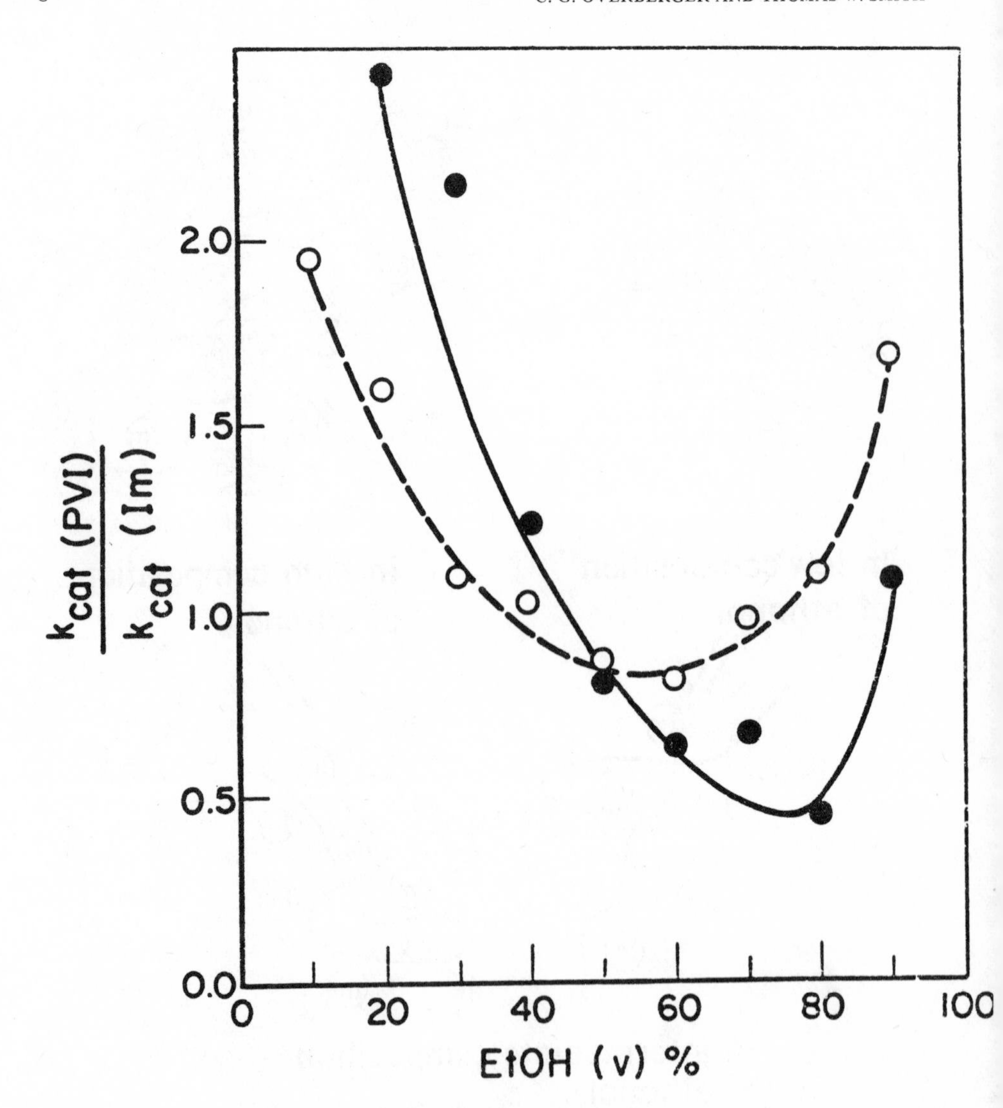

Figure 3. From Reference 26.

hydrolysis of S_{12}^- by monomeric imidazole. The kinetics were analyzed by assuming a Michaelis-Menten type mechanism, i.e., the intermediate formation of an apolar catalyst-substrate complex.

$$E + S \underset{k_{-1}}{\overset{k_1}{\rightleftharpoons}} E\cdot S \xrightarrow{k_2} E' + P_1 \xrightarrow{k_3} E + P_2 \tag{2}$$

$$v_o = \frac{k_2 E_o S_o}{S_o + E_o + K_m} \tag{3}$$

$$[E_o] > [S]_o, \ \bar{V}_m = k_2[S]_o \tag{4}$$

$$[E_o]/V_{obs} = K_m/\bar{V}_m + [E]_o/\bar{V}_m \tag{5}$$

$$[S]_o > [E]_o, \ \bar{V}_m = k_2[E]_o \tag{6}$$

$$[S]_o/V_{obs} = K_m/\bar{V}_m + [S_o]/\bar{V}_m \tag{7}$$

Interpretation of the kinetics was complicated by an autocatalytic effect and by the rate determining deacylation of an apparent acyl intermediate. Overberger and Okamoto have recently demonstrated that saturation phenomena can result from the accumulation of a polymeric acylimidazole intermediate. (28) They emphasized that it may be necessary to treat as little as 1% of

$$\nu_{ss} + \frac{\bar{V}_m [S]}{K_m + [S]}, \qquad \bar{V}_m = k_3[E]_o, \qquad K_m = k_3/k_{cat}$$

(8) (9) (10)

a kinetic run in determining the initial rate in instances where deacylation (k_3) is rate-determining.

Overberger, Glowaky and Vandewyer (27) have studied the effects of the acyl chain length in the substrate and the volume percent of water in aqueous-alcohol solvent systems on the rates of hydrolysis of a series of 3-nitro-4-acyloxybenzoic acid substrates (S_n^-) catalyzed by poly[4(5)-VIm]. In certain cases, particularly where the acyl-chain length in the substrate was seven carbons or more, deviations from pseudo-first-order kinetics were apparent. The kinetic data was again interpreted by assuming a Michaelis-Menten type mechanism. When the poly[4(5)-VIm]-catalyzed solvolysis of S_7^- and S_{12}^-, in <u>ca.</u> tenfold excess with respect to catalyst, was followed in propanol-water, (27) saturation behavior was observed for S_7^- in 15%(v) and 20%(v) propanol-water; the k_2 and k_m values obtained were 0.74 min^{-1} and 3.94 x 10^{-4}M, respectively, in 20% propanol-water, and 0.94 min^{-1} and 3.14 x 10^{-4}M, respectively, in 15% propanol-water. A somewhat bell-shaped plot of v_i <u>versus</u> (S_n^-) was observed for the poly-[4(5)-VIm]-catalyzed hydrolysis of S_7^- in 10% propanol-water as well as for the hydrolysis of S_{12}^- in 10, 15, and 20% propanol-water. The maximum of these curves occurred very near the critical micelle concentration (as determined by titration with pinacyanol chloride). This was taken as kinetic verification of the cmc values and the decrease of the reaction rate was attributed, perhaps erroneously, to progressive micellarization of the substrate.

The poly[4(5)-VIm], in ca. 10-fold excess with respect to substrate, catalyzed hydrolysis of S_2^-, S_7^- and S_{12}^- in 20% propanol—water all exhibited saturation kinetics.(27) The kinetic parameters obtained by assuming a Michaelis-Menten mechanism are listed in Table II.

Table II[c]
Influence of Catalyst Concentration on the Initial Esterolysis Rate[a]

Substrate	$k_2(min^{-1})$[b]	$K_m x10^4M$
S_2^-	0.15	2.96
S_4^-	0.15	2.86
S_7^-	0.086	4.04

(a) pH = 8.00, μ = 0.02; 26°; 20%(v) 1-propanol-water.
(b) The observed value has been corrected for protonated imidazole residues (α_1 = 0.94).
(c) Taken from Reference 27.

Anomalously, k_2 failed to increase and K_m failed to decrease with increasing chain length of the substrate. This data may be crucial in indicating that the saturation behavior heretofore observed with poly[4(5)-VIm] is at least in part covalent saturation due to accumulation of an acylimidazole intermediate.

Overberger and Glowaky (29) have studied, in depth, the acylation-deacylation behavior of poly[4(5)-VIm]. They were able to demonstrate, unequivocally, that the accelerative kinetic patterns typically observed in the poly[4(5)-VIm]-catalyzed hydrolysis of S_{12}^- and S_{18}^- in low alcohol water solvent systems were due to accumulation of acylated polymer which was more hydrophobic, thus a better catalyst for the hydrolysis of long-chain substrates. The accelerative kinetic pattern could be eliminated by increasing the rate of deacylation through the addition of the α-effect compound, hydroxylamine. In addition, the rate of deacylation was measured, first by isolating acylated polymer from a kinetic run and monitoring its rate of deacylation by following the rate of disappearance of the acylimidazole band at 245 nm in the uv. Secondly, the rate of deacylation of authentic poly(1-acyl-4-vinylimidazole)(PVIm-Ac_n), synthesized by free radical polymerization of the corresponding monomers, was monitored. Table III shows the rates of acylation and deacylation determined for S_2^-, S_7^-, S_{12}^- and S_{18}^-.

Overberger and Sannes (30) have determined the rate of the steady-state deacylation of PVIm-Ac_7 and PVIm-Ac_{12} prepared

Table III[f]

First-Order Rate Constants for Hydrolysis (K_{obs}) and Deacylation (K_{deacyl}) Reactions[a]

Substrate	k_{obs} (min^{-1})[b]	Intermediate	k_{deacyl} (min^{-1})[c]
S_2^-	0.022	PVIm-Ac	0.250
S_7^-	0.013	$PVIm\text{-}Ac_{17}$	0.263(0.251)[e]
S_{12}^-	0.090[d]	$PVIm\text{-}Ac_{12}$	0.041
S_{18}^-	0.500[d]	$PVIm\text{-}Ac_{18}$	0.006

(a) 40% Vol% ethanol-water, $\mu = 0.02$, TRIS = 0.02M, pH = 8.0, 26°.

(b) $[PVIm] = 5.0 \times 10^{-4}$ M, $[S_n^-] = 5 \times 10^{-5}$ M.

(c) Determined for >90% deacylation completed from Sephadex isolated samples.

(d) Accelerative kinetic behavior.

(e) Determined for >80% deacylation completed; from authentic sample.

(f) Taken from Reference 29.

during the poly[4(5)-VIm]-catalyzed hydrolysis of S_7^- and S_{12}^-, in *ca.* 10 to 20-fold excess substrate.

A typical plot of Abs *versus* time in excess substrate shows an initial burst (acylation) followed by a slower zero-order steady state deacylation. This, of course, is analogous to the situation in the α-chymotrypsin-catalyzed hydrolysis of *p*-nitrophenyl acetate. The mechanism for the poly[4(5)-VIm]-catalyzed hydrolysis of S_n^- can thus be described as shown in Scheme II. Whether or not an intermediate apolar polymer substrate complex is involved depends on the length of the acyl-chain of the substrate and the solvent composition.

Perhaps the clearest example of saturation kinetics in the hydrolysis of nitrophenyl esters by polymers of 4(5)-vinylimidazole was provided by Overberger and Pacansky (31) in studying the hydrolysis of long-chain nitrophenyl esters by copolymers of 4(5)-vinylimidazole and 1-vinyl-2-methyl-3-alkyl imidazolium iodides, (pIm + $C_n^+I^-$) where n denotes the chain length of the alkyl group (see Figures 4 and 5). In solvent composition of low ethanol content, the catalyzed hydrolysis of long-chain substrates proceeded by a Michaelis-Menten mechanism, exhibiting saturation phenomena in excess catalyst.

Increasing the chain length of the ester resulted in an increase in the catalytic rate of hydrolysis. Increasing the ethanol content decreased the catalytic activity and greatly reduced the apolar bonding between the catalyst and the substrate.

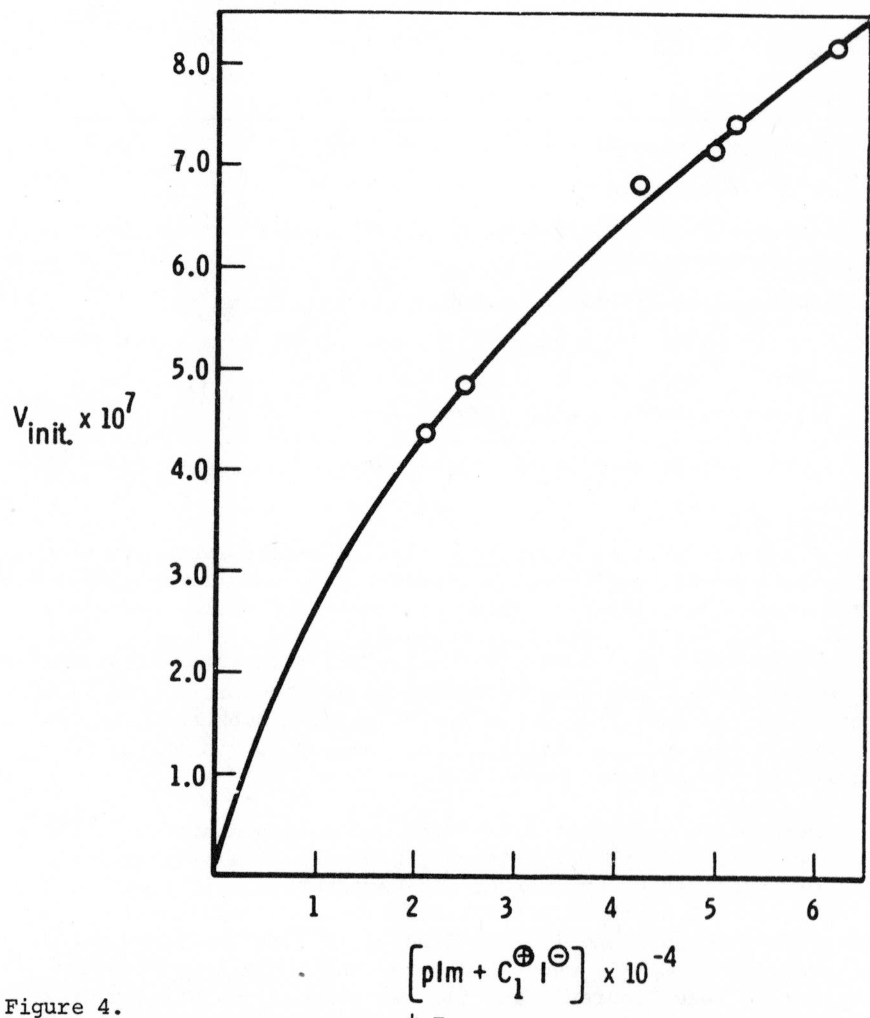

Figure 4.
Initial velocity *vs.* $[pIm + C_1^+I^-]$ for the hydrolysis of S_{12}^-, 34.7% ethanol(v), pH *ca.* 8;26°, [polymer]>[substrate]. From Reference 3.

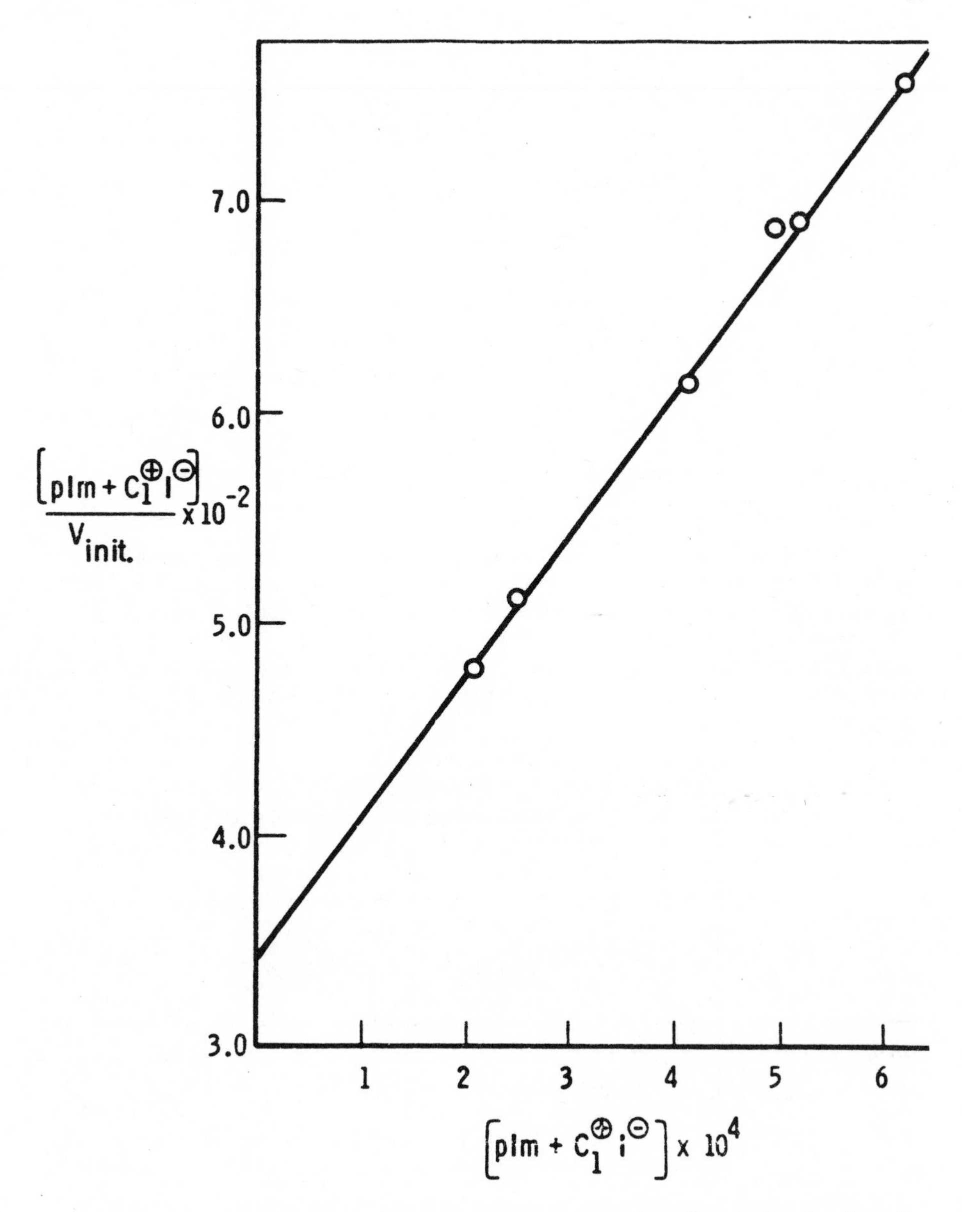

Figure 5. Modified Lineweaver-Burk Plot for the hydrolysis of S_{12}^- catalyzed by pIm + $C_1^+I^-$, 34.7% ethanol(v); pH <u>ca.</u> 8; 26°, [polymer]>[substrate]. From Reference 31.

Scheme II

$$\text{PVIm} + \text{S}_n^- \xrightarrow{k_{obs}} \text{P}^- + \text{PVIm-Ac}_n \xrightarrow{k_3} \text{Ac}_n\text{OH} + \text{PVIm}$$

$$E + S \rightleftharpoons E{\cdot}S \xrightarrow{k_2} E' + P_1 \xrightarrow{k_3} E + P_2$$

The effect of the chain length of the substrate, the chain length of the hydrophobic groups along the polymer backbone, and the ethanol-water composition on the extent of apolar interactions was ascertained by varying the length of the alkyl group in the catalysts and noting the effect on the rate of hydrolysis of S_n^- in varying ethanol-water compositions. (See Table IV).

Table IV[a]

Michaelis-Menten Kinetic Parameters for the Hydrolysis of Long-Chain Anionic Substrates Catalyzed by Copolymers Containing Pendant Imidazole and Long-Chain Cationic Groups at pH 8, 26°, μ = 0.02 [substrate]≃5x10^{-5}M

Ethanol (v)%	Polymeric Catalyst	Substrate	$\bar{V}_m \times 10^6$ M-min^{-1}	$K_m \times 10^4$ M	$k_2 \times 10^{12}$ (min^{-1})	M^{-1}min^{-1} k_2/K_m or k_{cat}
31.4	pIm+$C_1^+I^-$	S_{12}^-	2.3	1.5	4.6	307
34.7	pIm+C_1^+I	S_{12}^-	1.5	2.3	3.8	165
34.7	pIm+$C_4^+I^-$	S_{12}^-	2.4	2.6	5.0	192
40.0	pIm+$C_{16}^+I^-$	S_{12}^-	7.7	3.6	16.0	444
20.0	pIm+$C_{16}^+I^-$	S_7^-	8.1	3.8	15.0	395
40.0	pIm+$C_{10}^+I^-$	S_7^-	3.8	4.8	7.1	148
43.7	poly[4(5)-VIm]	S_{12}^-	- -	4.77	3.2	67
40	Im		- -	- -	- -	6.8

(a) Taken from Reference 31.

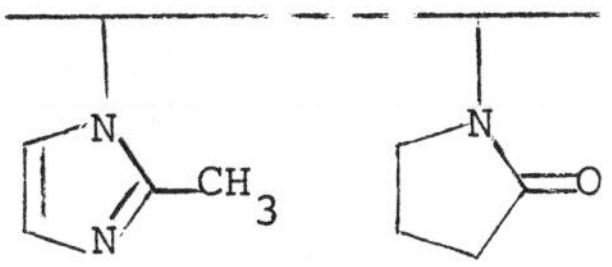

MVI-VP 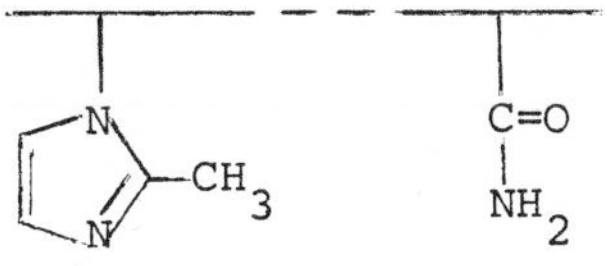 MVI-AA

Other workers have also recognized the importance of hydrophobic interactions in catalysis by polymeric species. Toyoki Kunitake has been very active since 1969 in studying the esterolytic activity of imidazole containing synthetic polymers.(11,15, 32-38) His first contribution in this area was a study of the hydrolysis of S_2^- by copolymers of 1-vinyl-2-methylimidazole (MVI) with N-vinylpyrrolidone (VP) and acrylamide (AA).(32) His kinetic studies were carried out at 30° and generally at pH 8.0 in 1.0M aqueous KCl. Although these polymeric catalysts were relatively ineffective catalysts, as compared to monomeric imidazole, the rate of the catalytic hydrolysis could be described by Michaelis-Menten kinetics, exhibiting saturation phenomena at high substrate concentrations.(32) Such a mechanism was given support by a subsequent study in which benzyl alcohol, 2,4-dinitrophenolate anion and dioxane were shown to competitively inhibit the VP-MVI copolymer-catalyzed hydrolysis of S_2^-.(15) It was determined that at less than 20% MVI content the kinetic pattern was independent of the copolymer composition, giving an average K_m value of 9.3 mM and an average k_2 value of 3.8 x $10^{-2}min^{-1}$ for the VP-MVI copolymers. The imidazole acrylamide copolymers were evidently less hydrophobic and more catalytically efficient, exhibiting K_m and k_2 values of 63 mM and 11 x $10^{-2}min^{-1}$, respectively. This is somewhat anomalous, because it would be expected that k_2 and K_m would increase or decrease in concert. Copolymers containing higher amounts of MVI proved to be less efficient catalysts, exhibiting K_m and k_2 values of 9.8 mM and 0.020 min^{-1}, respectively, for a copolymer containing 40% MVI units and 5.3 mM and 0.011 min^{-1}, respectively, for a copolymer containing 80% MVI units. Kunitake proposes a novel, if somewhat questionable, catalytic loop mechanism for the binding of substrate by the VP-MVI copolymers. He probably bases this proposal on the reported ability of poly(vinylpyrrolidone) to bind aromatic molecules in aqueous solution (32) and on the ability of such a model to explain the relative inefficiency of the catalysts containing higher amounts (40-80%) of MVI units in the copolymer. The catalytic loop is not proposed for the imidazole acrylamide copolymers.

In an effort to design more effective macromolecular catalysts, Kunitake and Shinkai prepared more hydrophobic polymers possessing 4(5)-substituted imidazole residues and sometimes acrylamidophenol (AP) units.(33)

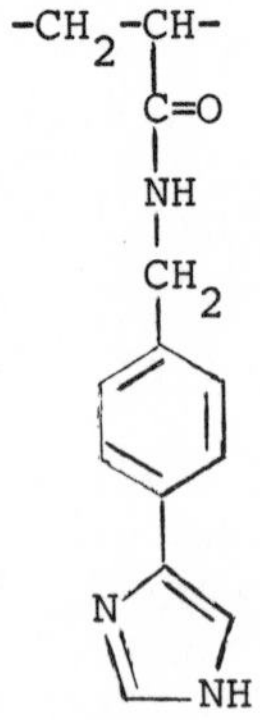

$-CH_2-CH(-C{=}O-NH-CH_2-C_6H_4-\text{imidazolyl})-$

Pl unit

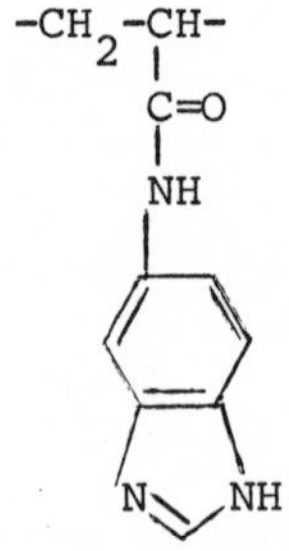

$-CH_2-CH(-C{=}O-NH-\text{benzimidazolyl})-$

VI unit

$-CH_2-CH(-C{=}O-NH-C_6H_4-OH)-$

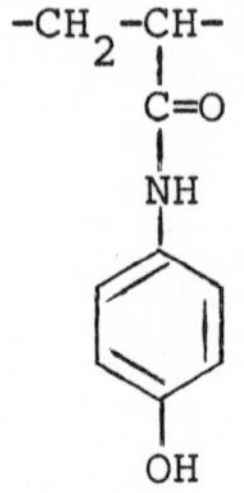

AP unit

These residues were incorporated into copolymers with vinylpyrrolidone (VP) or acrylamide (AA). Unfortunately, in their kinetic studies they chose a more difficultly hydrolyzable substrate, p-acetoxybenzoic acid (ABA). Thus, comparison with the previous studies of the hydrolysis of S_2^- is not possible. The kinetics of the hydrolysis of ABA by VP and AA copolymers containing BI, PI and/or AP units could be described by a Michaelis-Menten mechanism, again exhibiting saturation at high substrate concentrations.(33) The hydrolysis of ABA by the model compound [p-acetamidomethylphenyl-4(5)-imidazole](AcPI) followed second-order kinetics,(33) while species only slightly more hydrophobic such as [p-acetamidomethylnapthyl 4(5)-imidazole](N-yl-Im) and N-α-napthoylhistamine (N-oyl-Im) exhibited saturation kinetics.(34) These studies (33,34) generally included thermodynamic evaluations of the binding process and the "intracomplex reaction," (k_2).

Typical kinetic parameters are shown in Table V. It should be noticed that the polymers are typically no more efficient as catalysts than the monomeric imidazoles, further with these polymers again K_m is observed to decrease with no corresponding increase in the rate of turnover of substrate.

It was concluded from the thermodynamic studies that the decrease in free energy realized from substrate binding was mostly offset by the increase in the free energy of activation of the "intracomplex process." The interpretation of this result is that the hydrophobic interaction in the catalyst substrate complex becomes destroyed in the transition state of the intra-

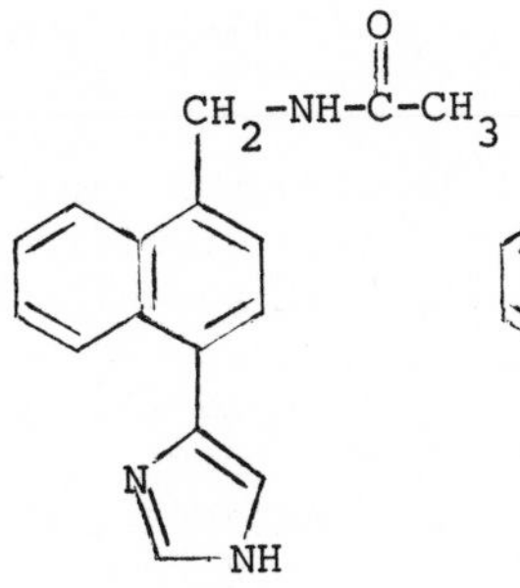

N-yl-Im

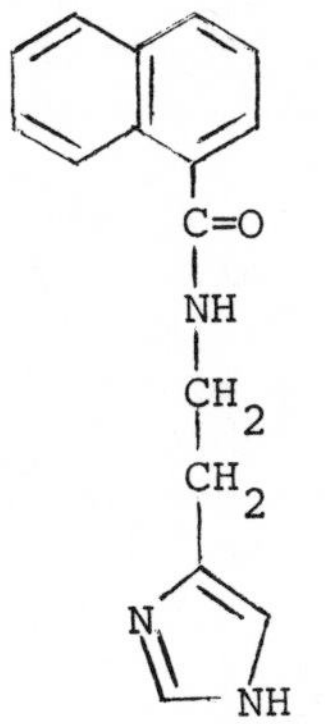

N-oyl-Im

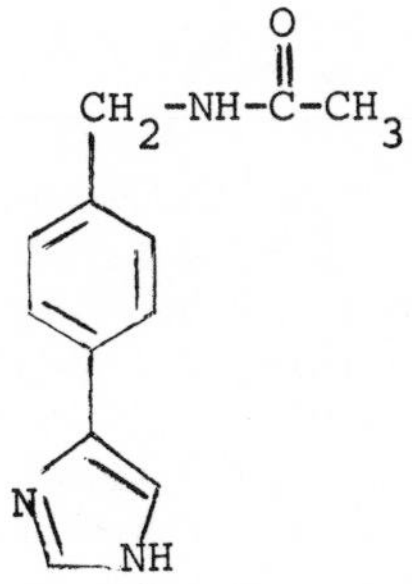

AcPI

Table V
Catalytic Hydrolysis with Polymer Catalysts[a]

Polymer	K_m,mM	k_2 $min^{-1}x10^3$	k_2/K_m $min^{-1}M^{-1}$
PI-AA (1.68% PI)[b]	83	19	0.23
PI-AA(8.91% PI)[c]	29	29	1.02
PI-VP(1.94% PI)[d]	19	13	0.66
PI-AP-AA(20% PI)[e] 1.8% AA	12	17	1.39
N-oyl-Im[e]	16.4	21	1.3
Nyl-Im[f]	13.9	54	3.9
AcPI[g]	- -	--	(k' = 0.27)

[a] At pH 8.0, 30°, in 1M aqueous KCl, total Im conc.

[b] 1.55 mM.

[c] 0.83 mM.

[d] 1.32 mM.

[e] 0.82 mM.

[f] 0.31 mM.

[g] 2.16 mM. Data taken from References 33 and 34.

complex reaction. Kunitake contends that the transition state is charged thus destabilized in a hydrophobic environment.(34)

Kunitake's most recent efforts have been directed towards preparing charged hydrophobic polymers which would stabilize a charged transition state in a hydrophobic region. In this regard the hydrolysis of the positively charged esters catalyzed by copolymers of PI and acrylic acid or methacrylic acid was studied.(35) The catalytic hydrolysis of phenyl esters, 3-acetoxy-N-trimethylanilium iodide (ANTI), and 3-nonanoyloxy-N-trimethylanilinium iodide (NNTI) proceeded according to Michaelis-Menten kinetics. The cationic substrates appear to be bound at the polymer catalytic sites by electrostatic and hydrophobic forces. NNTI showed a binding tendency greater than that of ANTI because of its enhanced hydrophobic character. The presence of the carboxylate group at the catalytic site facilitated the intracomplex reaction, thus binding ($1/K_m$) and esterolytic activity (k_{cat}) were observed to increase in concert.

Kunitake achieved similar effects in hydrolysis of the negatively charged ester S_2^-, by a copolymer of methylvinylimidazole partially quaternized with methyl iodide.(36) Figures 6 and 7 show curves for plots of the free energy of activation for this intracomplex reaction versus the molar free energy of the binding process for various uncharged and charged MVI copolymers. One should notice that the free energy of activation increases in a non-linear manner with increases in the molar free energy of the binding process in the curve for the neutral MVI containing polymers. With respect to the partially quaternized or protonated MVI containing polymers, one sees that the free energy of activation for the intracomplex reaction decreases almost linearly with the increase in the molar free energy of the binding process. This relation indicates that the substrate binding and intracomplex reaction rate are simultaneously enhanced or depressed. These results would imply that the hydrophobic nature, per se, of the catalytic site is not advantageous for the intracomplex hydrolysis. On the other hand, when the charged group exists at the predominantly hydrophobic catalytic site, the enhanced substrate binding due to additional hydrophobic interaction appears to result in better stabilization of the transition state of the intracomplex reaction.

I. M. Klotz and V. A. Kabanov and their associates have also been active in recent years in synthesizing hydrophobic polymers having pendant nucleophilic sites which are effective catalysts for the hydrolysis of activated phenyl esters.(10,37-44)

In 1969, Klotz and Royer (37) reported the results of a study in which lauroyl-PEI, prepared by the acylation of low molecular weight PEI with methyl laurate was utilized as a

Correlation of Substrate Bonding and Intracomplex Reaction

Unquaternized Polymers

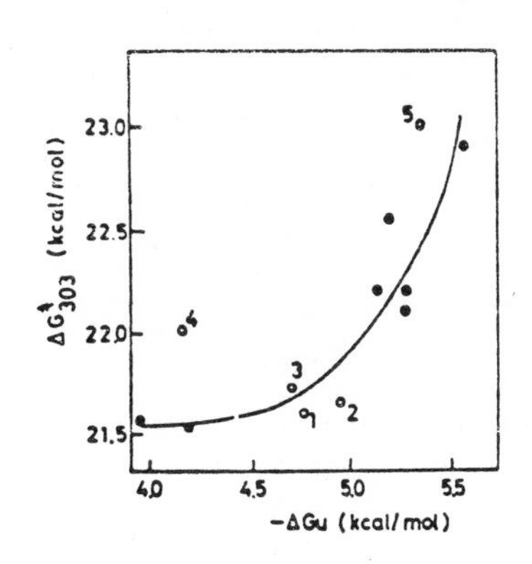

1) MVI-AP-AA-1

2) MVI-AP-AA-2

3) MVI-AP-AA-3

4) MVI-AP-AA-4

5) MVI-10

MVI-AA copolymers and

MVI-VP copolymer

$$E + S \underset{}{\overset{\Delta G_{\mu}}{\rightleftharpoons}} E\cdot S \quad \underline{\text{vs.}} \quad E\cdot S \xrightarrow{\Delta G^{\dagger}_{303}} E + P$$

Quaternized Polymers

G^{*}_{303} (kcal/mol)
24.0
23.5
23.0
5.0
5.5
6.0
AG_{u} (kcal/mol)

6) MVI-10 (pH 6.0)

7) MVI^{+}-1

8) MVI^{+}-2

9) MVI^{+}-3

10) MVI^{+}-4

11) MVI^{+}-5

12) MVI^{+}-AP-1

13) MVI^{+}-AP-2

14) MVI^{+}-AP-3

Figure 6 and 7. From Reference 36.

catalyst. The acylated PEI was a more effective catalyst towards nitrophenyl esters than was propyl amine or other poly(ethylene imines) of varying molecular weight. Marked selectivity of hydrolysis was observed in catalysis by lauroyl-PEI, p-nitrophenyl laurate being hydrolyzed 46 times faster than p-nitrophenyl acetate (See Table VI.)

Table VI
Lauroyl-Pei Catalyzed Hydrolysis of Nitrophenyl Esters
$k_{obs} \times 10^2 \text{ min}^{-1}$

Substrate / Catalyst	p-Nitrophenyl-acetate	p-Nitrophenyl-caproate	p-Nitrophenyl-laurate
Propyl amine	0.98	0.51	0.053
PEI-6	3.60	1.47	0.11
PEI-18	4.38	1.57	0.11
PEI-600	4.60	1.80	0.17
L(10%)PEI-6	15.2	68.1	698

Reaction conditions: pH 9.0, 0.02M Tris, 25°
93.3% H_2O/CH_3CN
Catalyst in 10-fold excess.

G. P. Royer and I. M. Klotz, J. Amer. Chem. Soc., 91, 5885 (1969).

Along a similar vein, Kabanov et al. (38) have reported on the esterolytic activity of lauryl- and benzyl-derivatized PEI. They observed that these polymers catalyzed the hydrolysis of p-nitrophenyl acetate and p-nitrophenyl stearate by a Michaelis-Menten type mechanism, exhibiting saturation in excess substrate. The kinetic parameters shown in Table VII indicate that both lauryl- and benzyl-derivatized PEI are more efficient catalysts than PEI itself. Further, a marked selectivity towards the hydrolysis of p-nitrophenyl stearate is exhibited.

In the previous communication, (37) Klotz and Royer suggested that the lauroyl-PEI polymer would provide a suitable framework for the introduction of catalytic functional groups to produce a macromolecule with even greater catalytic activity. This prediction was borne out in a 1971 communication by Klotz, Royer and Scarpa. (39) These workers introduced methylene imidazole and lauryl groups into a low molecular weight PEI by alkylation of the PEI with chloromethylimidazole and dodecyl iodide, respectively. This polymer proved to be an effective catalyst as compared to imidazole for the hydrolysis of p-nitrophenyl acetate. (See Table VIII). The absorbance versus time curve for the hydrolysis of p-nitrophenyl caproate, in excess substrate,

Table VII
Lauryl- and Benzyl-PEI Catalyzed Hydrolysis of Nitrophenyl Esters

	p-Nitrophenyl Acetate		p-Nitrophenyl Stearate	
Catalyst	k_{cat}/K_M $M^{-1}sec^{-1}$	$k_{cat}/K_M k_{OH^-}$	k_{cat}/K_M $M^{-1}sec^{-1}$	$k_{cat}/K_M k_{OH^-}$
PEI	1.0	0.12	$<10^{-2}$	$<10^{-3}$
$PEI\text{-}L_{0.48}$	3.6	0.45	10.1	0.78
$PEI\text{-}B_{0.66}$	7.9	0.98	26.0	2.0
α CT	3300	412	- -	8×10^7

Hydrolysis Conditions: pH 8.35, 25°
[PEI] = 6.6×10^{-5}M, [$PEI\text{-}L_{0.48}$] = 3×10^{-5}M, [$PEI\text{-}B_{0.66}$] = 1.95×10^{-6}M
Concn of Active Center.

O. V. Arkhangel'skaya, V. S. Pohezhetskii and V. A. Kabanov, Doklady Akad. Nauk. SSSR, 193, 525 (1970).

Table VIII
Hydrolysis of p-Nitrophenyl Acetate

Catalyst	k_{cat} $M^{-1}min^{-1}$
Imidazole	10
PEI(600)D(10%)-Im(15%)	2,700
α-Chymotrypsin	10,000

Irving M. Klotz, Garfield P. Royer and Ioannis S. Scarpa, Proc. Nat. Acad. Sci., 68, 263 (1971).

by lauryl-methylene imidazole-PEI shows an initial burst (acylation) followed by a steady-state zero-order region (deacylation). This is consistent with a two step reaction pathway, analogous to that of a hydrolytic enzyme such as α-chymotrypsin.

The utilization of derivatized poly(oxyethylene) as a catalyst is a possibility recently taken advantage of by Kabanov et al.(40) They report that poly(oxyethylene) whose terminal hydroxyl groups have been esterified with N-benzoyl histidine (POBH) is a more efficient catalyst toward the hydrolysis of p-nitrophenyl acetate than benzoyl histidine. The presence in the poly(oxyethylene) molecule of alternating sequences of ethylene units and oxygen atoms creates a unique hydrophilic-hydrophobic balance, providing for such varied properties as solubility in water and nonpolar hydrocarbons, an anomalous dependence of the solubility in water on the temperature and an ability for salting out from aqueous solutions. The kinetics of the hydrolysis of p-nitrophenyl acetate by POBH could be described by a Michaelis-Menten mechanism, exhibiting saturation in excess substrate.

Kabanov and his coworkers had also carried out numerous studies of the esterolytic activity of N-alkylated-poly(4-vinylpyridines). These polymers exhibited exceptionally high activity in the hydrolysis of activated esters, as compared to low molecular weight analogs. Unfortunately, Kabanov (45) has recently reported that these studies are in error due to impurities which were generated during the alkylation of the pyridine polymers and not removed during the work-up.

References:

1. C. G. Overberger, and J. C. Salamone, Accounts Chem. Res., **2**, 217 (1969).

2. H. Morawetz, Svensk Kem. Tidski., **79**, 309 (1967).

3. H. Morawetz, and J. A. Shafer, J. Phys. Chem., **67**, 1293 (1963).

4. S. Yoshikawa and O. K. Kim, Bull. Chem. Soc., Japan, **39**, 1729 (1966).

5. H. Morawetz and B. Vogel, J. Amer. Chem. Soc., **91**, 563 (1969).

6. H. Morawetz, J. Polym. Sci., **42**, 125 (1960).

7. C. L. Arcus, T. L. Howard and D. S. South, Chem. Ind. (London), 1756 (1964).

8. H. Morawetz, C. G. Overberger, J. C. Salamone and S. Yaroslavsky, J. Amer. Chem. Soc., **90**, 651 (1968).

9. N. Ise and F. Matusi, J. Amer. Chem. Soc., **87**, 296 (1965).

10. V. A. Kabanov, I. N. Topchieva, A. B. Solovéna and B. I. Kurganov, Vysokomol. Soedin., Ser. A, **14**(**8**), 1809 (1972). C.A. **77** 152786p (1972).

11. T. Kunitake and S. Shinkai, Bull. Chem. Soc. Japan, **44**, 3086 (1971).

12. H. Morawetz, "High Polymers. Vol. XXI," Interscience Publishers, New York, 1965, Ch. IX.

13. R. L. Letsinger, I. Klaus, J. Amer. Chem. Soc., 87, 3380 (1965).

14. C. G. Overberger, unpublished results.

15. T. Kunitake, F. Shimada and C. Aso, Makromol. Chemie, 126, 276 (1969).

16. C. G. Overberger, M. Morimoto, I. Cho and J. C. Salamone, J. Amer. Chem. Soc., 93, 3228 (1971).

17. C. G. Overberger, J. C. Salamone, I. Cho and H. Maki, Annals of the New York Academy of Sciences, 155, 431 (1969).

18. C. G. Overberger, T. St. Pierre, N. Vorcheimer, J. Lee and S. Yaroslavsky, J. Amer. Chem. Soc., 87, 296 (1965).

19. C. G. Overberger, T. St. Pierre, C. Yaroslavsky and S. Yaroslavsky, J. Amer. Chem. Soc., 88, 1184 (1966).

20. C. G. Overberger and C. M. Shen, Bioorganic Chemistry, 1, 1 (1971).

21. H. Ladenheim and H. Morawetz, J. Amer. Chem. Soc., 81, 4860 (1959).

22. R. L. Letsinger and T. J. Savereide, J. Amer. Chem. Soc., 84, 3122 (1962).

23. C. G. Overberger, R. Corett, J. C. Salamone and S. Yaroslavsky, Macromolecules, 1, 331 (1968).

24. C. G. Overberger and H. Maki, Macromolecules, 3, 214, 220 (1970).

25. C. G. Overberger, J. C. Salamone and S. Yaroslavsky, J. Amer. Chem. Soc., 89, 6231 (1967).

26. C. G. Overberger and M. Morimoto, J. Amer. Chem. Soc., 93, 3222 (1971).

27. C. G. Overberger, R. C. Slowaky and P.-H. Vandewyer, J. Amer. Chem. Soc., in press.

28. C. G. Overberger and Y. Okamoto, J. Polym. Sci., Polymer Chem. Ed., 10, 3387 (1972).

29. C. G. Overberger and R. C. Glowaky, J. Amer. Chem. Soc., in press.

30. C. G. Overberger and K. N. Sannes, unpublished results.

31. T. J. Pacansky, Ph.D. Thesis, The University of Michigan, 1972.

32. P. Molyneux and H. P. Frank, J. Amer. Chem. Soc., 83, 3169, 3175 (1961).

33. T. Kunitake and S. Shinkai, J. Amer. Chem. Soc., 93, 4247, 4257 (1971).

34. T. Kunitake and S. Shinkai, Bull. Soc. Chem. Japan, 43, 2581 (1970).

35. T. Kunitake and S. Shinkai, Makromol. Chemie, 151, 127 (1972).

36. S. Shinkai and T. Kunitake, Polymer Journal, 4(3), 253 (1973).

37. G. P. Royer and I. M. Klotz, J. Amer. Chem. Soc., 91, 5885 (1969).

38. O. V. Arkhangel'skaya, V. S. Psheyhetskii and V. A. Kabanov, Doklady. Akad. Nauk. SSSR, 193,525 (1970).

39. I. M. Klotz, G. P. Royer and I. S. Scarpa, Proc. Natl. Acad. Sci., USA, 68, 263 (1971).

40. A. B. Solov'eva, I. N. Topchieva, V. A. Kabanov and V. A. Kargin, Proc. Acad. Sci. USSR. Phys. Chem. Sec., 187, 524 (1969).

41. I. N. Topchieva, A. B. Solov'eva, B. I. Kurganov and V. A. Kabanov, Vysokomol. Soedin., Ser. A, 14(8), 1774 (1972). C.A. 78 152785n (1973).

42. I. N. Topchieva, A. B. Solov'eva, B. I. Kurganov and V. A. Kabanov, Vysokomol. Soedin., Ser. A, 14(4), 825 (1972). C.A. 78 83889k (1973).

43. Y. Birk and I. M. Klotz, Bioorganic Chemistry, 1, 275 (1971).

44. H. C. Kiefer, W. I. Congdon, I. S. Scarpa and I. M. Klotz, Proc. Natl. Acad. Sci., USA, 69, 2155 (1972).

45. V. A. Kabanov, Y. E. Kirsh, I. M. Papisov and V. P. Torchilin, Vysokomol. Soedin. Ser. B, 14(6), 405 (1972). C.A. 78 102304j (1973).

46. C. G. Overberger and Y. Okamoto, Macromolecules, 5, 363 (1972).

DISCUSSION SESSION

Discussion Leader- J. A. Moore

Fox: In your work, did you consider end-group effects?

Overberger: In all of the work that I have reported here on polymers with DP's over 20, we did not see any end group effects. Both azo-bis isobutyronitrile and benzoyl peroxide were used as initiators. When we are talking about the very small oligomers, dimers, or trimers, or even the monomer attached to two catalyst fragments, one can definitely see some steric interaction of the azo isobutyronitrile residues with substrates with charges; that is to say, the monomer or dimer exhibit less electrostatic binding than imidazole itself.

Smets: In your experiments with long-chain esters, what method was used to detect the reaction intermediate which passed through a maximum? Is this intermediate ES or ES' of you reaction scheme?

$$E + S \underset{k_{-1}}{\overset{k_1}{\rightleftharpoons}} ES \xrightarrow{k_2} ES' + P_1 \xrightarrow{k_3} E + P_2$$

Overberger: The intermediate whose concentration passes through a maximum is the acylated polymer, ES'. k_1 and k_{-1} constitute a fast pre-equilibrim and k_2 is fast in comparison to k_3, thus, ES' accumulates. The accumulation of acylated polymer, during a given kinetic run is reflected by an increase in the absorbance at ∿245 nm. However, in order to monitor the concentration of acylated imidazole residues, one must subtract out the varying contributions to the absorbance at 245 nm by the substrate(s) and the phenolate anion P_1, both of which having extinction coefficients much greater than that of an acylimidazole.

Consequently, the rate of deacylation, k_3, was determined independently on acylated polymer which was separated from S, P, and P_2 by gel-permeation chromatography and on authentic polymer prepared by the polymerization of the corresponding 1-acyl-4-vinylimidazole.

The observed rate of hydrolysis of long-chain substrates in excess catalyst, k_{obs}, was taken as the rate of acylation. Knowing the rate of acylation and the rate of deacylation, the curve reflecting the concentration of ES' _versus_ time was calculated.

Smets: In what respect is the Michaelis-Menton reaction scheme applicable to the kinetics of your system?

Overberger: The Michaelis-Menton reaction Scheme,

$$E + S \underset{k_{-1}}{\overset{k_1}{\rightleftharpoons}} E \cdot S \xrightarrow{k_2} ES' + P_1 \xrightarrow{k_3} E + P_2$$

appears to be applicable to the poly-4(5)-VIm catalyzed hydrolysis of long-chain esters (C_{12} or greater) in buffer systems of low alcohol content, ca. ≤30% (v). However, there is a problem here in that if one is not careful in monitoring only the very initial part of the reaction (<2%) the kinetics become complicated by the rate-determining k_3 step and the accelerative kinetic behavior which accompanies accumulation of acylated polymer. In studies of the esterolytic activity of N-alkylated 4- and 5 vinylimidazoles we have been able to circumvent this problem.

Salamone: There has been considerable interest in recent years in the study of gel-entrapped enzymes. Have you considered the incorporation of 4(5)-vinylimidazole in a hydroxyethyl methacrylate matrix?

Overberger: We have not attempted the incorporation of poly-4(5)-vinylimidazole in a hydroxyethyl methacrylate matrix. One possibility might be incorporation of the monomer and polymerization within the matrix.

Moore: You have invoked "hydrophobic interactions" as an explantion for the concentration of substrate on the reactive areas of yourcatalysts but couldn't this tendency be explained by varying solubility of the substrate in the bulk of solution and in the solvent held in the vicinity of the polymer?

Overberger: If I understand your question correctly, you are indicating that local variation of solubility parameters along the hydrocarbon coil might be a way of explaining the increased activity of the substrate with long alkyl side chains. I am not quite certain what that kind of solubility would mean in these dilute solutions and how one could distinguish that concept of solubility from the general aspect of hydrophobic interactions. I suggest that hydrophobic interactions are precursors in the sense of any localized increased solubility.

INTERIONIC REACTIONS IN POLYELECTROLYTE SOLUTIONS

Norio Ise

Department of Polymer Chemistry
Kyoto University
Kyoto, Japan

Recently a large number of publications dealing with various types of reactions in the presence of polyelectrolytes have appeared.[1-3] In our laboratory, "catalytic" influences of polyelectrolytes on interionic reactions were studied. In this paper, we wish to review the progress of our work in the two years since the publication of an earlier review article.[4]

Aquation reactions of Co-complexes were systematically investigated.[5] The reactions proceeded between cationic complexes

$$Co(NH_3)Br^{+2} + M^{+n} + H_2O \rightarrow Co(NH_3)_5H_2O^{+3} \qquad [M^{+n}: Ag^+, Hg^{+2}, Tl^{+3}] \quad (1)$$

and metal ions, and anionic polyelectrolytes were quite effective in enhancing the reactions. Figure 1 gives the acceleration factor k_2/k_{20} [k_2 and k_{20} are the second-order rate constants in the presence of polyelectrolytes and in their absence, respectively] as a function of the polyelectrolyte concentration. At lower polyelectrolyte concentrations, polyphosphate (PP) was most effective and polystyrene sulfonate (PSt) least effective. Polyethylene sulfonate (PES) was intermediate in effectiveness. This is probably due to the difference in the charge density of the polyelectrolytes. Figure 2 shows the acceleration factor for the three inducing cations. The order of the factor is $Tl^{3+} \simeq Hg^{2+} \gg Ag^+$. Because of partial hydrolysis of Tl^{3+}, at least half of the Tl ions exist in the form of $Tl(H_2O)_5OH^{2+}$, so that Tl^{3+} was as effective as Hg^{2+}.

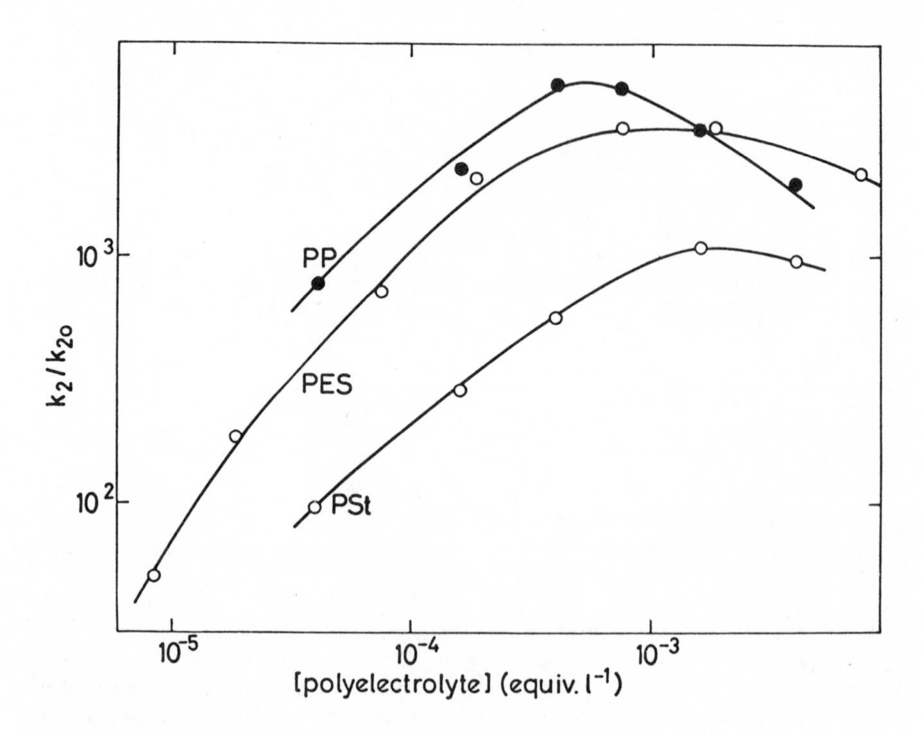

Figure 1. Dependence of the acceleration factor for the Hg^{2+} induced aquation on polyelectrolyte concentration at 25°C. $[Co(NH_3)_5Br^{2+}] \simeq 6 \times 10^{-5}$ M, $[Hg^{2+}] \simeq 5 \times 10^{-4}$ M, $k_{20} = 8.67\ M^{-1} \cdot sec^{-1}$. Taken from Ref. 5.

Figures 1 and 2 suggest that the most important factor in determining the acceleration of the reaction under consideration is the electrostatic forces between polyelectrolyte ions and the reactant ions. Addition of inert salts such as $NaClO_4$ weakens the electrostatic forces, resulting in smaller acceleration. This is shown in Figure 3, which indicates that the acceleration factor becomes smaller monotonically with increasing $NaClO_4$ concentration.

Anion-anion reactions which were studied[6] were:

$$Cl-CH_2-CO_2^- + S_2O_3^{-2} \longrightarrow {}^-O_3S_2-CH_2-CO_2^- + Cl^- \quad (2)$$

$$Cl-C_6H_2(NO_2)_2-CO_2^- + OH^- \longrightarrow HO-C_6H_2(NO_2)_2-CO_2^- + Cl^- \quad (3)$$

$$\left(O_2N-C_6H_3(CO_2^-)-S\right)_2 + CN^- \longrightarrow {}^-S-C_6H_3(CO_2^-)(NO_2) + SCN-C_6H_3(CO_2^-)(NO_2) \qquad (4)$$

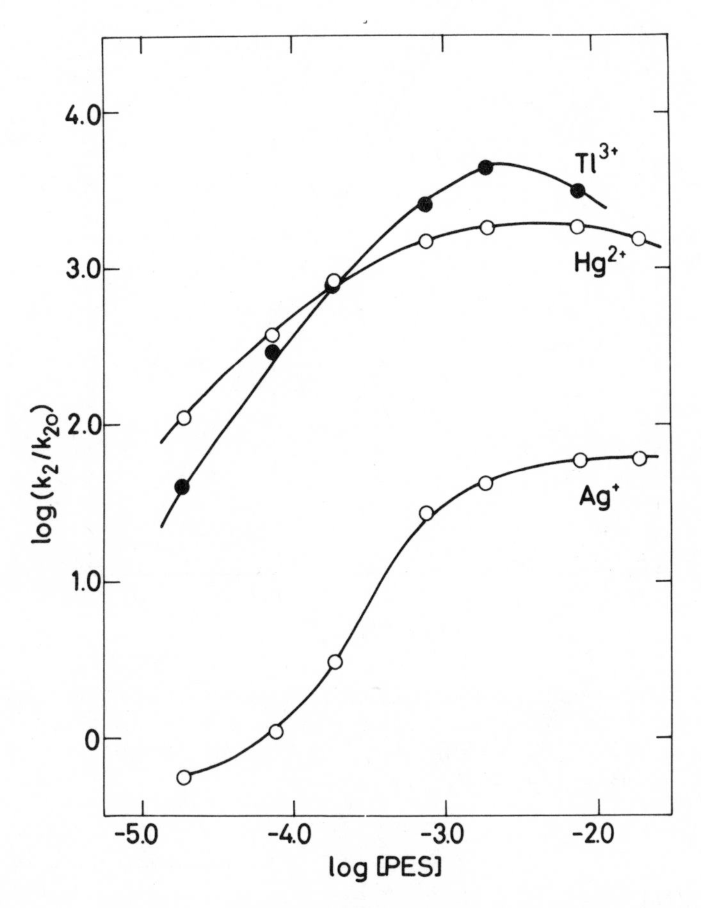

Figure 2. Comparison of the polyelectrolyte influence on the induced aquations at 20°. $[Co(NH_3)_5Br^{2+}] \simeq 6 \times 10^{-5}$ M, [inducing cation] $\simeq 1 \times 10^{-3}$ equiv. l^{-1}, $[HClO_4] \simeq 0.05$N. Taken from Ref. 5.

These reaction systems were accelerated by the cationic polyelectrolytes shown below:

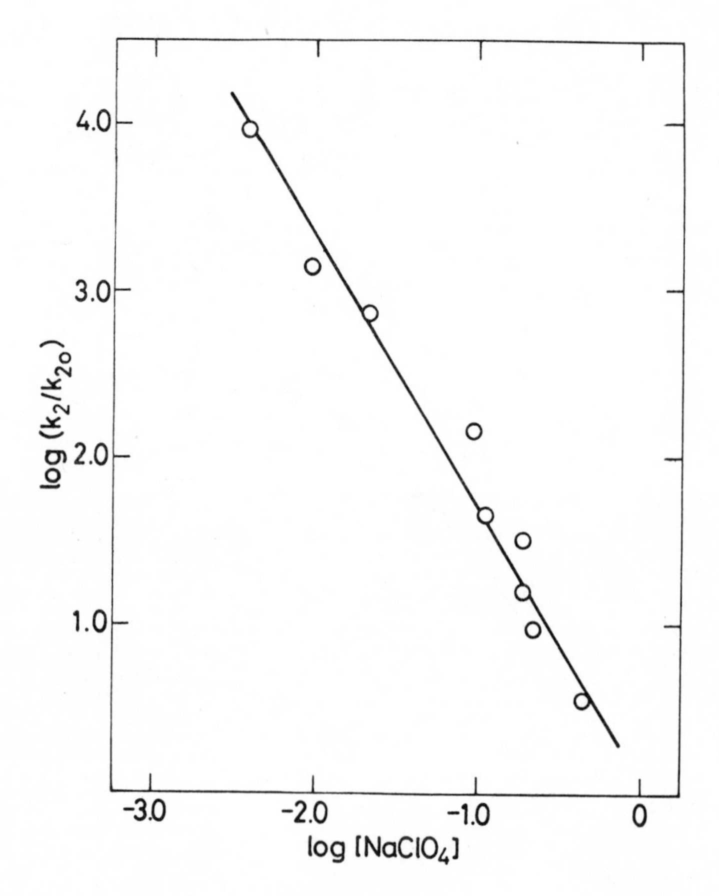

Figure 3. Dependence of the polyelectrolyte influence on simple salt concentration at 30°. $[Co(NH_3)_5Br^{2+}] \simeq 6.7 \times 10^{-5}$ M, $[Hg^{2+}] \simeq 1.3 \times 10^{-4}$ M, $[HClO_4] \simeq 1.9 \times 10^{-3}$ M, $[PSt] \simeq 3.8 \times 10^{-4}$ equiv. l^{-1}. Taken from Ref. 5.

	R_1	R_2
DM	CH_3	CH_3
DE	C_2H_5	C_2H_5
MBz	CH_3	CH_2-ϕ
MNBz	CH_3	CH_2-ϕ-NO_2
NAP	CH_3	CH_2-α napthyl

Reaction (2) was enhanced to about the same extent (the maximum acceleration $\simeq 10^2$) by any of these polycations whereas the acceleration of reactions (3) and (4) became larger with increasing hydrophobicity of the polycations. This result shows the important role of the hydrophobic interactions between the reactant and polymer "catalysts,"* confirming earlier observations by Sakurada.[1]

In the reaction systems under consideration, the hydrophobic interactions accelerated the reactions. The electrostatic interactions are long-ranged whereas the hydrophobic ones are short-ranged. This basic difference is expected to manifest itself in reaction systems in which the effect of steric hindrance is important. In other words, hydrophobic interactions decelerate the reactions. An example is the alkaline fading reaction of phenolphthalein.[8] As seen in Table I, addition of less hydrophobic polyelectrolytes such as DE, C_2PVP, and C_3PVP enhanced the reaction. However, a highly hydrophobic polyelectrolyte, namely C_4PVP, had practically no influence and BzPVP, which is much more hydrophobic, retarded the reaction, or more exactly, stopped it completely at a polyelectrolyte concentration of 1.05×10^{-3} equiv. l^{-1}. It is believed that the attractive interaction between the substrate anions and the polycations is so strong, or that the substrates are bound so tightly by the polycations that the OH^- attack on the anions is hindered.

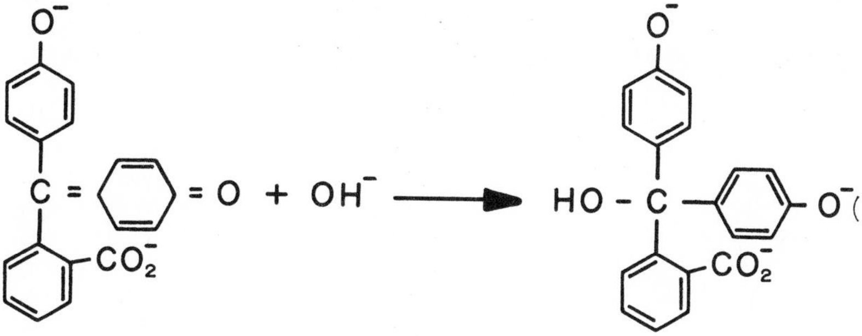

$$C = \langle \rangle = O + OH^- \longrightarrow HO - C - \langle \rangle - O^- \qquad (5)$$

(O⁻ and CO_2^- substituents on the rings, as drawn)

*Klotz and his associates[7] demonstrated that the hydrolysis of 2-hydroxy-5-nitrophenylsulfate can be enhanced by a factor of 10^{12} in the presence of a derivative of polyethylenimine containing dodecyl and methyleneimidazole groups. This is the largest acceleration factor reported for synthetic polymers. Furthermore, it is interesting to note that this polymer was shown to be more effective than a naturally occurring enzyme, arylsulfatase.

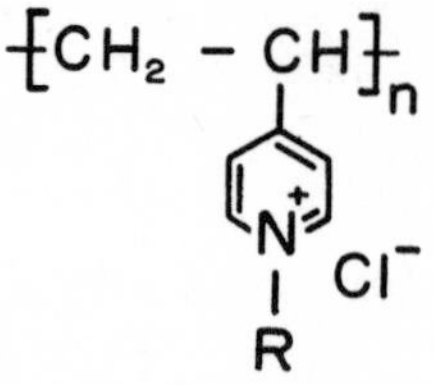

$$\left[CH_2 - CH\right]_n \text{ (4-pyridinium, } N^+\text{–R, } Cl^-)$$

	R
C_2PVP	C_2H_5
C_3PVP	C_3H_7
C_4PVP	C_4H_9
BzPVP	CH_2-ϕ
$C_{16}BzPVP$	$C_{16}H_{33}$ (5%), CH_2-ϕ (95%)

Table I. Polyelectrolyte Influence on the Alkaline Fading Reaction of Phenolphthalein at 25°.[a]

Polyelectrolyte	Concn. equiv. l^{-1}	k_2 M^{-1} sec^{-1}
None	0	0.015
DE	5.26×10^{-5}	0.018
	2.63×10^{-4}	0.024
	1.05×10^{-3}	0.041
C_2PVP	5.26×10^{-5}	0.020
	2.63×10^{-4}	0.037
	1.05×10^{-3}	0.081
C_3PVP	5.26×10^{-5}	0.017
	2.63×10^{-4}	0.021
C_4PVP	5.26×10^{-5}	0.017
	2.63×10^{-4}	0.017
BzPVP	3.26×10^{-5}	0.015
	2.63×10^{-4}	0.0064
	1.05×10^{-3}	0
NaPSt	1.05×10^{-3}	0.015
	7.90×10^{-3}	0.015

[a] [PP] = 4.21×10^{-5} M, [NaOH] = 1.05×10^{-1} M.

Electron-transfer reactions between Co-complexes and various reductants are also subject to the influence of polyelectrolytes.[9] We studied both inner-sphere and outer-sphere electron-transfer reactions in the presence of anionic polyelectrolytes.[10,11,12]

$$Co(III)L_6^{n+} + Ru(NH_3)_6 \rightarrow Ru(NH_3)_6^{3+} + Co^{2+} \quad (6)$$

$$Co(III)L_5X^{n+} + V^{2+} \rightarrow Co^{2+} + 5L + X^- + V^{3+}(III) \quad (7)$$

$$Co(III)L_5X^{n+} + Fe^{2+} \rightarrow Co^{2+} + 5L + X^- + Fe^{3+} \quad (8)$$

L: ligand

The results are shown in Figures 4, 5 and 6. It is seen that the reactions are largely accelerated, even in the presence of a fairly large amount of a foreign salt. The acceleration factor itself does not depend on the ligand whereas the rate constant in the absence of the polyelectrolytes (k_{20}) does vary with the ligand. It is not the ligand but the valency of the CO-complex which determines the acceleration factor. This is in agreement with an earlier observation by Gould,[13] and indicates that the important factor in determining k_2/k_{20} is the electrostatic interaction between the polyelectrolyte and the reactants in the present systems.

In the preceding paragraphs, interionic reactions between similarly charged ionic species were shown to be accelerated by addition of oppositely charged macroions.* The next question is how interionic reactions between oppositely charged ionic species are affected by polyelectrolytes. Thus, we investigated a most simplified case:[14]

$$NH_4^+ + OCN^- \rightleftharpoons (NH_2)_2CO \quad (9)$$

As seen in Table II, the addition of anionic and cationic polyelectrolytes and surfactants retarded the reaction. Simple electrolytes also showed a retarding action, which however is much smaller than that caused by polyelectrolytes.

---

*This statement is correct with the exception of the fading reaction of phenolphthalein. It is to be remembered that, in this system, hydrophobic interactions are involved, in addition to electrostatic forces.

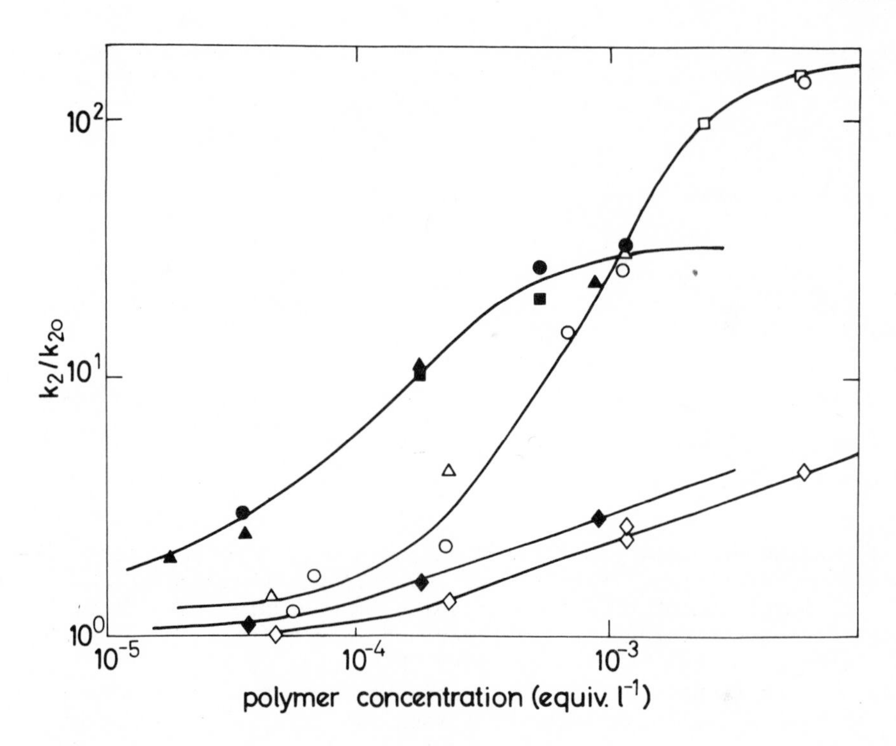

Figure 4. Polyelectrolyte influence on electron-transfer reactions between Co-complexes and $Ru(NH_3)_6^{2+}$ at 25°. [NaCl] = 0.2 M, [HCl] = 2×10^{-4} M, [Co-complex] ≃ 5×10^{-5} M, $[Ru(NH_3)_6^{2+}]$ ≃ 5×10^{-4} M.

PES	PSt	Co-complex	k_{20} (M^{-1} sec^{-1})
○	●	$Co(en)_2pyCl^{2+}$	9.3×10^2
△	▲	$Co(en)_2H_2OCl^{2+}$	3.5×10^2
□	■	$Co(en)_2NH_3Cl$	3.6×10
◇	◆	$Co(en)_2Cl_2^+$	4.4×10^3

Taken from Ref. 10.

In the formation of urea, only electrostatic interactions are involved. Recent work on the alkaline fading reactions of triphenylmethane dyes, which contain hydrophobic groups, revealed much more complicated polyelectrolyte influence.[8] Figures 7 and 8 show the second-order rate constant for ethyl violet and brilliant green. Addition of quaternized polyvinylpyridine and CTABr unexpectedly enhanced the reaction whereas anionic polyelectrolytes and surfactants retarded the reaction. It is apparent that the more hydrophobic the added electrolytes, the larger the acceleration

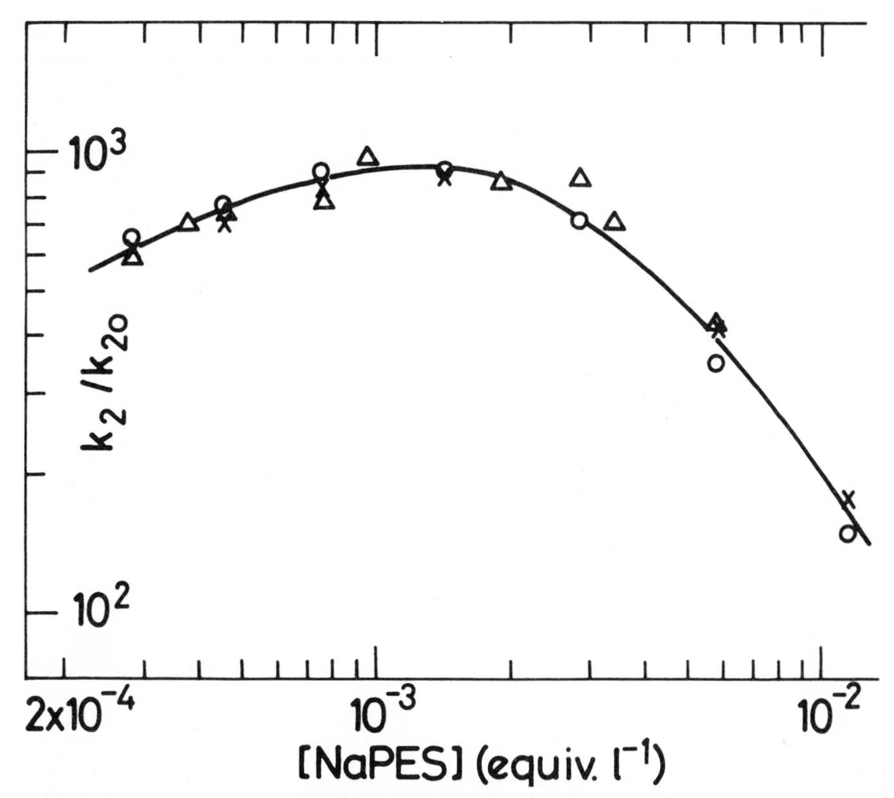

Figure 5. Polyelectrolyte influence on electron-transfer reactions between Co-complexes and Fe^{2+} at 25°. $[Fe^{2+}] \simeq 1.5 \times 10^{-3}$ M, [Co-complex] $\simeq 3 \times 10^{-5}$ M, $[H^+] \simeq 10^{-3}$ M.

O: $Co(NH_3)_5Cl^{2+}$ Δ: $Co(NH_3)_5Br^{2+}$

X: $Co(NH_3)_5N_3^{2+}$

Taken from Ref. 12.

$$R_1-C_6H_4-C(C_6H_4R_2)=C_6H_4=\overset{+}{N}R_3 + OH^- \rightleftharpoons R_1-C_6H_4-C(OH)(C_6H_4R_1)-C_6H_4-NR_3 \quad (10)$$

Dye	R_1	R_2	R_3
Ethyl violet	$N(C_2H_5)_2$	$N(C_2H_5)_2$	$(C_2H_5)_2$
Crystal violet	$N(CH_3)_2$	$N(CH_3)_2$	$(CH_3)_2$
Brilliant green	H	$N(C_2H_5)_2$	$(C_2H_5)_2$
Malachite green	H	$N(CH_3)_2$	$(CH_3)_2$

Table II. Polyelectrolyte Influence on the Ammonium Cyanate-Urea Conversion at 50° [a]

Electrolyte	Conc. (equiv. l^{-1})	$k_2 \times 10^2$ (M^{-1} min^{-1})
None	--	3.24
NaPAA[b]	0.020	3.02
	0.056	2.35
	0.070	2.25
	0.112	1.81
	0.150	1.37
	0.273	1.23
DE	0.100	1.47
NaCMC (0.68)[c]	0.100	2.58
NaCMC (0.96)[c]	0.100	2.14
NaLS[d]	0.100	1.85
CTABr[e]	0.100	2.24
$CaCl_2$	0.111	2.68
NaCl	0.0444	3.24
	0.111	2.85
	0.200	2.46
	0.300	2.19

[a] $[NH_4OCN]$ = 0.1025 M.
[b] Polyacrylate sodium.
[c] Sodium salt of carboxymethylcellulose with the degree of substitution indicated in the brackets.
[d] Sodium lauryl sulfate.
[e] Cetyltrimethylammonium bromide.

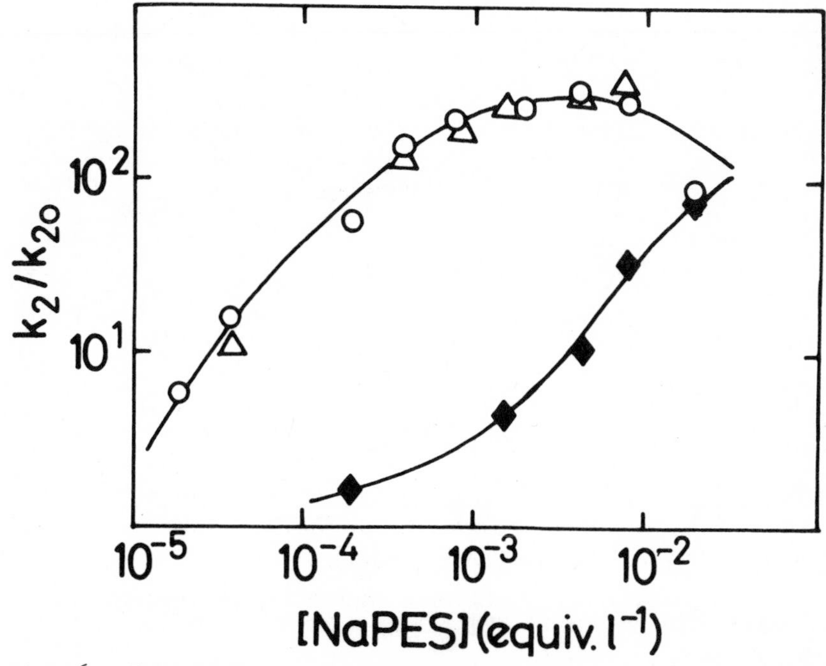

Fig. 6. Polyelectrolyte influence on electron-transfer reactions between Co-complexes and V^{2+} at 25°.

O: $Co(NH_3)_5N_3^{2+}$ Δ: $Co(NH_3)_5Br^{2+}$

♦: $Co(en)_2Cl_2^+$

Taken from Ref. 11.

and the deceleration. The rate enhancement indicates the important role of hydrophobic interactions; in the presence of NaPSt, hydrophobic and Coulombic interactions are at work simultaneously between the dye cations and the macroanions. The macroanions repulse the hydroxyl ions by Coulombic forces. For quaternized polyvinylpyridine, the electrostatic attraction between OH^- and the macrocations, and the hydrophobic attraction between the dye cations and the macrocations enhanced the reaction.

The investigations mentioned above clearly show that interionic reactions can be markedly influenced by polyelectrolyte additions. If we observe an acceleration, there can be two possibilities; (1) The forward reaction step is accelerated more than the backward step and (2) the backward step is appreciably more hindered than the forward step. The experimental data so far accumulated do not allow an answer to be given to this question.

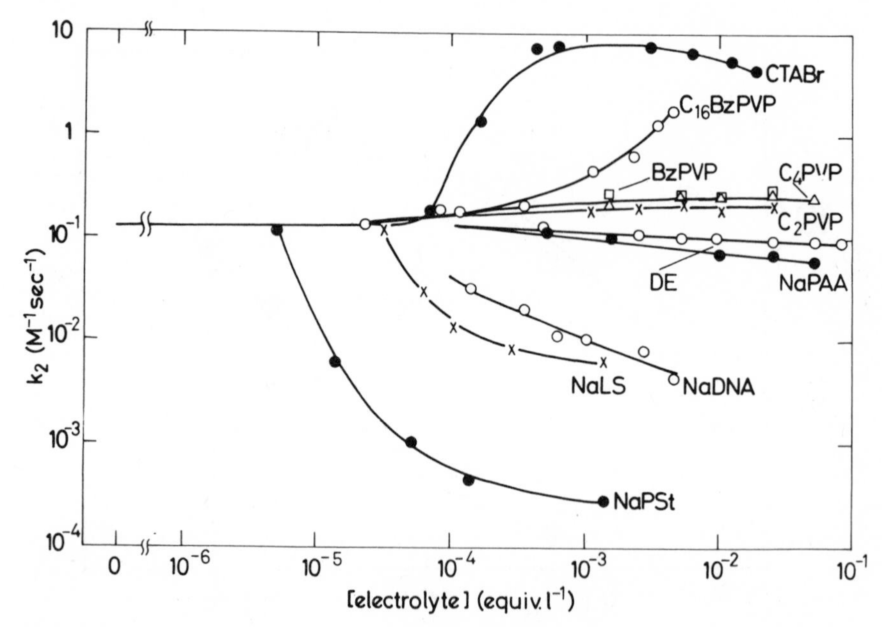

Figure 7. Polyelectrolyte influence on the fading reaction of ethyl violet (EV) at 30°. [EV] = 1.05×10^{-5} M, [OH^-] = 1.05×10^{-2} M. Taken from Ref. 8.

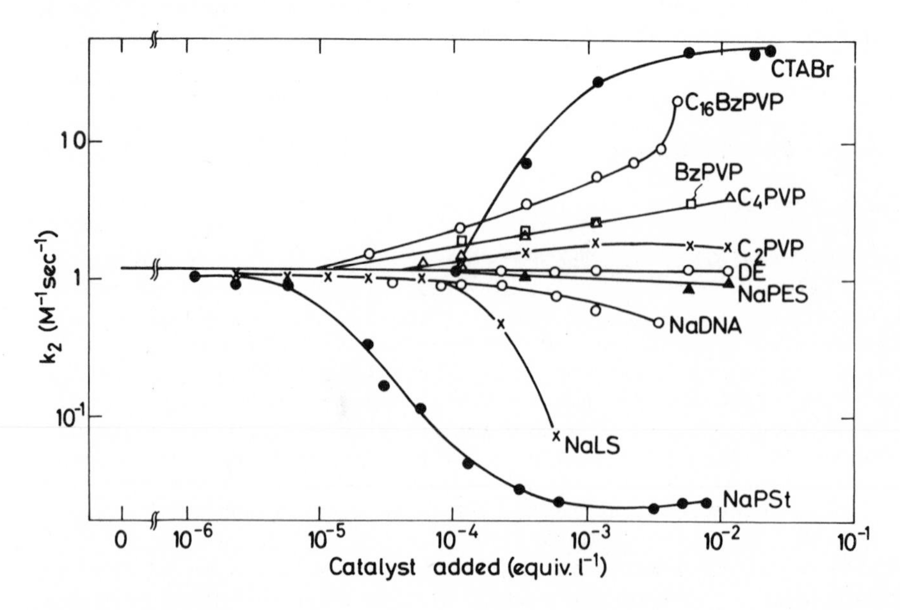

Fig. 8. Polyelectrolyte influence on the fading reaction of briliant green (BG) at 30°. [BG] = 1.25×10^{-5} M, [OH^-] = 3.49×10^{-2} M.

Table III. Interactions between Polyelectrolytes and Reactants.

Reactants \ Polyelectrolyte	Quaternized Polyvinylpyridine	Polystyrene Sulfonate	Polyethylene Sulfonate
OH^-	Coulombic attraction	Coulombic repulsion	Coulombic repulsion
Dye^+	hydrophobic attraction (Coulombic repulsion	hydrophobic and Coulombic attraction	Coulombic attraction
Polyelectrolyte Influence	Acceleration	Deceleration	Small deceleration

It is therefore of interest to study equilibrium reactions in the presence of polyelectrolytes. Thus we studied the reaction:[15,16]

$$\text{(pyridinium-}CH_2CH_2\text{-adenine, } Cl^-\text{)} + CN^- \underset{\overleftarrow{k}}{\overset{\overrightarrow{k}}{\rightleftharpoons}} \text{(4-cyano-1,4-dihydropyridine-}CH_2CH_2\text{-adenine)} \quad (11)$$

Since the forward step is a reaction between oppositely charged ionic species, we may expect deceleration by polyelectrolytes. This was actually the case, as is shown in Fig. 9. The ratio of the second-order rate constants ($\overrightarrow{k}/\overrightarrow{k}_0$) in the presence and absence of polyelectrolytes was smaller than unity. Quite interestingly, the polyelectrolyte influence on the equilibrium constant K or K/K_0, fell on the same curve for each polyelectrolyte. Since the equilibrium constant is equal to $\overrightarrow{k}/\overleftarrow{k}$, where $\overleftarrow{k}$ is the second-order rate constant of the backward step, the fact that $K/K_0 = \overrightarrow{k}/\overrightarrow{k}_0$ simply indicates that the backward step was not influenced at all by the polyelectrolytes. This is most reasonable because the

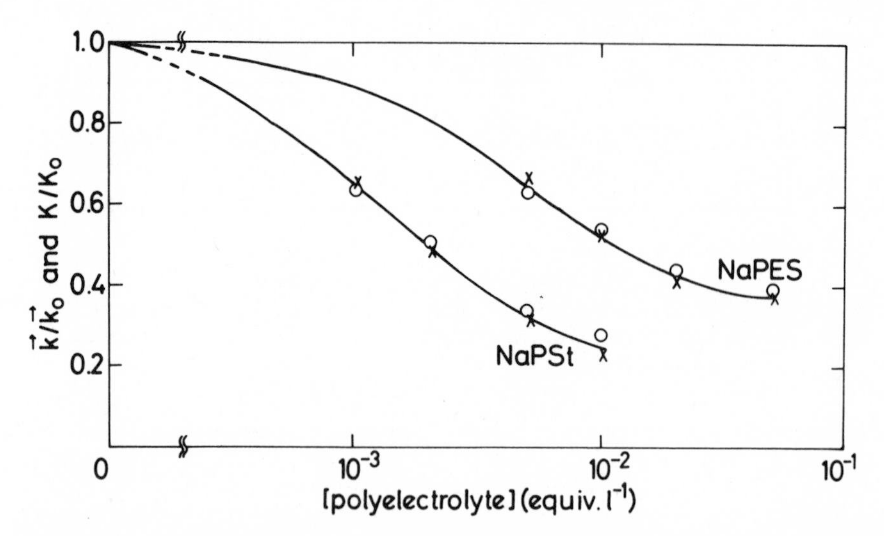

Figure 9. Comparison of the polyelectrolyte influences of NaPES and NaPSt on the forward rate constant with those on the equilibrium constant at 25°. [NAA] = 1.98×10^{-4} M, [KCN] = 4.04×10^{-3} M, [KOH[= 1.0×10^{-3} M.

$$O = \vec{k}/\vec{k}_0 , \qquad X = K/K_0$$

Taken from Ref. 16.

reactants are electrically charged and may be subjected to the strong influence of polyelectrolytes whereas the product is uncharged and cannot be affected to the same extent as the charged reactants.

Different influences on the two elementary processes were also observed for the following reactions.

(12)

$$\mathrm{Ni^{2+} + Murexide^{-}} \underset{\overleftarrow{k}}{\overset{\overrightarrow{k}}{\rightleftharpoons}} \left[\mathrm{O{=}C\langle{}^{NH-C(=O)}_{NH-C(\cdots O)}\rangle C{=}N{=}C\langle{}^{C(=O)-NH}_{C(\cdots O)-NH}\rangle C{=}O \;\; (O\cdots Ni\cdots N\cdots Ni\cdots O)} \right]^{+} \qquad (13)$$

Figures 10 and 11 give the rate constants of the forward and backward steps of reaction (12).[17] Addition of cationic polyelectrolytes increased $\overrightarrow{k}$, as can be expected for reactions between like-charged species. On the other hand, the backward rate constant

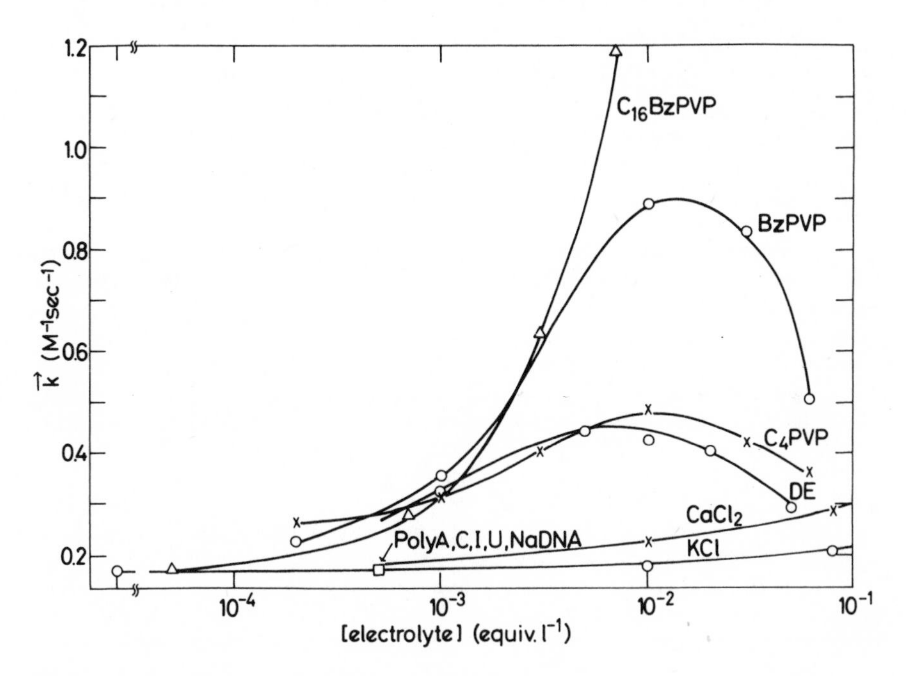

Figure 10. Polyelectrolyte influence on the forward rate constant of the β-NAD^+-CN^- reaction at 25°. [β-NAD^+] ≃ 1.5×10^{-4} M, [KCN] ≃ 7×10^{-3} M, [KOH] ≃ 10^{-3} M.

Taken from Ref. 17.

was decreased except for C_{16}BzPVP. Though the reason for this deceleration is not clear, the two elementary steps were affected by polyelectrolyte in different ways. The forward rate constant ratio k/k_0 of reaction (13) was found to be 0.03 at a polystyrene sulfonate concentration of 5×10^{-3} equiv. l^{-1};[18] in other words, the reaction was retarded. This result is quite reasonable

change by polyelectrolyte addition if only the accumulation of reactants is responsible for the polyelectrolyte acceleration. Any change in $\Delta H^{\neq}$ would suggest that the reaction mechanism itself is modified. Thus the variation of the thermodynamic quantities demonstrated above is evidence showing definitely the inappropriateness of the accumulation effect.

We are not asserting that reactant ions are not accumulated by macroions. As a matter of fact, there exists experimental evidence which shows that a relatively large portion of gegenions are "associated" with macroions.[29] The inhomogeneity in the gegenion distribution is most reasonable, but the same phenomenon can also be expected for simple electrolyte solutions. The difference in the inhomogeneity in polyelectrolyte and simple electrolyte solutions is a matter of degree. According to Debye and Hückel,[30] we have the so-called ionic atmosphere structure; the (local) concentration of gegenions around a central ion is higher than the bulk concentration. Nevertheless, it seems that the solution properties of simple electrolytes, including the primary salt effect, have never been discussed in terms of the local ionic concentrations.

From the foregoing discussion, we conclude that the "accumulation" effect is not an appropriate explanation for the observed polyelectrolyte influence, and, even if so, an additional factor or factors must be introduced to account for the changes in the thermodynamic quantities. Our basic attitude is, and should be, such that simple electrolytes and polyelectrolytes have to be dissected according to a unified point of view, the fundamental difference being that the repeating units of high molecular weight compounds are connected to each other by covalent bonds. In this connection, it should be remembered that physical chemistry shows that interionic reactions can be accelerated or decelerated by an increase in the (bulk) concentrations of simple electrolytes. This phenomenon has been called the primary salt effect. According to Bronsted,[31] we have

$$k_2/k_{20} = f_A \cdot f_B/f_X \quad (15)$$

for a bimolecular reaction $A+B \rightleftarrows X \rightarrow C+D$ (X = the critical complex) where k_2 is the second-order rate constant, the subscript 0 denotes the rate constant at zero ionic strength, and f is the single-ion activity coefficient. Since the f's are calculable by using the Debye-Hückel Theory[30] if only simple electrolyte ions are involved, Eq. (15) can be written as

$$\log_{10}k_2 = \log(kT/h)K^{\neq} + 1.018\ Z_A\ Z_B\ I^{1/2} \quad (16)$$

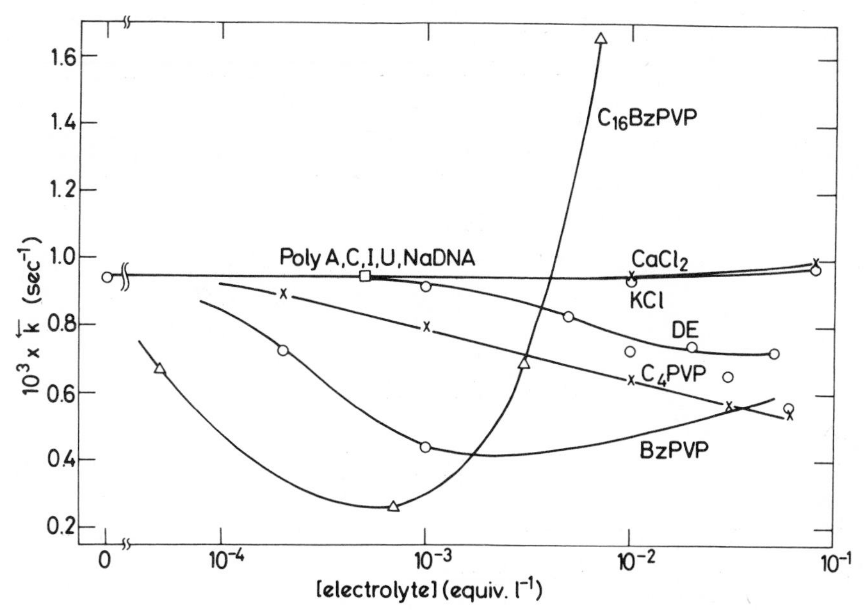

Figure 11. Polyelectrolyte influence on the backward rate constant of the β-NAD$^+$ - CN$^-$ reaction at 25°. [β-NAD$^+$] ≃ 1.5 × 10^{-4}M, [KCN] ≃ 7 × 10^{-3} M, [KOH] ≃ 10^{-3} M.

Taken from Ref. 17.

because the reaction occurs between oppositely charged ionic species. The equilibrium was shifted toward the reactand side; K/K_0 was 0.03, indicating that the backward reaction was not influenced at all.

The elementary steps of all three reactions were undoubtedly affected in different proportions. It follows from this very fact that polyelectrolytes should not be called "catalysts," because a catalyst is a substance which affects the forward and backward reaction processes in the same proportion, according to Ostwald.[19] Since no study has been reported on the "catalytic" effect of polymers in general on the elementary steps, the use of the term "polymer catalysis" is not justified.

Now we turn to the thermodynamic aspect of polyelectrolyte influences. It should be pointed out that the polyelectrolyte acceleration of the reactions between oppositely charged ionic reactions was caused mostly by decreases in the enthalpy and entropy of activation ($\Delta H^{\neq}$ and $\Delta S^{\neq}$). This tendency is in sharp contrast with that[20,21] found for the "catalytic" influence of simple electrolytes, or for the primary salt effect. Table IV clearly shows this contrast for the bromoacetate-thiosulfate reaction. Table V

Table IV. Thermodynamic Quantities of the Bromoacetate-Thiosulfate Reactions (25°).

Electrolytes	Concn. 10^2 equiv. l^{-1}	$\Delta H^{\neq}$ (Kcal mol^{-1})	$\Delta S^{\neq}$ (eu)	$\Delta G^{\neq}$ (Kcal mol^{-1})
		Polyelectrolyte		
---	---	15.7	-15.4	20.3
PEI·HCl[a]	0.6	13.2	-19.3	19.0
	1.2	12.1	-21.5	18.5
		Simple Electrolytes[b]		
----	--	15.3	-15.4	20.3
$BaCl_2$	12	15.6	-14.5	19.9
$LaCl_3$	9	17.5	-6.2	19.3

[a]Hydrochloride of polyethylenimine.
[b]Taken from Ref. 20.

gives thermodynamic data for the aquation reactions of $Co(NH_3)_5Br^{2+}$. Except for the Tl^{3+}-induced aquation reactions, the addition of polyelectrolytes decreased both $\Delta H^{\neq}$ and $\Delta S^{\neq}$, resulting in a decrease in $\Delta G^{\neq}$. Obviously, the polyelectrolyte acceleration is due to a decrease in $\Delta H^{\neq}$ (not due to an increase in $\Delta S^{\neq}$).

We note that the following reaction systems were also accelerated by polyelectrolytes as a result of a decrease in $\Delta H^{\neq}$; hydrolysis of various types of esters,[22,23] solvolysis of p-nitrophenylacetate,[24] spontaneous hydrolysis of dinitrophenylphosphate dianion,[25] aquation reaction of $Co(NH_3)_5Cl^{2+}$ induced by Hg^{2+},[26] outer-sphere electron transfer reactions between $Co(en)_2NH_3Cl^{2+}$ and $Ru(NH_3)_6^{2+}$ (or V^{2+}),[9] outer-sphere electron transfer reactions between Co-complexes and $Ru(NH_3)_6^{2+}$ (or V^{2+}),[10,11] alkaline fading reaction of triphenylmethane dyes,[8] template polymerization of acrylate monomer,[27] substitution reaction of dinitrochlorobenzoate by OH^-,[6] reactions of dithiobisnitrobenzoate with CN^-,[6] and reactions of dinitrofluorobenzene with aniline, dipeptides, and amino acids.[28]

Thus, our generalization is that the polyelectrolyte acceleration[*] is due to decreases in H (or the activation energy) and in $\Delta S^{\neq}$. This seems to be a very important point. It has been

---

[*]An exceptional case will be discussed later.

Table V. Thermodynamic Quantities of the Aquation Reactions of $Co(NH_3)_5Br^{2+}$ (25°).

Electrolytes	Concn.×10^4 equiv. l^{-1}	$\Delta H^{\neq}$ (Kcal mol^{-1})	$\Delta S^{\neq}$ (eu)	$\Delta G^{\neq}$ (Kcal mol^{-1})
	Ag^{2+} (7.5×10^{-4} M)			
---	--	14.4	-15	18.9
NaPSt	0.3	12.3	-19	18.1
	3.0	3.3	-41	15.5
	30.0	12.6	-13	16.6
$NaNO_3$		20.8	+8	18.4
	Hg^{2+} (5×10^{-4} M in $HClO_4$, 0.01N)			
---	---	14.1	-7	16.2
NaPES	0.185	9.5	-12	13.0
	1.85	6.3	-18	11.7
	18.5	4.5	-23	11.4
NaPP	4.00	5.0	-20	11.1
NaPSt	4.00	9.2	-11	12.4
$NaClO_4$	200	14.5	-5	16.0
	Tl^{3+} (10^{-3} M_x in $HClO_4$, 0.01N)			
---	----	10.2	-23	17.1
NaPES	1.85	8.9	-15	13.4
	7.40	9.7	-11	12.8
	14.8	9.9	-9	12.6
	18.5	9.9	-9	12.5

claimed[3] that the addition of polyanions in reactions between cationic species accumulates the reactants around the polyanions so that the collision factor (A) between the reactants should be raied. If this is really true and were the only factor affecting the reaction rate, $\Delta S^{\neq}$ should increase by polyelectrolyte addition, because these quantities are related by the equation

$$A = e\left(\frac{kT}{h}\right)\exp(\Delta S^{\neq}/R) \qquad (14)$$

where k, T, h, and R have the usual meanings. However this was not the case. We, furthermore, point out that $\Delta H^{\neq}$ should not

where for aqueous solutions at 25°, $K^{\neq}$ is the equilibrium constant between X and the reactants, Z is the valency and I is the ionic strength. According to Eq. (16) interionic reactions between similarly charged ionic species can be accelerated when the ionic strength is raised, whereas those between oppositely charged ionic species can be decelerated. These tendencies have actually been found for various reaction systems,[31] and the agreement of the experimental data with Eq. (16) has been most satisfactory.[31]

For polyelectrolytes which brought about much large acceleration for reactwons between similarly charged ionic species and much larger decelerations for those between oppositely charged ionic species, Eq. (16) cannot be applied since the Debye-Hückel Theory is not valid. However, the basic concept proposed by Bronsted, or Eq. (15), can be shown to be valid as seen from Table VI, which demonstrates satisfactory agreement between the polyelectrolyte influence observed and that calculated from the activity coefficient data. This agreement definitely implies that the observed polyelectrolyte influence is basically the primary salt effect.

Table VI. Estimation of Polyelectrolyte Influence on Urea Formation (50°)[a]

[NaPAA] equiv.l^{-1}	$\log(k_2/k_{20})$ obsd.	$\log(k_2/k_{20})$ calcd.
0	0	0
0.056	-0.140	-0.145
0.070	-0.158	-0.174
0.11	-0.253	-0.251
0.22	-0.421	-0.408

[a]For details of the calculation, see Ref. 14.

The calculation of k_2/k_{20} for the formation of urea was possible because the critical complex has no electric charge so that f could be assued to be unity in the presence and absence of the polyelectrolyte. For interionic reactions between similarly charged ionic species, however, k_2/k_{20} cannot be calculated because the critical complex is more highly charged than the reactants and its f_X cannot be estimated because of the lack of a complete theory. Thus, discussion for this reaction type is qualitative; the existing data of the activity coefficients of both polyelectrolytes and simple electrolytes show a general trend that the mean activity coefficient decreases with increasing electrolyte concentration at low concentrations.[32] In other words, we have the following relation between the interaction parameter β_{ij}'s ($= \partial \ln \gamma_i / \partial m_j$, γ_i is the mean activity coefficient of the component i, m_j is the concentration of the component j) for ternary systems containing water (component 1), polyelectrolyte (component 2) and simple electrolyte (component 3). This relation implies that addition of higher valence electrolytes including polyelectrolytes lowers the mean activity coefficient of simple electrolytes. Since the mean activity coefficient is the geometrical average of the single-ion activity coefficient (f), we may expect the same concentration dependence for f_X as for the mean activity coefficient. Thus, polyelectrolyte addition would cause lowering of f_A and f_B and an even larger decrease in f_X because of the higher valance of the critical complex than the reactants. Thus, $f_A \cdot f_B / f_X$ or k_2/k_{20} increases by addition of polyelectrolytes. In other words, acceleration results.

As an exceptional case, we found that the Tl^{3+}-induced aquation reaction of $Co(NH_3)_5Br^{2+}$ is accelerated because of an increase in $\Delta S^{\neq}$. (See Table V.) It is interesting to inquire what causes the difference between Tl^{3+} and Hg^{2+}. As is shown in Table VII, the ionic radius of the trivalent Tl ion is smaller than that of the divalent Hg ion, causing a much higher electrostatic potential at the surface of the former ion. It is quite acceptable that water molecules in the hydration shell are more strongly bound by Tl^{3+} than by Hg^{2+}, as is reflected in the specific exchange rate constant of hydrated water molecules (k_c).[33]

Table VII. Comparison of Hg^{2+} and Tl^{3+}

	ionic radius (A)	k_c (sec^{-1})
Ag^{+}	1.13	-
Hg^{2+}	1.12	~ 10^{+9}
Tl^{3+}	1.05	~ 10^{+5}

The strong hydration is furthermore supported by Tanbe's experiments, which showed that the water molecule in the aquo-complex origination from the hydration shell of Tl^{3+} if this cation was used as an inducing agent whereas it came from solvent for the Hg^{2+}-induced case. On the basis of this fact it is tempting to suggest that the Hg ions can be easily dehydrated in the vicinity of macroions whereas the Tl ions cannot. The dehydration for the Hg^{2+} case should raise the entropy of the reactants. On the other hand, the entropy of the reactants for the Tl^{3+} case would not be changed by polyelectrolyte addition. Once the desolvation of Hg^{2+} takes place, practically no water molecules can, nor must be released from the critical complex whereas the degree of freedom of water molecule or molecules in the critical complex must be increased in the case of Tl^{3+} if the reaction is to proceed. As shown in Fig. 12, therefore, the $\Delta S^{\neq}$ value for the Hg^{2+} case should be lowered by electrolyte addition whereas that for the Tl^{3+} case should increase. This is exactly what was observed.

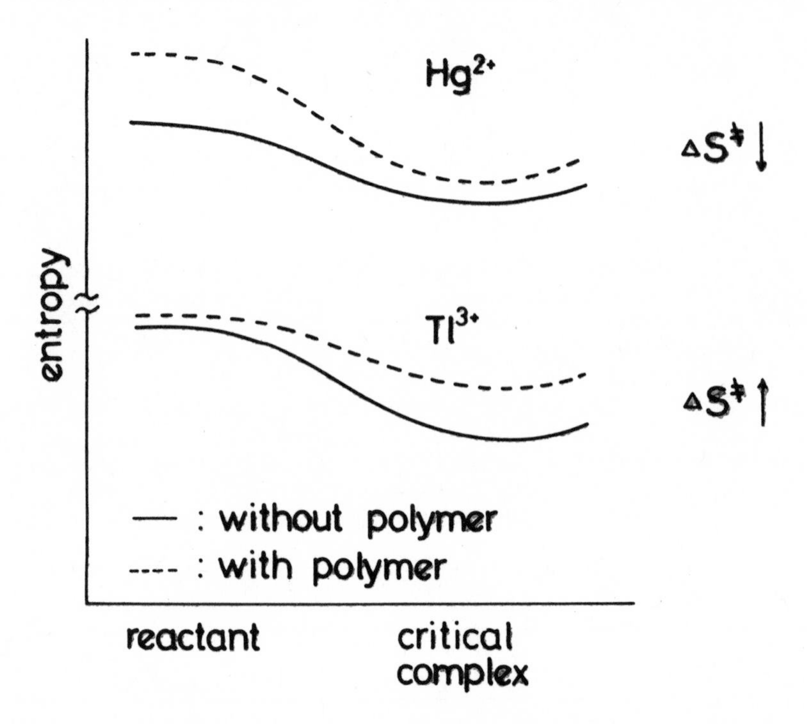

Figure 12. Desolvation and entropy.

If the desolvation of the cations occurs, the closest distance of approach between the cations and macroions should become smaller. Naturally, this change results in stronger interactions, giving rise to smaller values for f_A or f_B and even much smaller values for f_X than in the absence of dehydration.

It would be reasonable to expect $\Delta H^{\neq}$ to decrease if desolvation of the reactant takes place. This is the case for Hg^{2+}. Since Tl^{3+} is not dehydrated, on the other hand, $\Delta H^{\neq}$ is practically not affected, as shown in Table V.

Thus is seems plausible that the Bronsted relation is a useful one for the interpretation of the polyelectrolyte influence on interionic interactions, if combined with the desolvation hypothesis. At present, the hypothesis is not directly substantiated by experiments. Intensive study is in progress in our laboratory to obtain experimental proof.

REFERENCES

1. I. Sakurada, Pure Appl. Chem., 16, 263 (1968).
2. C. Overberger and J. C. Salamone, Accounts Chem. Res., 2, 217 (1969).
3. H. Morawetz, Accounts Chem. Res., 3, 354 (1970).
4. N. Ise, Adv. Polymer Sci., 7, 536 (1970-1971).
5. N. Ise and Y. Matsude, J.C.S. Faraday Trans. I, 69, 99 (1973).
6. T. Ueda, S. Harada, and N. Ise, publication in preparation.
7. H. C. Kiefer, W. I. Congdon, I. S. Scarpa and I. M. Klotz, Proc. Natl. Acad. Sci. U.S., 68, 263 (1971).
8. T. Okubo and N. Ise, J. Amer. Chem. Soc., 95, 2293 (1973).
9. R. C. Patel, G. Atkinson, and E. Baumgartner, Bioinorganic Chem. in press.
10. S. Kungi and N. Ise, to be published.
11. C. Kim and N. Ise, to be published.
12. M. Shikata and N. Ise, to be published.
13. E. S. Gould, J. Amer. Chem. Soc., 92, 6797 (1970).
14. T. Okuko and N. Ise, Proc. Roy. Soc., A327, 413 (1972).
15. N. Ise and T. Okubo, Nature, 242, 605 (1973).
16. T. Okubo and N. Ise, J. Amer. Chem. Soc., 95, 4031 (1973).
17. T. Okubo and N. Ise, publication in preparation.
18. S. Kunugi and N. Ise, publication in preparation.
19. W. Ostwald, Phys. Z., 3, 313 (1902).
20. V. K. LaMer and M. E. Kamner, J. Amer. Chem. Soc., 57, 2662 (1935).
21. R. C. Patel, Ph.D. Thesis, Boston University, 1969.
22. I. Sakurada, Y. Sakaguchi, T. Ono, and T. Ueda. Chem. High Polymers, Japan, 22, 696 (1965).
23. K. Suzuki and Y. Taniguchi, presented at the 21st Annual Meeting of the Society of Polymer Science, Japan, Osaka (1972).
24. C. G. Overberger, T. St. Pierre, C. Yaroslavsky, and S. Yavoslavsky, J. Amer. Chem. Soc., 88, 1184 (1966).
25. T. Ueda, S. Harada, and N. Ise, Polymer J., 3, 476 (1972).
26. H. Morawetz and B. Vogel, J. Amer. Chem. Soc., 91, 563 (1969).
27. J. Furguson and S. A. O. Shah, Europ. Polymer J., 4, 343 (1968).
28. T. Ueda, S. Harada, and N. Ise, publication in preparation.
29. J. R. Huizenga, P. H. Grieger, and F. T. Wall, J. Amer. Chem. Soc., 72, 2636 (1950).
30. P. J. Debye and E. Hückel, Physik. Z., 24, 185 (1923).

31. J. N. Bronsted, Z. physik. Chem., 103, 169 (1922).
32. See, for example, W. J. Moore, "Physical Chemistry," Prentice-Hall, Inc., Englewood Cliffs, N.J., Third Edition, p. 369.
33. For concentration dependence of the mean activity coefficients of polyelectrolytes, see for example Ref. 4.
34. M. Eigen, Pure Appl. Chem., 6, 105 (1963).

DISCUSSION SESSION

Discussion Leader - J.A.Moore

Moore: Most of your data are reported for Na^+ as the gegenion. Would there be any difference in your results if the gegenion were Cs^+?

Ise: No. As far as our results are concerned, we observed no differences among gegenions. We tried only Na^+,K^+,Cs^+.

Salamone: Do you believe that the primary salt effect is responsible for all the calytic effects in polyelectrolytes?

Ise: Yes, to a first approximation, I believe that the primary salt effect is responsible for the observed accelerations and or decelerations. However, I don't think it is almighty.

Salamone: You don't think that the accumulation process plays a role?

Ise: Some sort of accumulation can take place, but it is not the only factor. If it were, then one should not observe any change in ΔH.

Salamone: In one of your slides the rate of reaction between two inorganic species went through a maximum as the concentration of added polyelectrolyte was increased. Prof. Morawetz has interpreted this as being due to the accumulation effect.

Ise: Our data for the activity coefficients indicate that changes in these values cause the changes in the observed rates.

Moore: Prof. Morawetz has observed that self quenching of the fluorescence of the uranyl ion, in the presence of a polyelectrolyte, occurs at concentrations of the uranyl ion much lower than in the absence of the polyelectrolyte. This is indicative of the accumulation of ions on the polymer. It seems, therefore, that both accumulation and activity effects can be operative.

Ise: Yes, of course. Some sort of accumulation should take place. What I mean is that accumulation does not necessarily cause an increase in the collision frequency of the reacting species.

Moore: That is true only when the reacting species are very tightly bound to the macroion.

Ise: Yes. In those cases the decrease in ΔH is responsible for the acceleration.

Smets: I am interested in the "fading reaction". When you use a polyanion, you see a decrease in rate?

Ise: Yes. We observed a deceleration.

Smets: The addition of a polyanion to positivily charges dyes usually causes a appreciable shift in λ_{max} because of association. How did you follow the fading reaction?

Ise: We used a stopped-flow technique and also followed the complete spectrum using a rapid scan technique.

Bamford: In the work we did on the polymerization of sodium acrylate in the presence of poly(ethyleneimine) the rate of polymerization went through a maximum and became quite low in the presence of excess imine. Do you think that the primary salt effect is the controlling factor there?

Ise: Yes, but unfortunally, the activity coefficient of the acrylate ion was not measured in that system so I am not positive.

St. Pierre: Is it possible, in your system, to demonstrate competitive inhibition by adding extraneous salts?

Ise: Yes. If you add simple, indifferent salts. In that case the acceleration factor itself becomes smaller with increasing concentration of salt, because of its effect on the activity coefficient.

Salamone: Are saturation phenomena consistent with the theory of the primary salt effect?

Ise: I don't know if such effects have been observed in the case of polyectrolyte catalysis.

Salamone: They are very well known in micellar catalysis.

Ise: That is a very different system because the micelles are in a state of flux, constantly being destroyed and reformed.

Moore: If such a saturation effect were observed in a polyelectrolyte system, that would be difficult to reconcile with the theory of the primary salt effect.

Ise: Before we can say that, the activity data must be accumulated for the system.

Klesper: How do you measure the activity coefficients?

Ise: Electrochemical methods, i.e. a concentration cell. The second in the isopiestic method in which the vapor pressure of the solvent is measured and that of the solute is caculated from the Gibbs- Duhem equation.

Klesper: How do the experimental values compare with the calculated values?

Ise: Quite unfortunately, we have no complete theory for polyelectrolyte solutions which would permit the necessary calculations.

POLYMERS AS CATALYSTS FOR POLYMERIZATION REACTIONS

C.H. Bamford
Department of Inorganic, Physical and Industrial Chemistry, University of Liverpool, England.

I should like to make a few remarks about our work on polymers as catalysts for processes of different types from those mentioned by Prof. Overberger, namely polymerization reactions. As far as I am aware, the first example was discovered by Ballard and myself in 1956[1] and involved the polymerization of N-carboxy-α-amino acid anhydrides (NCAs(I)). These monomers polymerize in the presence of bases with elimination of CO_2 and formation of poly α-amino acids; the reactions initiated by primary and secondary amines mainly follow the route shown in equation (1), although there are other processes occurring which need not concern us here.

$$R'NH_2 + \underset{\text{(I)}}{\begin{array}{l} R^2HC - C{=}O \\ \quad | \qquad\;\; O \\ \;\; HN - C{=}O \end{array}} \longrightarrow R'NH\text{-}COCHR^2NH_2 + CO_2 \qquad (1)$$

$$R'NH(COCHR^2NH)_nH + (I) \longrightarrow R'NH(COCHR^2NH)_{n+1}H + CO_2$$

When the secondary base polysarcosine dimethylamide (II) is used as initiator for the polymerization of DL-plenylalanine NCA ($R^2{=}\phi CH_2$) in nitrobenzene solution at 25°, polymerization is much faster than expected (e.g. 100times) and has unusual kinetic features.

$$Me_2N(COCH_2NMe)_nH$$

(II)

Thus the rate of polymerization increases with the degree of polymerization of the initiator (n) up to n=20, approximately (Fig.1). At high concentrations of NCA the rate tends to become independent of the concentration of NCA. The reaction is retarded by addition of poly (DL-phenylalanine). Observation[1] shows that polysarcosine chains are able to catalyse the polymerization, but subsequent investigations revealed that this is only possible when the initiating base groups are attached to the ends of the chains. These and other observations led to the conclusion that the NCA molecules become "adsorbed" on to the polymer chains by virtue of hydrogen-bond formation and are then favorably situated for attack by the terminal base group, equation (2)

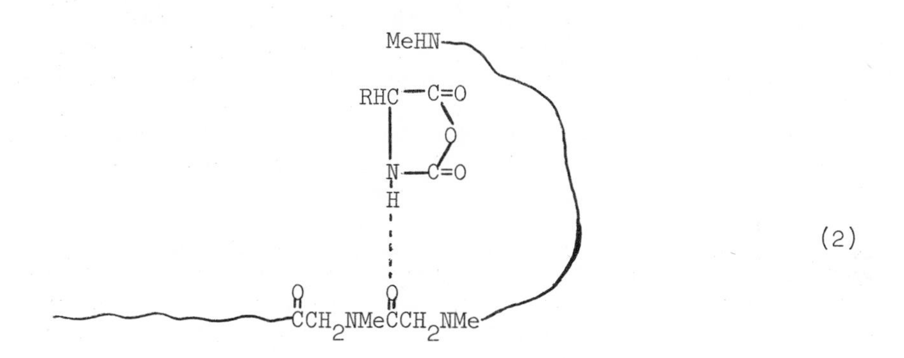

(2)

Subsequently, hydrogen-bonding of this type was confirmed by in-infrared observations[2]. A simple kinetic scheme based on these ideas was shown to account for the experimental results reasonably well. Kinetic analysis allowed the velocity coefficient k_2 for attack of an absorbed NCA molecule by the terminal base group to be estimated as a function of its position,i; the latter is the number of adsorption sites obtained by counting along the polysarcosine chain from the terminal base. It appears (Fig.2) that k_2 peaks rather sharply near n=7, a result which must reflect the the conformational properties of the polysarcosine chains carrying adsorbed NCA molecules under the experimental conditions. We called the phenomenon responsible for these catalytic properties, the "chain effect".

The chain effect is clearly selective in the sense that it depends on the presence in the NCA molecule of an unsubstituted NH group. It would therefore be anticipated that in the poly sarcosine -initiated polymerization of a mixture of N-substituted and unsubstituted NCAs, the latter would polymerize rapidly by the chain effect mechanism, while the former would subsequently react in conventional fashion. The resulting copolymer should therefore have some block-like character. This was tested[3] by examining the copolymerization of sarcosine and γ ethyl L glutamate NCAs (in the latter R^2=CH_2-CH_2-CO_2Et) initiated by polysarcosine dimethylamide. Measurments of the optical rotatory dispersion of the final copolymer were consistent with the view that about 30% of the glutamate residues were in the α helical conformation, which could only occur if a block of these residues occurred in the copolymer. Structures of the type shown in III must therefore be present in the latter

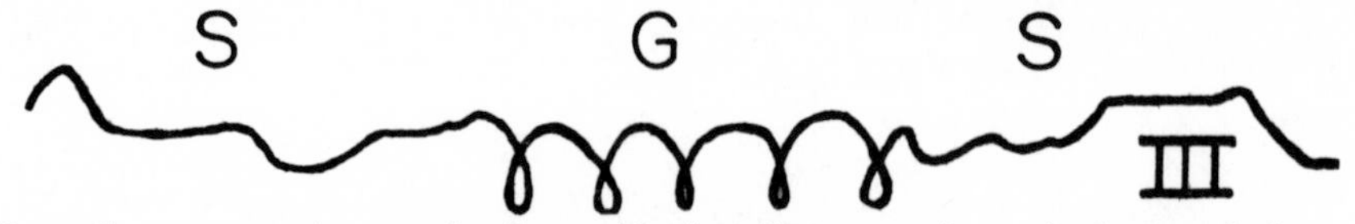

Here S represents randomly coiled polysarcosine chains and G,α helical polyglutamate chains. Copolymers of the same overall composition prepared from mixtures of the NCAs by initiation with a non-polymeric base gave no indications of block character. Evidently the chain effect is capable of introducing some degree of order into copolymers prepared from monomer mixtures.

We have also observed related phenomena in free-radical polymerization[4]. The reaction studied was the polymerization of acrylic acid in the presence of polyethylenimine in acetone-water solution. Fig.3 shows the initial rate of photoinitiated polymerization of acrylic acid (M) at 25^0as a function of the total base molar concentration of polyethylenimine [T]o. [M]o was constant at 0.1 mole/l. At this low concentration no significant rate of polymerization is observed with [T]o=0. However, with increasing [T]o the rate rapidly increases,passing through a maximum near the point at which [T]o/[M]o=1. The maximum rate in Fig.3 is high, corresponding to 30% conversion per minute showing that the presence of polyethylenimine can bring about a great enhancement in rate. These observations are consistent with a "template" polymerization, the monomer being bound to the polyethylenimine chains. (the template) and there polymerized. We believe that the high rates arise from high local concentrations of adsorbed monomer. At low [T]o the template chains are effectively saturated with monomer, so that the rate increases with increasing [T]o. On the other hand, for [T]o/[M]o>1, there is insufficient monomer to maintain saturation and the continuity of monomer on the template is interrupted. Further, the presence of a non-polymerizable acid, which completes with acrylic acid for the template sites,reduces the rate of polymerization. Tetraethylene pentamine has little, if

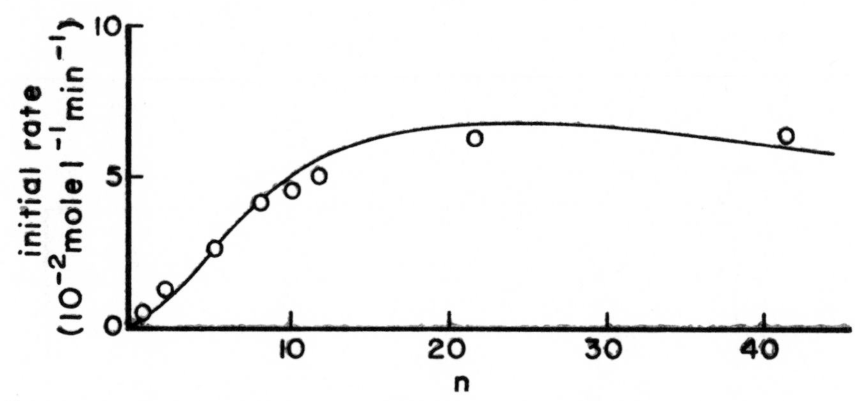

Fig. 1. Polymerization of DL-phenylalanine NCA initiated by polysarcosine dimethylamides with different degrees of polymerization. Nitrobenzene solution; temperature 15^0; $[M]_o$=0.100 mole /l., [II]=5.4x10^{-3}mole/l.;___, theoretical curve.

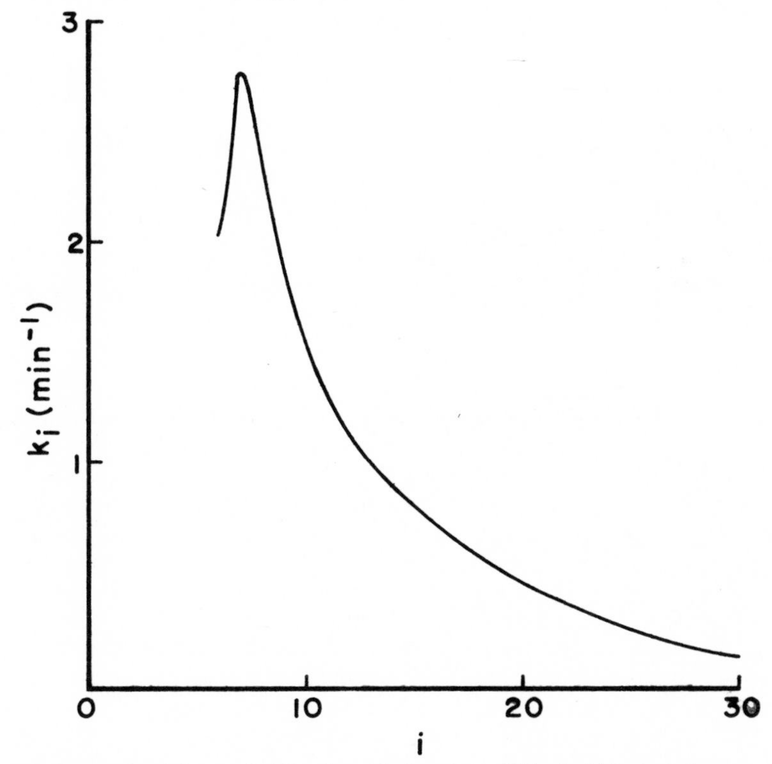

Fig. 2. Polymerization of DL-phenylalanine NCA initiated by polysarcosine dimethylamide. Nitrobenzene solution; temperature 15^0. The velocity constant k_i is plotted as a function of the position i of the sarcosine residue in the polysarcosine chain.

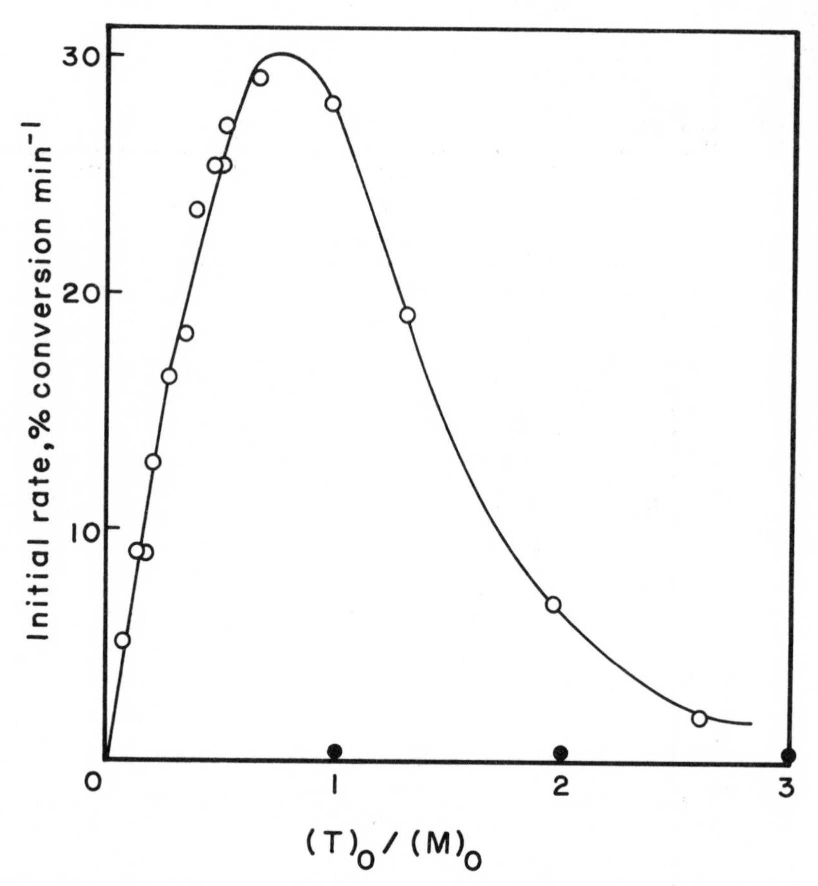

Fig. 3. _O Dependence of initial rate of polymerization on base molar concentration of polyethylenimine $[T]_o$ at 25^0. _● Reaction in the presence of tetraethylene pentamine. ($[T]_o$=0). Acrylic acid concentration $[M]_o$=0 1 mole/l. ABIN ($6x1o^{-3}$ mole/l. as photosensitizer;λ in the range3650-3663Å.

any, catalytic activity (Fig.3). The effects described are independent of the type of free-radical initiator. The systems were homogeneous throughout the reaction.

In the experiments described, the monomer was bound to the template by polar forces. Recently[5] we have examined a species of template polymerization involving covalently bound monomer. The system was based on polyvinylmethacrylate, IV, prepared by reaction of polyvinyl alcohol with methacrylyl chloride; the product was not completely substituted.

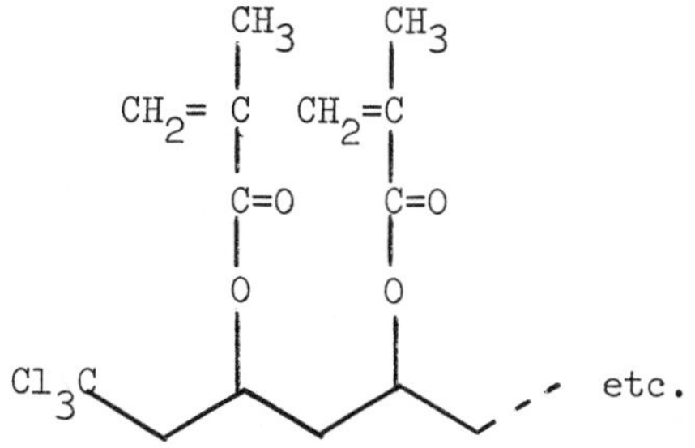

$$\begin{array}{ccccc} & CH_3 & & CH_3 & \\ & | & & | & \\ CH_2= & C & CH_2= & C & \\ & | & & | & \\ & C=O & & C=O & \\ & | & & | & \\ & O & & O & \\ & | & & | & \\ Cl_3C\diagdown\diagup & & \diagdown\diagup & & \diagdown \cdots \text{ etc.} \end{array}$$

(IV)

A terminal $-CCl_3$ group was provided to ensure initiation from one end of the monomer sequence; photochemical initiation was effected (λ=436nm) with the aid of manganese carbonyl, $Mn_2(CO)_{10}$[6]. Measurements in ethyl acetate solution clearly revealed enhanced rates from template action. We endeavored to determine the rate of propagation along the template chains by the rotating sector technique. It appears that the mean half-life of a propagating radical has a minimum value of 1 ms.

These studies demonstrate the catalytic activity of polymers in different types of polymerization processes and they hold out the hope that , with sufficient ingenuity, it may be possible to synthesize polymers with desired structural features by template reactions. However, much work will be required before that aim is achieved.

REFERENCES

1. D.G.H.Ballard and C.H.Bamford, Proc. Roy. Soc. 236A 384 (1956)

2. C.H.Bamford and R.C. Price, Trans. Faraday Soc. 61 2208 (1965)

3. C.H.Bamford, H. Block and Y. Imanishi, Biopolymers,4 1067 (1966)

4. C.H.Bamford and Z. Shiiki, Polymer 9, 596 (1968)

5. C.H.Bamford and A. Wrobel, unpublished observation

6. C.H.Bamford, P.A. Crowe and R.P.Wayne, Proc. Roy. Soc. 284A, 455 (1965)

DISCUSSION SESSION

Discussion Leader- J. A. Moore

Challa: A stereoregulating effect exerted by polymer chains rather than a kinetic effect is found for the stereospecific replica polymerization of MMA on tactic PMMA[1,2,3]: The driving force is the tendency of isotactic and syndiotactic PMMA to form stable stereocomplexes.

1. G. Challa, J. Polymer Sci. A1, 10, 1031 (1972)

2. ibid. 11, 989, 1003,1013 (1973)

3. G. Challa, Polymer 14,171 (1973)

Moore: The ability to initiate chains selectively at one end, rather than randomly along the chain, is a prime requisite for the formation of a perfect ladder polymer. This represents a very significant advance in the control of polymerization processes.

THE REACTION OF POLYVINYLAMINE WITH p-NITROPHENYL ACETATE

T. St. Pierre, G. Vigee and A. R. Hughes

Chemistry Department
University of Alabama in Birmingham
Birmingham, Alabama 35293

We have heard Professor Overberger describe the factors which account for the enhanced reactivity of polymers. For instance, the imidazole catalyzed hydrolysis of esters is more efficient for the imidazole group of poly (4(5)-vinylimidazole) than imidazole itself when, in the case of the polymer, favorable electrostatic, hydrophobic or catalytic factors are operative.

I will describe a polymer reaction where two of these factors should reasonably lead to enhanced reactivity but, in fact, do not. In the reaction of simple amines (A) with esters (E), the rate of phenol (P) formation is given by

$$d[P]/dt = k_o + k_a [A][E] + k_{ac} [A]^2[E] + k_{bc} [OH^-][A][E].$$

The term which is second order in amine concentration makes a significant contribution to the overall rate. In the catalyzed hydrolysis of esters with imidazole there is little or no contribution of the corresponding term. The aminolysis of p-nitrophenyl acetate by polyvinylamine should maximize the amine catalyzed aminolysis term. Polyvinylamine was chosen also for its unique titration behavior which has been suggested is due to amine-amine interaction.

The synthesis of polyvinylamine outlined by R. Hart[1] was employed in this work. We are presently exploring alternatives to this multistep synthesis. M. Sugiura[2] has reported the synthesis of polyvinylamine from polyacrylamide via the hofmann rearrangement but we have not been able to duplicate this synthesis in our laboratory. We prefer, in principle, the Schmidt route to polyvinylamine starting with polyacrylic acid. Under the acidic conditions of this reaction the amine function, as it is formed, is protonated which cancels its ability to react as a nucleophile in side reactions and also stabilizes it against oxidation. Furthermore, since it is possible to prepare stereoregular polyacrylic acid this synthesis could lead eventually to stereoregular polyvinylamine. However, our best conversion of acid function to amine function is about 80%. This does not compare favorably with the 94% conversion of amide to amine reported for the Hofmann reaction.

The reaction of polyvinylamine with p-nitrophenyl acetate was studied under pseudo first order reaction conditions, $[PVA]_o \gg [E]_o$, at constant ionic strength, adjusted 1.0 M with KCl, and at constant pH. The reaction was found to be first order in polyvinylamine and the rate increases with increasing pH. A plot of the observed rate constant, corrected for the hydroxide reaction, against the fraction of amine (α) present as the free base gives a second order rate constant of 4.4 ± 0.3 M min^{-1}. The plot was based on a kinetic pK_a of 9.7. There was no evidence for amine catalyzed aminolysis in the polyvinylamine reaction and the rate constant was less than you would expect for a simple amine of comparable basicity.

We then examined the possibility of introducing a favorable hydrophobic interaction between the polymer and ester. This was accomplished by promoting a tight coil conformation in the process of forming Ni^{2+} complexes with the amine functions of the polymer. The coiling effect has been demonstrated in the viscosity work of P. Teyssie[3]. Our UV spectral data and titration data show the dependence of complex formation on pH. There can be little doubt that even at low Ni^{2+} concentration that there is chelation which promotes coil formation.

The rate of reaction of polyvinylamine-nickel complex with p-nitrophenyl acetate deviates from first order kinetics with increasing polymer complex concentration.

We assumed a preequilibrium between polymer complex and ester which then reacts to give product. This data was accommodated by the standard saturation kinetics treatment. After applying the correction for the fraction of amine-free base at each of the pH's where data was collected, we calculate an association constant and first order rate constant of 40 M^{-1} and 0.65 $min.^{-1}$ respectively. This does not represent a significant rate enhancement.

While the ester is attracted to the polymer it seems that the polymer domain is not a favorable reaction environment. It is possible that the ester is buried in the hydrophobic portions

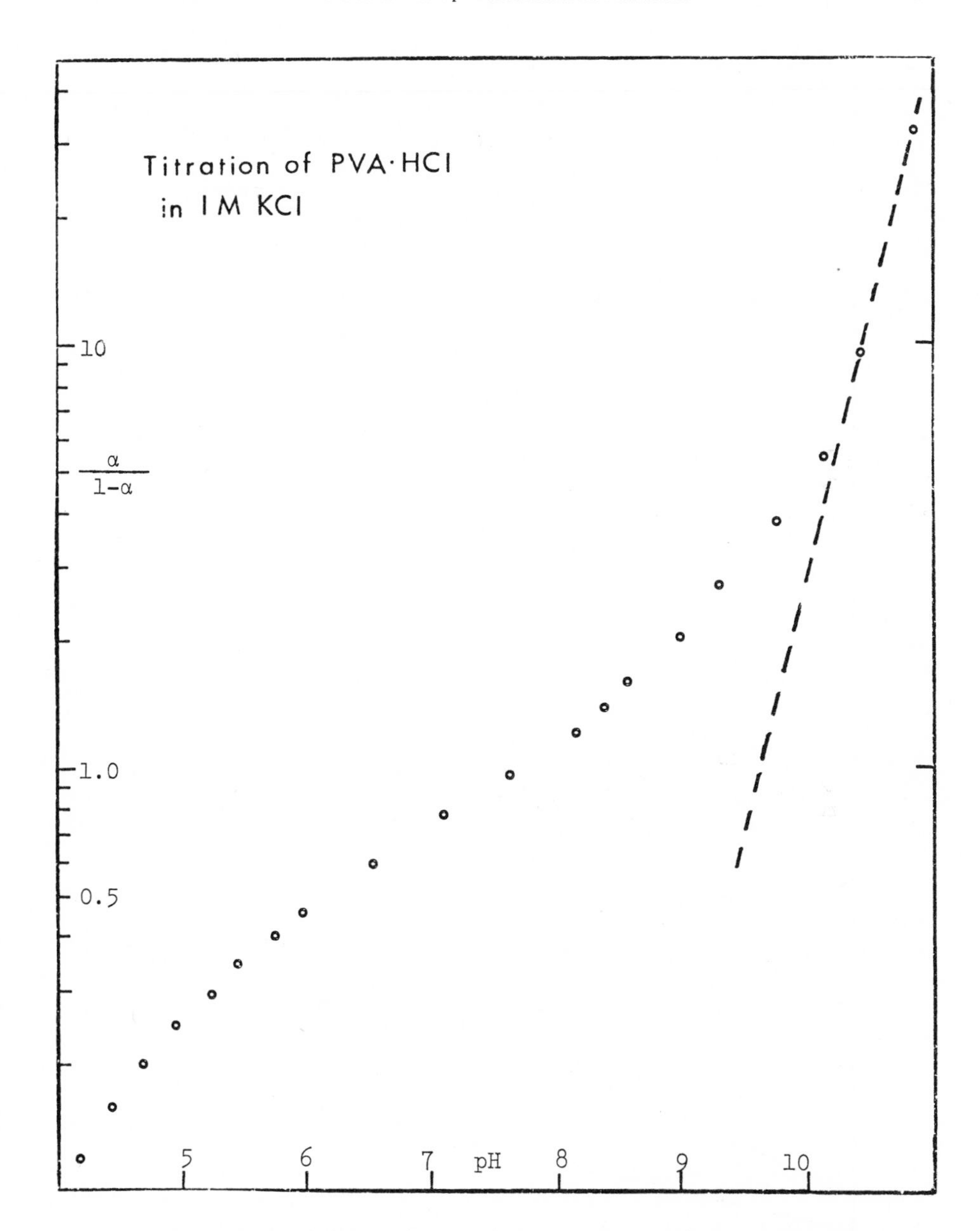

Fig. 1. The titration of polyvinylamine hydrochloride with KOH in IN KCl. The fraction of amine free base is defined as α. The straight line is for reference only and has a slope of one.

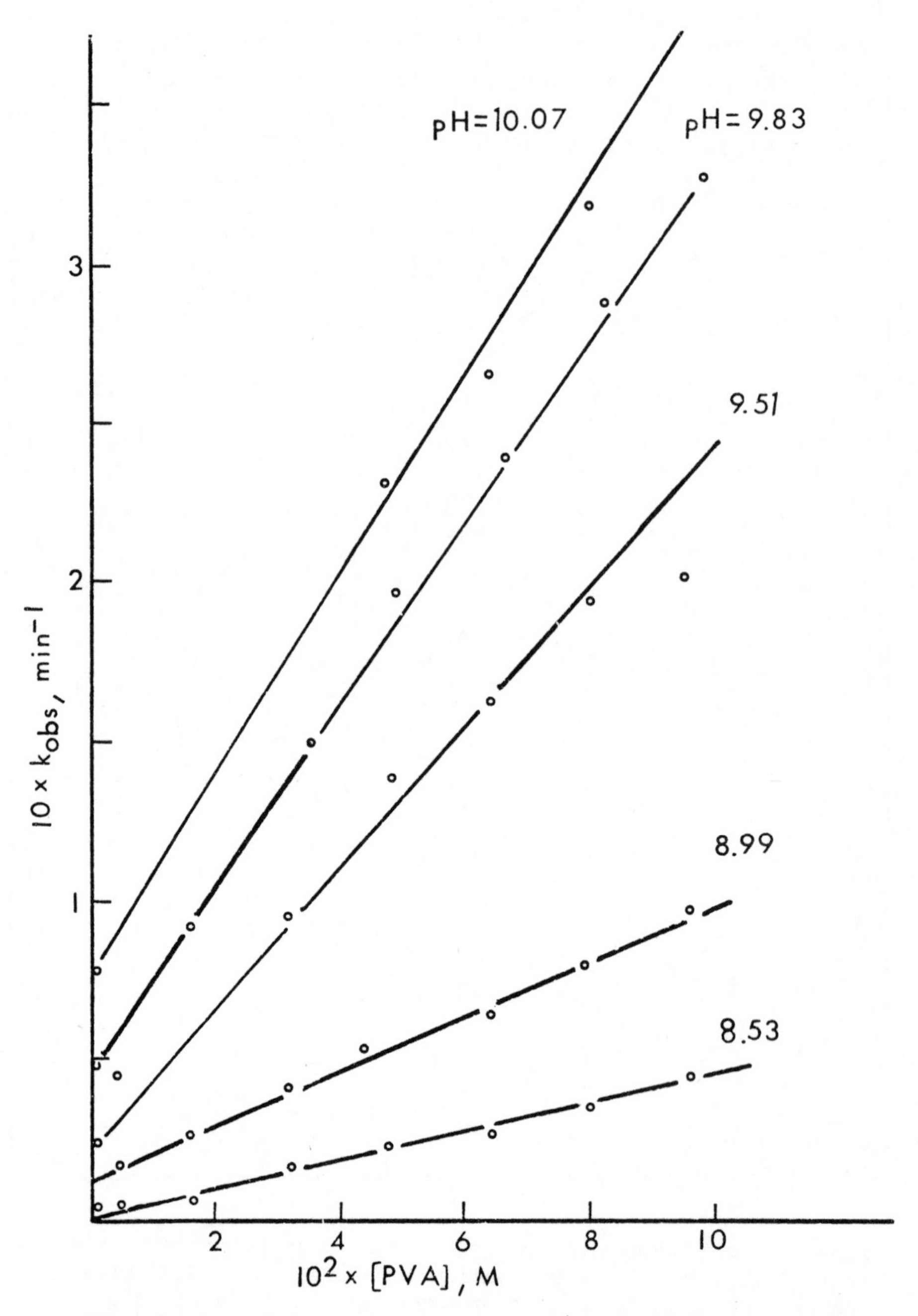

Fig. 2. The observed rate constant for the reaction of PVA and p-nitrophenyl acetate as a function of total PVA concentration and pH.

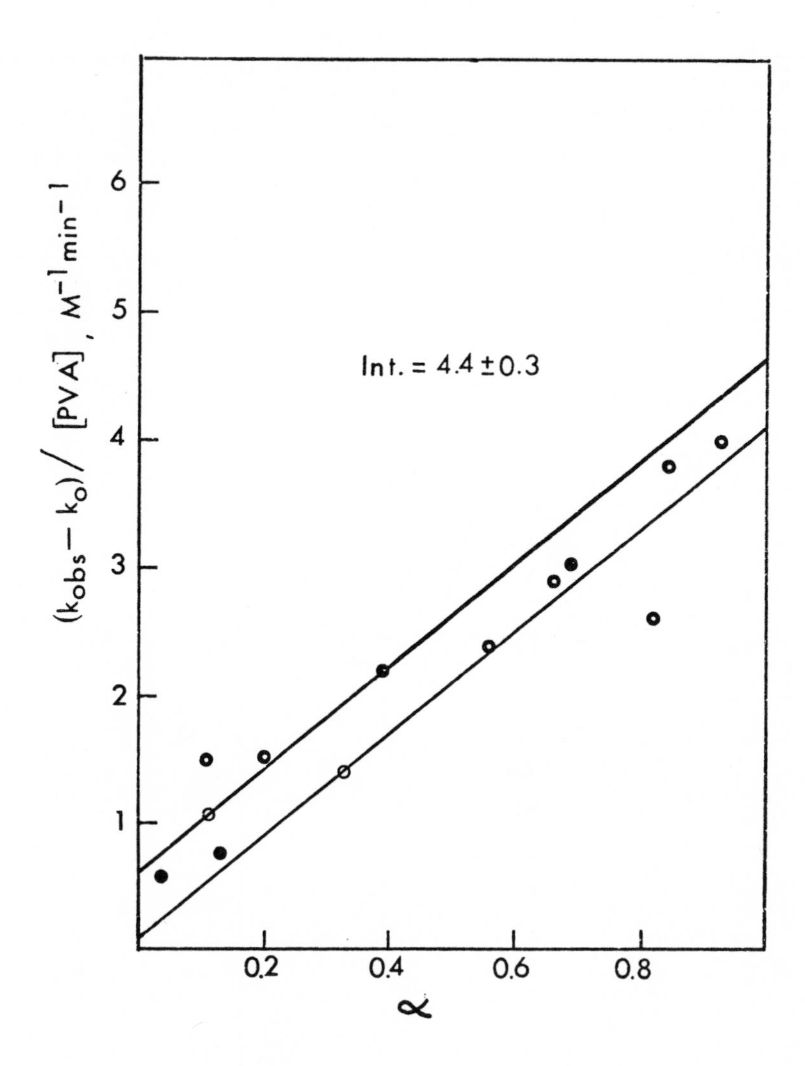

Fig. 3. The corrected apparent second order rate constant as a function of the fraction of amine present as the free base.

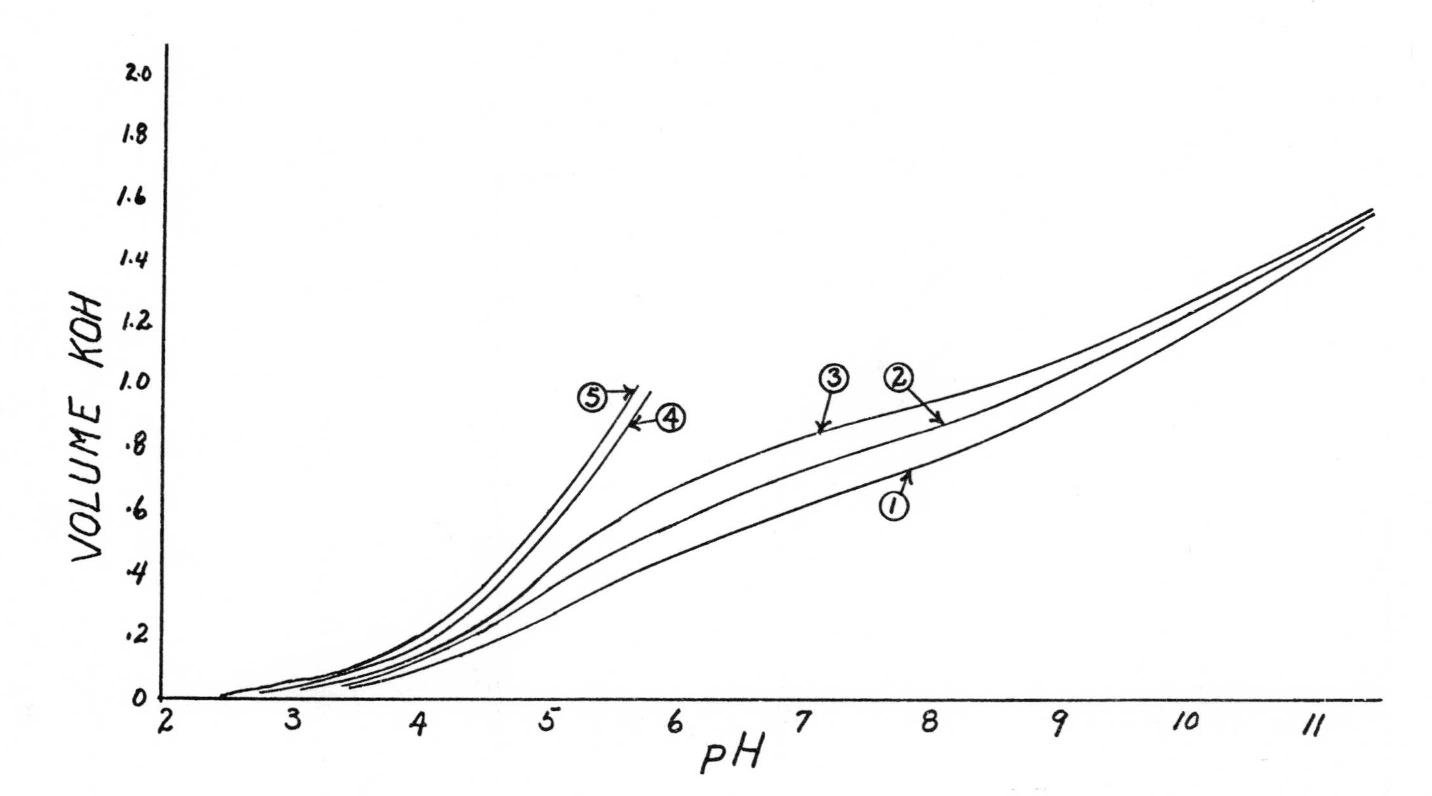

Fig. 5. Titration of PVA·HCl with KOH in IM KCl with various amounts of Ni^{2+} present [PVA]= 0.026 M. Nickel/amine=0 curve 1; 0.16 curve 2; o.3; o.6 curve4, and 1.0 curve 5.

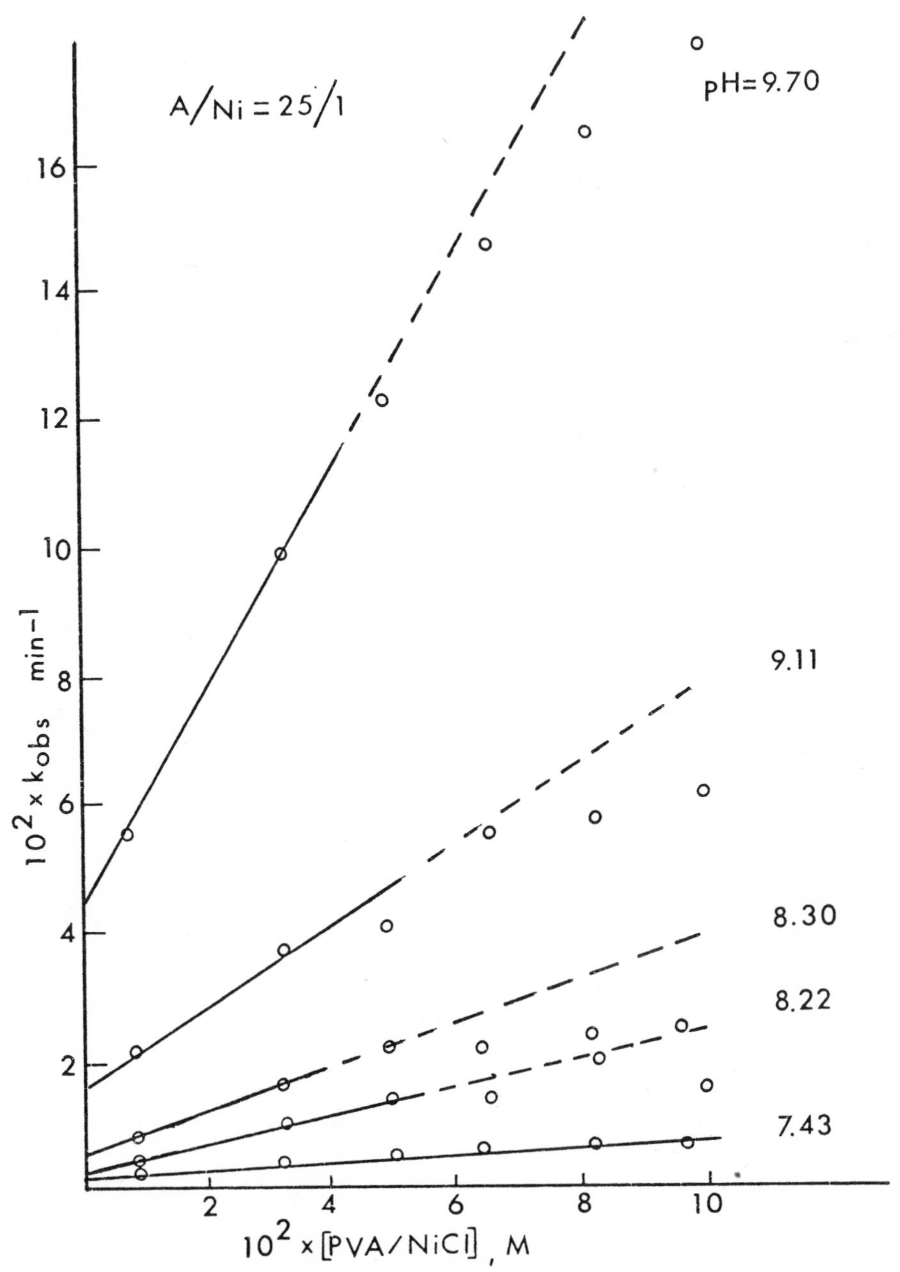

Fig. 6. A first order plot of the observed rate constant as a function PVA/NiCl concentration for the various pH's indicated.

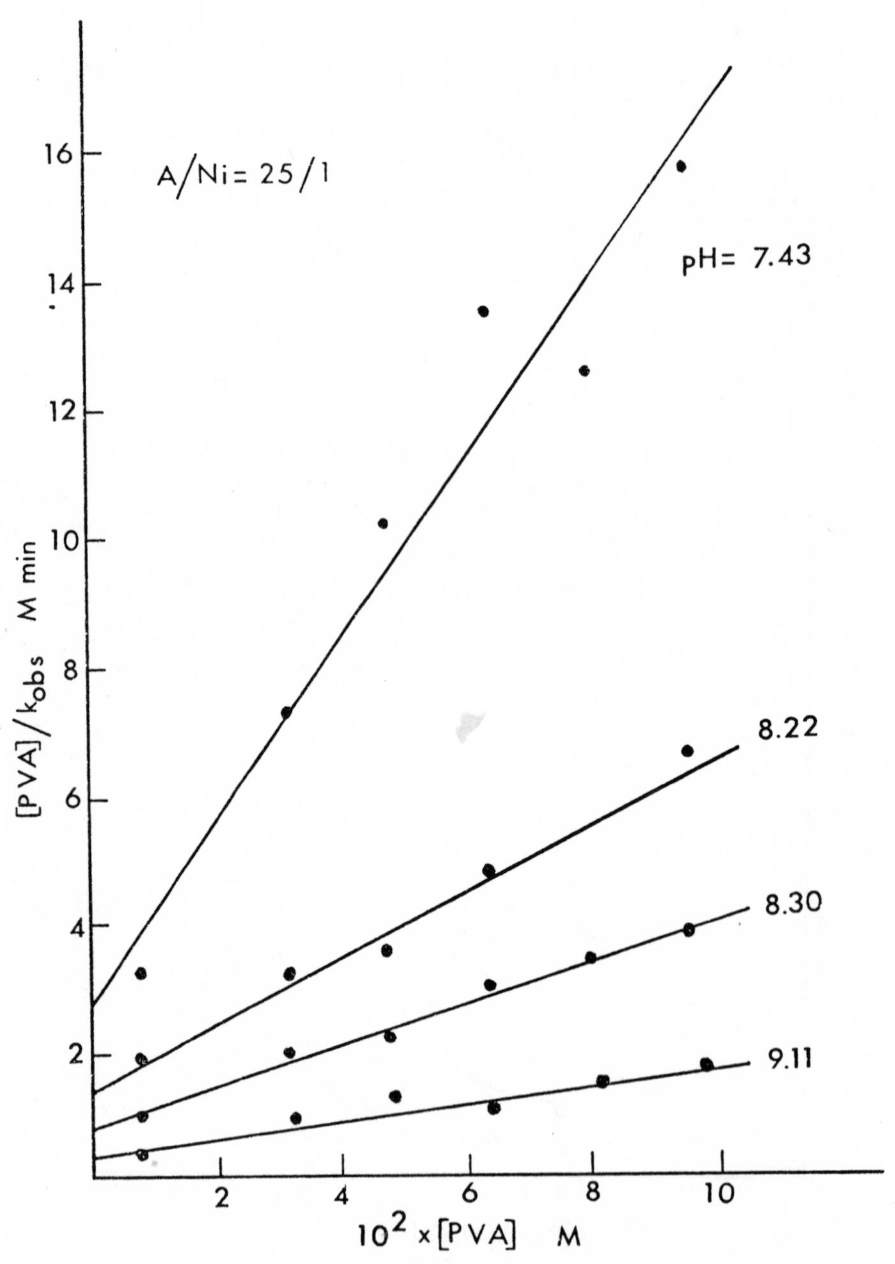

Fig. 7. A plot of [PVA]/k_{obs} versus [PVA]. The intercept equals 1/K k and the slope equals 1/k_{cat}.

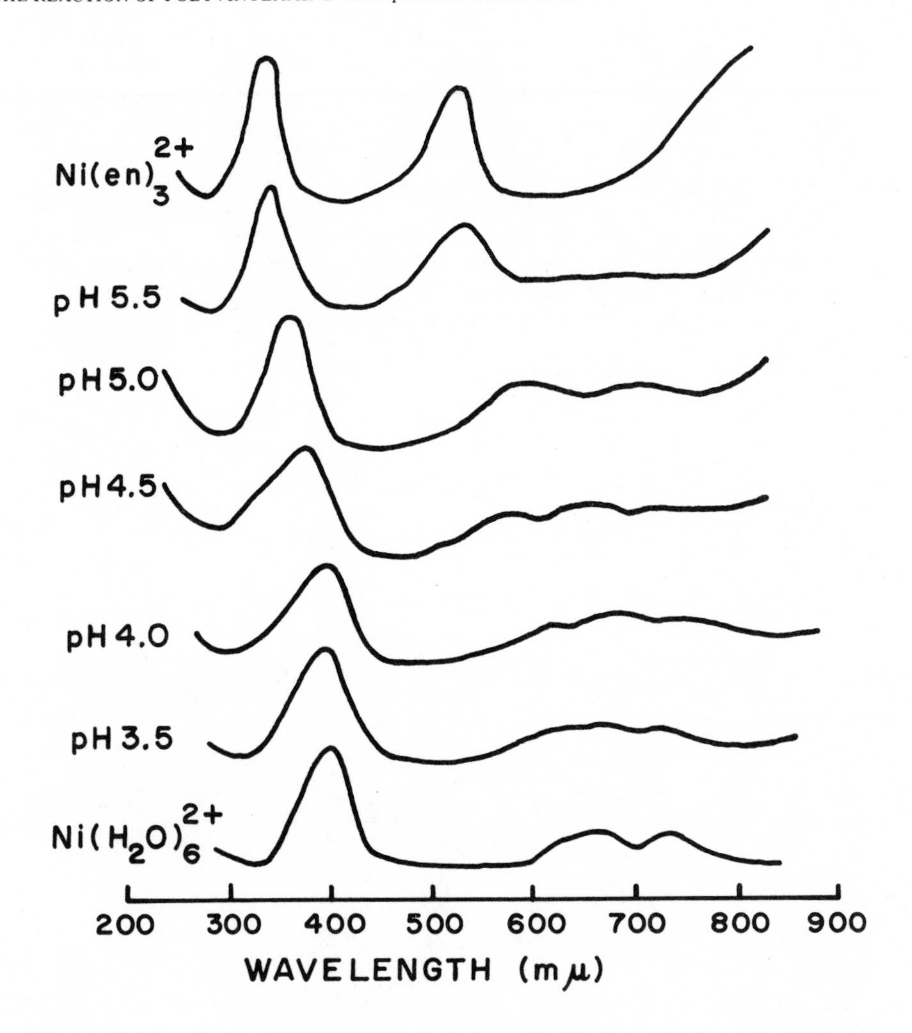

ABSORPTION SPECTRA OF Ni-PVA AT pH 3.5-5.5 IN 1.0 M KCl SOLUTION

of the polymer where the amine functions are less available for reaction. Even if you discount this possibility, the reaction with the polymer complex is a reaction in as environment less polar than water. For a reaction such as the aminolysis of an ester which is neutral in the reactant state and a zwitter ion in the transition state, this change in reaction environment will have an adverse effect on the rate of reaction. It may be as Dr. Cho has suggested to me, that the full benefit of hydrophobic interactions will not be achieved except in a micelle-like arrangement brought about by the reaction of the polymer with an ester of a long chain acid and a charged phenol.

REFERENCES

1. R. Hart, J. Polymer Sci. 29, 629 (1958).

2. M. Sugiura, et. al. Kogaku Zasshi, 72, 1926 (1969)

3. P. Teyssie et. al. Macromol. Chem., 84, 51 (1965).

DISCUSSION SESSION

Discussion Leader - M. Lewin

Carraher: I am not sure I understand fully your reaction conditions.

St. Pierre: The reaction was run in water solution with the ionic strength adjusted to 1. Each kinetic run was done at constant pH. The purpose of adding Ni ion, in the ratio of amine groups to Ni ion of 25:1, was to promote coiling and thus to foster favorable interaction between polyvinylamine and p-nitrophenyl acetate.

Carraher: The amine groups will form rather strong complexes with Ni under your conditions. I would expect it to inhibit the reaction by formation of a "package" which actually ties up amine groups and inhibit the reaction.

St. Pierre: If we had used a high ratio of Ni to amine groups there would be no reason to expect a favorable influence, but in the low ratio used I would expect to see some favorable effect.

Daly: Did you determine the concentration of free amino groups which are readily accessible?

St. Pierre: I did not. Is there a convenient method for assaying the available amine functions?

Daly: You might titrate with a relatively large reagent which would be very reactive but would only attack the accessible amine groups and would not attack the hindered amines.

Moore: One way to modify your polymer to introduce some more hydrophobic groups would be to alkylate some of the amines with a long chain aldehyde and then reduce the imine formed. These groups could cluster and interact with substrates such as long chain esters.

St. Pierre: This technique has been used by some workers. They used a copolymer to incorporate hydrophobic groups. We wanted, however, to keep our system as simple as possible.

Unidentified: How do you interpret the titration behavior of PVA and PEI?

St. Pierre: Katchalsky proposed that there are two types of ionizable groups, one which involves hydrogenbonded amine groups and one which involves free groups.

Smets: What is the slope of the Hasselback diagrams in one of your slides, $^{1 - \alpha}/_{\alpha}$ versus pH? What was the pK? It should be the pK of the compound.

St. Pierre: The curve has a break in it so it is difficult to answer your question. The pH at half neutralization is about 7.5 for poly(vinylamine).

Smets: That is extremly low.

St. Pierre: Correct, in fact,the kinetic pKa is much closer to 9.5. If we extrapolate from the high end of the titration curve, it too goes to 9.5.

Smets: What was the yield of the Schmidt reaction on poly(acrylic acid)?
St. Pierre: We were able to convert approximately 80% of the carboxyl groups into amine groups.

Smets: Does that figure include lactam groups which may form?

St. Pierre: We titrated the product and compared the resulting curve with similar data obtained from the starting acid and poly-(vinylamine)(from an alternative route). The curves are entirely different and copolymer cross-over from poly(acrylic acid) to poly(vinylamine) gives approximately 80% amines functions. Under these strongly acidic conditions one would expect that each amine function would be protonated. Lactam formation is therefore not as likely to occur as it is in the Hoffman reaction.

Harwood: Earlier workers encountered crosslinking in the polymeric Hoffman reaction because isocyanates were intermediates. Did you have the same results?

St. Pierre: Although many side reactions may occur, we have not found any evidence for crosslinking in the Schmidt or the Hoffman reaction.

FUNCTIONALIZATION OF POLYMERS

Paolo Ferruti

Istituto di Chimica Industriale del Politecnico,
Sezione Chimica Macromolecolare e Materiali,
Milan, Italy.

Considerable interest is being focused at present on synthetic polymers bearing reactive functional groups. By "reactive functional groups" I mean chemical functions enabling the polymer to chemically react with, or to exert a definite chemical action on one or more components of a surrounding medium, either in artificial or in biological systems. This definition is not absolute, since any chemical group may undergo chemical reactions. However, a polymer containing "reactive chemical groups" is usually prepared to be, and expected to act as a reactive substance and not merely as inert material.

Synthetic polymers have in fact been considered for a very long time as materials, intended to replace naturally - occurring substances such as cellulose, wood, wool, silk, etc. in many practical applications. Hence, emphasis has been laid mainly on the chemical inertness, mechanical properties, and processability of synthetic polymers under study. The most notable exception to this view occurs in the case of ion - exchange resins.

More recently, however, an increasing number of chemists have started to realize the tremendous potentiality of purpose - tailored synthetic polymers as "chemically active" substances, for instance, macromolecular catalysts, enzyme carriers, and macromolecular drugs.

From a preparative point of view, the most important problem facing a macromolecular chemist working in this field is how to attach the appropriate chemical function or appropriate set of different functions to a macromolecular backbone.

Basically, this can be achieved with two principal methods. The first is to polymerize or copolymerize suitable monomers already containing the desired functions in their molecule.

The second method involves performing one or more modification reactions, on a suitable macromolecule thus introducing the desired functions on a pre-synthesized polymer. This second method we shall call "functionalization of polymers".

It is apparent that the first method, simple as it may seem, cannot be universally applied because of its inherent limitations. First, the desired function, or one of the desired functions if several are required, may interfere with the polymerization process. Secondly, synthesis of the starting monomers may be difficult or even virtually impossible, especially if complex structures are required. In these cases, it may be necessary to functionalize pre-existing polymers of a suitable structure.

Grafting polymerization may be also considered as a special way of functionalizing pre-existing polymers. I do not however intend to deal with this here.

Equally, I shall not be concerned with the functionalization of natural macromolecules, even though many of the following considerations might be extended to apply to these.

If a functional group cannot be conveniently introduced as such by polymerization onto a polymer backbone, there are three synthetic ways of obtaining the desired end product.

The first entails attacking a conventional polymer by a suitable reagent. An example of this procedure is the chloromethylation of polystyrene in order to introduce $-CH_2Cl$ groups as a starting point for further modification reactions in the ion - exchange resin field.

The second method involves preparing and polymerizing monomers containing chemical groups which, though inert toward the polymerization processes, may be easily converted into the desired functions by a modification reaction performed on the resulting polymers. This method has been widely applied, for instance, in synthesis of polymers containing phenolic and sulphydrilic groups.

The third method consists of synthesizing first macromolecules with chemical functions able to react selectively and quantitatively with, for instance, hydroxyl or amino groups giving ester or amide bonds. By a subsequent reaction step performed with alcohols or amines bearing the desired groups attached as substituents, including such high-molecular weight substances as natural proteins, multifunctional polymers may ultimately be obtained. This method is being investigated in our Laboratory with a view to developing synthetic means of obtaining multifunctional polymers with complex structures to be tested for their biological applications. It has also been used to obtain natural enzymes bonded to a polymeric matrix, thus stabilizing the enzymes themselves while retaining most of their catalytic activity.

These three synthetic methods will be discussed separately.

FUNCTIONALIZATION OF CONVENTIONAL SYNTHETIC POLYMERS

The introduction of reactive functions into conventional polymers has been the subject of much scientific and technical research in the past. I will not attempt to make a review of the subject, but only briefly to summarize the scope and the limitations of this procedure.

When dealing with the functionalization of conventional synthetic polymers, some shortcomings are immediately apparent. In most cases, selectivity of the reaction is poor. This means that unwanted groups may be introduced, through side-reactions which are difficult to control. The site of attack may also be hard to control. Furthermore, crosslinking and chain scission are often likely to occur.

This is best exemplified by the functionalization of polyethylene and related polyolefines. In this case, most reactions such as chlorination or chlorosulfonation take place through a radical mechanism. This leads to partly degraded and crosslinked products (1). Furthermore, it is usually difficult to find inert solvents which are capable of dissolving both the polyolefines and the reaction products. This poses serious problems when the reaction has to be performed in solution (2).

Most reactions on polyolefines have been carried out in the heterogeneous phase, often by contacting the polymers with reagents in the gaseous phase. In these case, permeability is the most important factor governing the reaction. If the reaction is performed on crystallizable polymers these are attacked preferentially since permeability is higher in the amorphous regions. This leads in many cases to an oddly - distributed functionalization of the starting material (3).

The situation improves somewhat when it comes to functionalization of other synthetic polymers, for instance vinylaromatic polymers, such as polystyrene. In this case, reactions such as sulfonation, chloromethylation, nitration etc have been performed, using techniques which do not basically differ from what is normally indicated in the organic chemistry of aromatic substitutions. Most of these reactions have been performed on crosslinked polymers in a swollen state, with a view to obtaining ion - exchange resins. Since several reviews have appeared on this topic (4), I do not think worthwhile discussing it any further. Another very important use of crosslinked functionalized resins, mostly based on polystyrene, is the well-known solid-phase peptide synthesis.

In the electrophilic substitutions of poly-vinylaromatics,the main chain is usually considered to be a *para* orienting substituent. However, it is obviously impossible to separate the *o*, *m*, and *p*-substituted isomeric units, as well as disubstituted units if present. Hence, the structure of the products resulting from a substitution reaction on poly-vinyl aromatics is often somewhat irregular.

From the foregoing considerations, it follows that the functionalization of conventional polymers may prove unsuitable, if multifunctional polymers with a strictly controlled structure are to be obtained. In this case, the second and the third main synthetic routes previously defined appear to be more appropriate. These will be dealt with in more detail in the following sections.

FUNCTIONALIZATION OF POLYMERS BY MEANS OF CHEMICAL TRANSFORMATION OF THE POLYMERIZATION PRODUCTS OF SUITABLE MONOMERIC PRECURSORS

This method has been widely used to synthesize multifunctional polymers designed for special applications. Its main advantages are that both the number of reactive functions introduced into the polymer, and the absence of unwanted groups deriving from side reactions can be easily achieved, provided the "precursor" units have been properly chosen, so to be able to react quantitatively in the desired way.

The introduction of each functional group leads to special problems, as would be expected. Hence, we have selected some relevant examples as a basis for further discussion. These examples will be individually discussed.

Polyvinylphenols

All three monomeric vinylphenols have been prepared (5-9). In particular, the para isomer was prepared by Dale and Hennis and by Carson and his colleagues by saponification of 4-benzoyloxystyrene (7) or 4-acetoxystyrene (8), and, more recently, by Overberger and colleagues (9) by decarboxylation of 4-hydroxycinnamic acid (Scheme 1). Several studies on the direct homo- or copolymerization of vinylphenols have appeared, mainly centered on p-vinylphenol. The first experiments were performed with radical initiators, but they gave only low molecular weight polymers (10). A phenolic function, in fact, is well known to interfere with radical polymerization, as well as with most reactions proceeding through a radical mechanism, and a monomer like a vinylphenol has to be expected to be self-poisoning in such a process. Better results have been more recently obtained by Kato (11) with cationic initiators. Unfortunately, the phenol nucleus is strongly activated toward nucleophilic alkylation, and consequently direct polymerization of vinylphenols by cationic initiators often yields polymers whose structures are not fully polyvinylic, but may partly arise from the reaction between the vinyl group and the phenol nucleus. The same considerations hold good when dealing with hydroxyalkylstyrenes or dihydroxystyrenes such as vinylhydroquinone (12) and related compounds (13).

On the other hand, high molecular weight polymers or copolymers of hydroxy-substituted styrenes may be of interest in several

Scheme 1

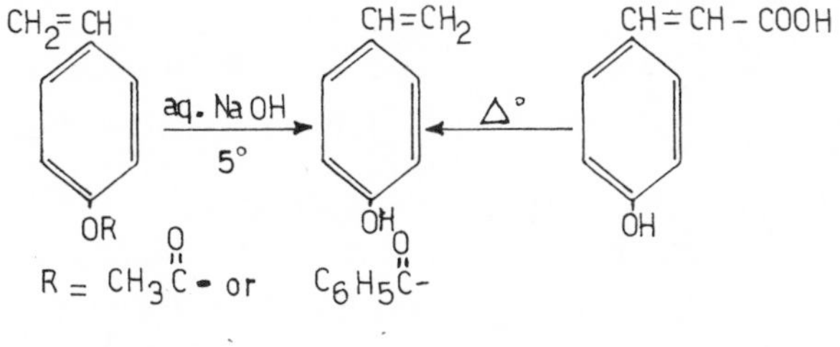

$CH_2{=}CH$ $CH{=}CH_2$ $CH{=}CH{-}COOH$

aq. NaOH, 5° →   ← Δ°

OR   OH   OH

$R = CH_3\overset{O}{\overset{\|}{C}}-$ or $C_6H_5\overset{O}{\overset{\|}{C}}-$

fields. They can be expected, for instance, to act as polymeric stabilizers in very much the same way as non-macromolecular phenols.

Multifunctional polymers containing both imidazole and phenolic groups have also been shown to have enzyme-like activity in catalyzing at pH $\geqslant$ 9 the hydrolysis of nitrophenyl esters (14). Finally, as we shall see later, a phenolic group may be the starting point for further reactions leading to the introduction of other chemical groups.

We first obtained linear, high molecular weight poly-4-vinylphenol (15) and poly-3-vinylphenol (16) and some poly-3-alkyl-4-vinylphenols (17) by radical polymerization of the corresponding acetoxy derivatives, followed by hydrolysis of the acetoxy groups (Scheme 2). The same procedure has been followed by other Authors

Scheme 2

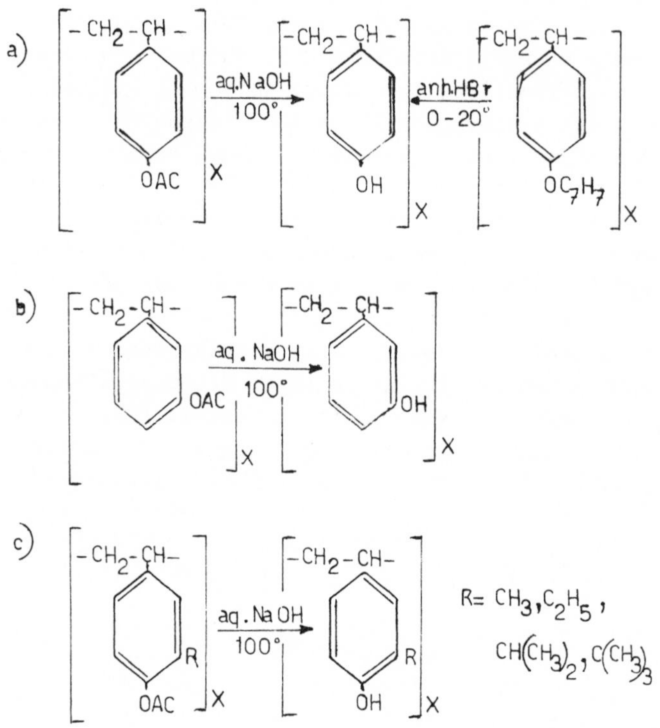

a) $[-CH_2-CH(C_6H_4OAc)-]_x \xrightarrow[100°]{aq.NaOH} [-CH_2-CH(C_6H_4OH)-]_x \xleftarrow[0-20°]{anh.HBr} [-CH_2-CH(C_6H_4OC_7H_7)-]_x$

b) $[-CH_2-CH(C_6H_4OAC)-]_x \xrightarrow[100°]{aq.NaOH} [-CH_2-CH(C_6H_4OH)-]_x$

c) $[-CH_2-CH(C_6H_3R\text{-}OAC)-]_x \xrightarrow[100°]{aq.NaOH} [-CH_2-CH(C_6H_3R\text{-}OH)-]_x$ R= $CH_3, C_2H_5, CH(CH_3)_2, C(CH_3)_3$

in synthesizing poly-vinylhydroquinones (18). Hydrolysis occurs quite easily in water, under an inert gas atmosphere, and with sodium or potassium hydroxides, the product gradually dissolving in the alkaline medium. Alternatively, hydrolysis may be carried out homogeneously in aqueous dioxane in the presence of mineral acids.

More recently, we have developed another synthetic route to poly-4-vinylphenol, which may possibly be applied to alkyl-substituted vinylphenols as well. This procedure (Scheme 2) involves debenzylation with anhydrous hydrobromic acid of poly-4-vinylphenylbenzylether, obtained by polymerization of the corresponding monomer (19). This route may be preferable to that described previously if high molecular weight polymers or copolymers of 4-vinylphenol are to be prepared with catalytic systems for which 4-acetoxystyrene does not appear to be suitable. Unlike 4-vinylphenylbenzylether, 4-acetoxystyrene has in fact so far yielded unsatisfactory results with most cationic or anionic initiators.

Tertiary Amino-Polymers and Their Derivatives

Either aliphatic or cycloaliphatic tertiary amino polymers may be extensively used in different applications as materials or (more generally) as reagents. Some tertiary polyamines and their N-oxides may also be used in pharmacology. For example, a number of these polymers are endowed with preventive action against silicosis (20) while others show antiheparinic (21) or antimetastatic (29) activity.

We have performed systematic research on the synthesis of tertiary amino polymers (22), mainly because we were interested in their pharmacological applications.

I shall now consider only the preparative methods involving the introduction of tertiary amino groups in to pre-synthesized polymeric chains, as well as some transformations of these groups, by further reaction, into related chemical functions.

Tertiary poly-allylamines and polyvinylamines. Since tertiary poly-allylamines and poly-vinylamines are structurally among the most simple of the conceivable tertiary amino polymers we were most interested in their preparation.

Direct synthesis of both classes of polymers by polymerization of the corresponding monomers is not feasible. Simple N-vinyl aliphatic or cycloaliphatic amines are not even known with certainty in a pure state, although we have demonstrated that in some cases gem--diamino compounds may be isolated, which are in a dissociation equilibrium between secondary amine and N-vinylamine (23). We also found that under certain conditions these compounds react as vinylamines, both in simple organic reactions (24) and in copolymerization reactions, e.g. with acrylonitrile (25) (Scheme 3). On the other hand, it is known that allylamines, though easily prepared, do not homopolymerize readily to high molecular weight polymers.

We obtained satisfactory results by reducing the carbonyl groups of poly-N,N-dialkyacrylamides (Scheme 4) (22). The reducing agent was excess of lithium aluminium hydride dissolved in dioxane, N-methylmorpholine or anisole, at a reaction temperature of 100 to 150°. This agent had previously been used with fairly good results to reduce other polymers (26). Little chain degradation occurred during the reaction. Pure poly-allylamines with intrinsic viscosities up to 1 dl/g could be obtained.

All the poly-allylamines obtained by us were amorphous by X-rays, even the modified stereoregular poly-acrylamides. This might be due to lack of stereoregularity from racemization that occurred during

Scheme 3

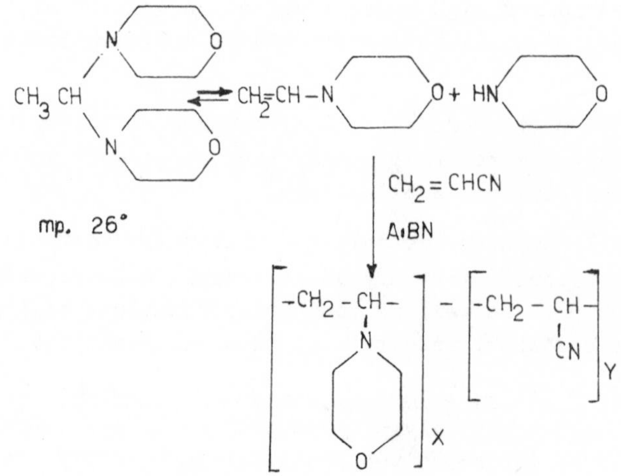

CH_3CH (bis-morpholino) ⇌ $CH_2{=}CH{-}N$ (morpholine) + HN (morpholine)

mp. 26°

$CH_2{=}CHCN$

AIBN

$[-CH_2-CH(N\text{-morpholino})-]_X - [-CH_2-CH(CN)-]_Y$

Scheme 4

$$\left[-CH_2-\underset{\underset{NR_2}{|}}{\underset{CO}{\underset{|}{CH}}}-\right]_x \xrightarrow[100^\circ-150^\circ]{LiAlH_4} \left[-CH_2-\underset{\underset{NR_2}{|}}{\underset{CH_2}{\underset{|}{CH}}}-\right]_x$$

modification. It is well known that stereoregular vinyl polymers, especially those containing a carbonyl group in the α position at the tertiary carbon atom, may lose stereoregularity by heat treatment with basic catalysts (27).

Similar results were obtained from the reduction of poly-N--vinylacetamides, thus obtaining high-molecular weight poly-dialkylvinylamines.

Poly-dialkylaminomethylphenols. Polymers containing both tertiary amino and phenol groups may have interesting properties both in the pharmaceutical field, like other tertiary amino polymers, and as chelating agents or as macromolecular stabilizers.

We found that when high molecular weight poly(2-alkyl-4-vinylphenols) were allowed to react in a suitable solvent, such as dioxane, with an excess of formaldehyde and a secondary amine, a smooth Mannich reaction ensued and poly(2-alkyl-4-vinyl-6(dialkylamino-methyl)phenols) were obtained in very high yields (Scheme 5) (28). The degree of substitution, as indicated by elemental analysis, was essentially complete. Under similar conditions, poly(4-vinylphenol) gave poly(2,6-bis(dialkylaminomethyl)-4-vinylphenols) (Scheme 5).

Further modifications of tertiary amino polymers. Further transformations may be induced in order to convert tertiary amino polymers into derivatives which can be used in the pharmaceutical field. For instance, the tertiary amino groups may be easily transformed with peracetic or perbenzoic acids into the corresponding N-oxides (Scheme 6). N.-oxide polymers of various kinds are able to prevent completely the onset of silicosis as indicated in a number of papers published by Schilpköter, Brockaus et al., and by ourselves (20).

Scheme 5

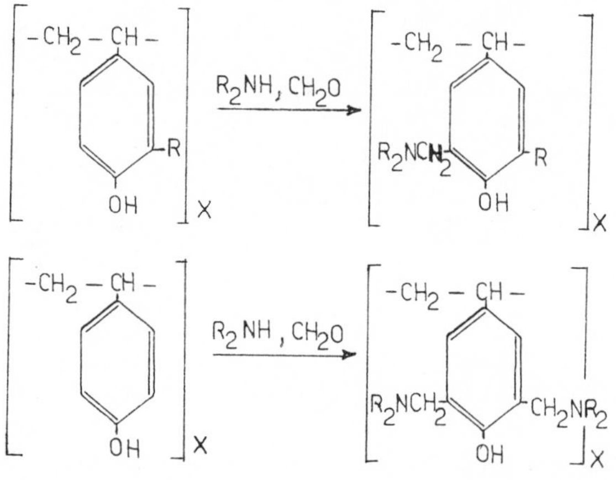

$\left[-CH_2-CH- \text{(4-hydroxy-3-R-phenyl)} \right]_x \xrightarrow{R_2NH,\ CH_2O} \left[-CH_2-CH- \text{(3-}R_2NCH_2\text{-4-hydroxy-5-R-phenyl)} \right]_x$

$\left[-CH_2-CH- \text{(4-hydroxyphenyl)} \right]_x \xrightarrow{R_2NH,\ CH_2O} \left[-CH_2-CH- \text{(3,5-bis(}CH_2NR_2\text{)-4-hydroxyphenyl)} \right]_x$

Scheme 6

$\left[-CH_2-CH(-R-NR^1R^2)- \right]_x \xrightarrow[15^\circ-50^\circ]{RCO_3H} \left[-CH_2-CH(-R-NR^1R^2\rightarrow O)- \right]_x$

$\left[-CH_2-CH(-R-NR^1R^2)- \right]_x \xrightarrow{CH_2-O,\ CH_2-C=O} \left[-CH_2-CH(-R-{}^{+}NR^1R^2-CH_2CH_2COO^-)- \right]_x$

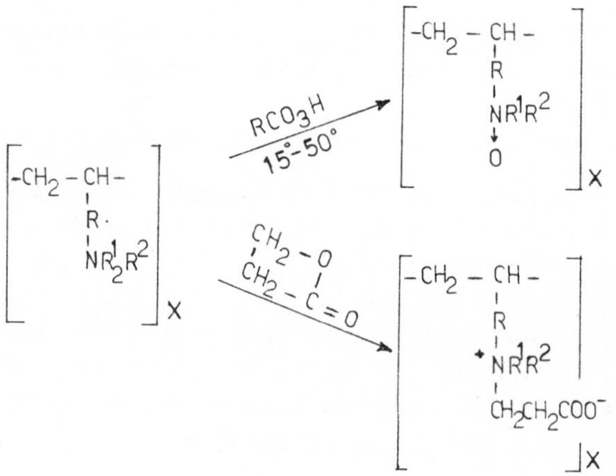

Furthermore, we have found that high molecular weight partially N-oxidized poly-allyldietylamine is a good antagonist of heparin (21)

and we are currently developing a set of heparin-complexing agents based on tertiary amino polymers and their N-oxides, some of which have been patented and may come into clinical use in the near futu - re (21).

An interesting reaction may be quantitatively performed on tertiary amino polymers, to give poly betaines (Scheme 6). This can be achieved by the use of β -propiolactone. The resulting poly--betaines are under study as antimetastases agents following the in-dications of a previous paper of ours, indicating that some water--soluble synthetic polymers having zwitterionic properties show con-siderable activity against cancer cell dissemination and metastases (29).

Polyampholites

A polyampholite may be defined as a polymer which in water sol-ution at pH near neutrality bears positive and negative charges in the same time. The poly-betaines described previously are an example of polyampholites. However, polyampholites are a larger class of poly-mers, since every polymer bearing acidic and basic groups of suffi-cient strength may be considered a polyampholite.

Polyampholites may be prepared, for instance, by copolymeriza-tion of pairs of monomers bearing basic and acidic groups respecti-vely. In several instances, however, polyampholites have been pre-pared by chemical reactions performed on suitable macromolecular precursors. Among the most interesting examples is the work of Neu-feld and Marvel (30) who obtained polymers by hydrogen-transfer po-lyaddition of α , ω -dithiols to 3,6-diallyl-2-piperidone, and then hydrolized them (Scheme 7), and that of Panzik and Mulvaney (31) who cyclopolymerized N-2-phenylallylacrylamides, and then hydrolized the resulting polymers (Scheme 8).

In a research on antimetastases polymers, we needed polyamphol-ites bearing positive charges in the main chain, and negative char ges in the side substituents. We solved this problem in the follo-wing way (32).

First of all, we successfully copolymerized Schiff bases with acrylonitrile by radical initiators. This, as far as I know, is the first instance of the use of Schiff bases as monomers or co-monomers in radical polymerizations. Secondly, we hydrolyzed the -CN groups

Scheme 7

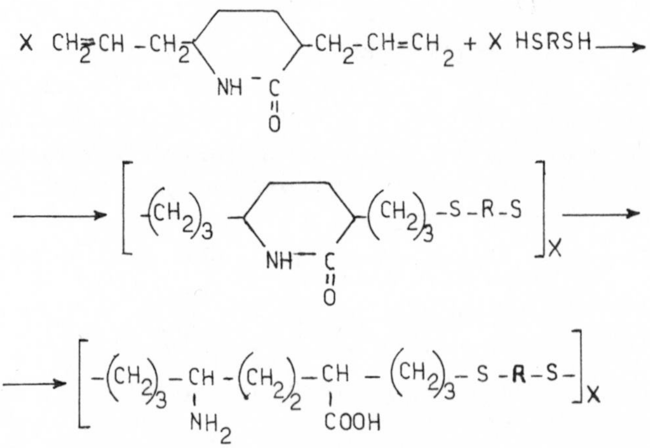

$$x\ CH_2{=}CH{-}CH_2{-}\underset{NH-C(=O)}{\bigcirc}{-}CH_2{-}CH{=}CH_2 + x\ HSRSH \longrightarrow$$

$$\longrightarrow \left[-(CH_2)_3 - \underset{NH-C(=O)}{\bigcirc} -(CH_2)_3-S-R-S \right]_x \longrightarrow$$

$$\longrightarrow \left[-(CH_2)_3 - \underset{NH_2}{CH} - (CH_2)_2 - \underset{COOH}{CH} - (CH_2)_3 - S - R - S - \right]_x$$

Scheme 8

$$CH_2{=}C(C_6H_5){-}CH_2{-}N(R){-}C({=}O){-}CH{=}CH_2 \longrightarrow \left[-CH_2-C(C_6H_5)< \begin{matrix} CH_2-CH- \\ CH_2-N(R)-C{=}O \end{matrix} \right]_x \longrightarrow$$

$$\longrightarrow \left[-CH_2 - \underset{\underset{NHR}{|}}{\underset{CH_2}{C(C_6H_5)}} - CH_2 - \underset{COOH}{CH} - \right]_x$$

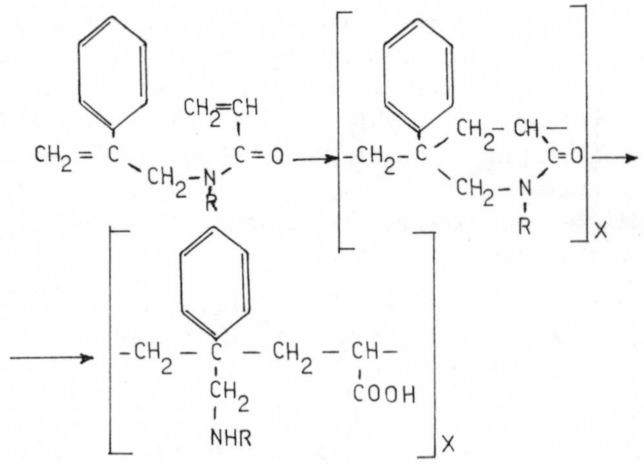

and finally obtained the desired polyampholites (Scheme 9). These proved to be water-soluble at temperatures above 35°, and are currently being studied for possible pharmacological activity.

Conclusions

I have only given some examples of chemical functions which may be introduced in polymers by means of chemical transformations of the polymerization products of suitable monomeric precursors. These, however, may constitute a basis for some conclusions about this synthetic route for the functionalization of polymers.

First of all, it may be pointed out that all the successful syntheses have been performed when simple modification reactions have been involved. By successful syntheses I mean those which occur quantitatively, without any extensive degradation of pre-existing polymeric chains and without the introduction of chemical functions other than those the operator is concerned with. By simple reactions, I mean reactions involving only one or two macroscopic reaction steps, and only one or two reagents in addition to the substrate.

Secondly, all these reactions involve only a single chemical group of the polymer at once. Reactions involving two or more chem-

Scheme 9

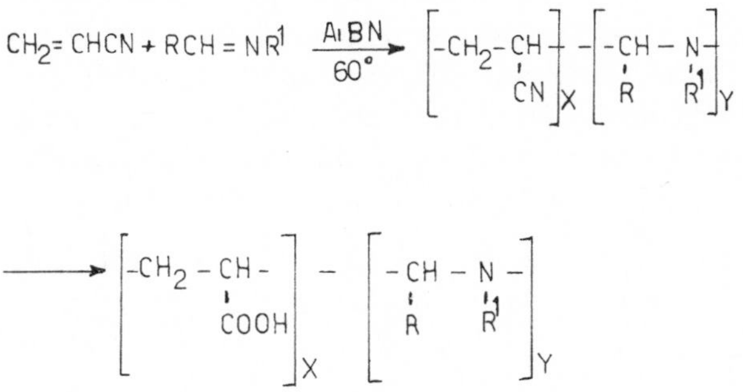

$$CH_2{=}CHCN + RCH{=}NR^1 \xrightarrow[60^\circ]{AIBN} \left[-CH_2-\underset{CN}{CH}-\right]_X - \left[-\underset{R}{CH}-\underset{R^1}{N}-\right]_Y$$

$$\longrightarrow \left[-CH_2-\underset{COOH}{CH}-\right]_X - \left[-\underset{R}{CH}-\underset{R^1}{N}-\right]_Y$$

ical groups at once attached to the same macromolecular chain have in fact been demonstrated both on theoretical and practical grounds (see for instance the Flory's approach on this subject (33)) to be incapable of proceeding quantitatively, except in specific cases.

On the other hand, the precursors functions must be sufficiently inert to allow synthesis of the monomer and its polymerization to proceed satisfactorily. They must also be able to be transformed specifically into the desired functions in a way that meets the requirements outlined previously.

From the foregoing considerations, it is apparent that it is not easy to plan a suitable precursor for some functions to be introduced on a polymer. It would be rather difficult, for instance, to plan a monomeric precursor in order to introduce pendant vinyl groups into a preformed polymer chain quantitatively and without any side-reactions. This is even more evident if a set of functions has to be introduced into the same monomeric unit of a given polymer, as may occur when dealing with polymers to be tested in the pharmaceutical field. In such cases other methods for the functionalization of polymers must be sought. This will be the subject of the following section.

FUNCTIONALIZATION THROUGH POLYMERIC PRECURSORS CONTAINING REACTIVE FUNCTIONS ABLE TO COUPLE WITH SUITABLE CHEMICAL GROUPS

The basic principle of this method lies in preparing first macromolecules with chemical functions able to react selectively and quantitatively with particular chemical groups giving stable covalent bonds, such as hydroxyl or amino groups giving ester or amidic bonds. Multifunctional polymers may ultimately be obtained by a subsequent reaction step performed with chemical compounds containing the desired set of chemical functions in addition to the groups to be involved in the coupling reaction.

The coupling functions may be introduced in the precursor polymer either by a modification reaction performed on a pre-synthesized polymeric matrix usually of a conventional type, or by direct homo- or copolymerization of the corresponding monomers.

Both methods have been utilized. It may be observed that with the former method it is usually difficult to achieve a clean and straightform synthesis of multifunctional polymers. The introduction of reactive functions such as isocyanato-, diazo-, isocyanide groups and similia, into a preformed polymeric matrix can only be performed by a quantitative and selective onestep process in a very limited number of cases. On the contrary, the Latter method is not affected by these difficulties.

The choice between these two methods clearly depends on the aim of the operator. This particular kind of functionalization has two major fields of application, namely the binding of enzymes to macromolecular matrices, and the preparation of polymers with specific functions to be used as active substances in the biomedical and pharmaceutical fields.

The chief goal of chemists working in the former field, is to be able to bind a conveniently large amount of an enzyme to a polymeric matrix, preferably of low cost and endowed with good mechanical properties as well as chemical inertness in the enzyme's working conditions. Hence, they often prefer the first of the routes outlined previously. Binding between the protein and the polymeric matrix is usually performed through the amino, the hydroxyl or the carboxyl groups of the protein. Several devices have been studied in order to introduce chemical functions into conventional polymers, able to couple with these groups. A few examples of results achieved in this field will be given.

Isliker (34) converted a carboxylic acid ion exchange resin into a polymeric acid chloride. This reaction was not quantitative, since it was later discovered that a considerable amount of anhydride groups were also present. This author subsequently treated the resin with alkaline solutions of proteins, thus fixing the proteins themselves to the resin (Scheme 10a). Grubhofer and Schleith (35) nitrated, reduced and diazotized polystyrene, and coupled this product with albumin and different enzymes (Scheme 10b). Brandemberger (36) synthesized crosslinked poly-isocyanatostyrene from poly-aminostyrene, and also succeeded in binding enzymes with this method (Scheme 10 c). A superior method was developed by Manecke (37) who prepared a crosslinked copolymer of acrylic acid and 3-fluorophenylacrylamide. Nitration of the phenyl nucleous afforded a derivative which was able to bind several enzymes (Scheme 10d).

More recently, many devices have been developed in order to bind enzymes to polysaccharides or synthetic polyamides. For instance, an interesting method to functionalize nylon has been described by Goldstein (38).

Under certain conditions, a 4-component condensation occurs, between an amine, an aldehyde, an isocyanide and a carboxylix acid,

Scheme 10

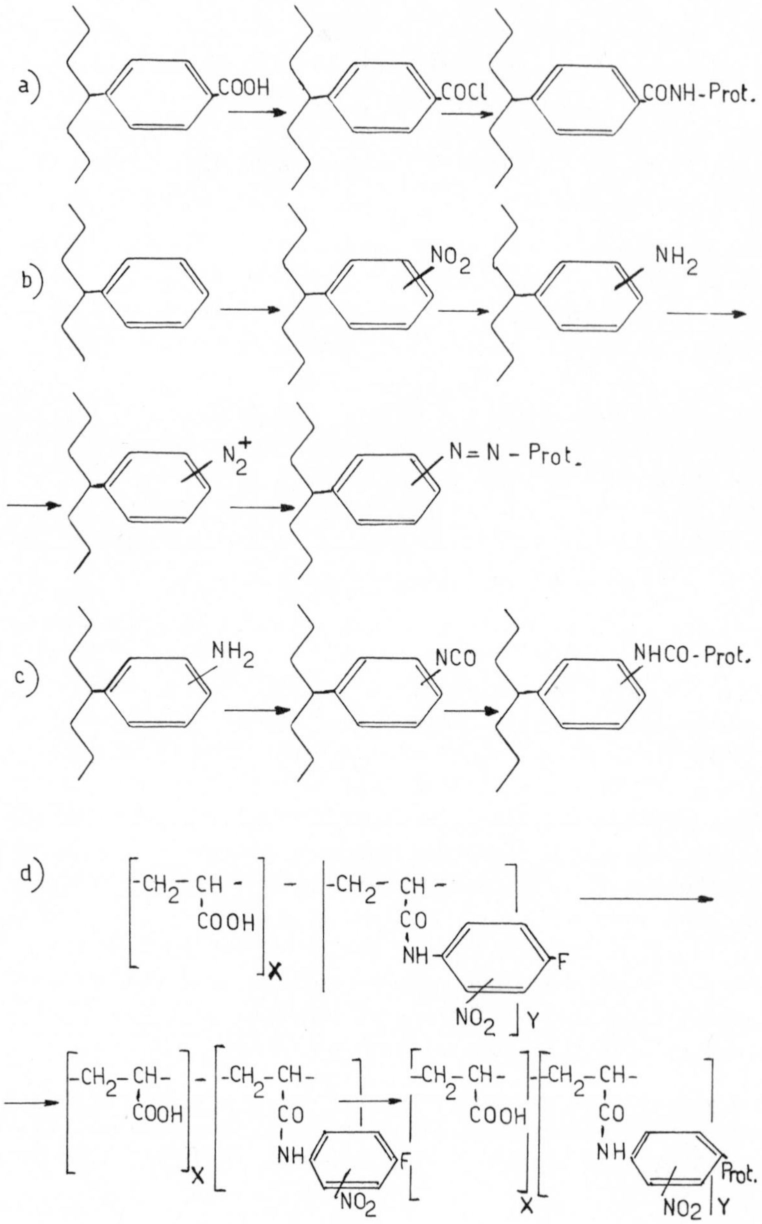

a) COOH → COCl → CONH-Prot.

b) → NO_2 → NH_2 → → N_2^+ → N=N - Prot.

c) NH_2 → NCO → NHCO-Prot.

d) $[-CH_2-CH(COOH)-]_X-[-CH_2-CH(CONH-C_6H_3(NO_2)F)-]_Y$ → $[-CH_2-CH(COOH)-]_X-[-CH_2-CH(CONH-C_6H_3(NO_2)F)-]$ → $[-CH_2-CH(COOH)-]_X[-CH_2-CH(CONH-C_6H_3(NO_2)Prot.)-]_Y$

giving a diamide. This reaction could be performed on partially hydrolyzed nylon, and by using a diisocyanide at this stage, pendant isocyanide groups could be introduced while practically restoring the original molecular weight of the nylon matrix (Scheme 11). Furthermore, other chemical functions could be introduced into the newly-formed side chains, or the chemical amino or carboxyl groups of the protein itself could be involved in this reaction, thus achieving direct binding of the protein to the matrix.

Covalent binding of enzymes to polymeric matrices has also been achieved in some instances by utilizing synthetic polymers in which the coupling functions were introduced during the polymerization process by the use of suitable monomers, i.e., by following the second of the two main routes outlined previously. For example, Ken-

Scheme 11

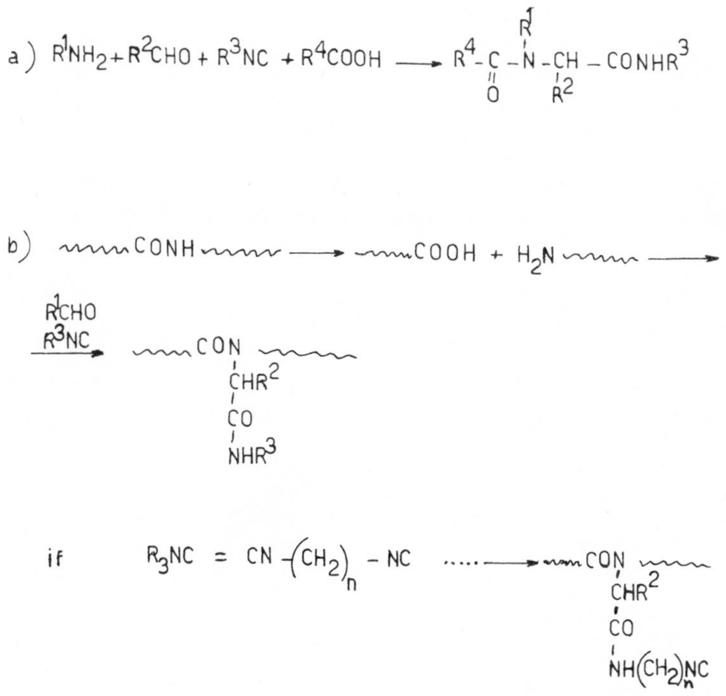

a) $R^1NH_2 + R^2CHO + R^3NC + R^4COOH \longrightarrow R^4\text{-}C(=O)\text{-}N(R^1)\text{-}CH(R^2)\text{-}CONHR^3$

b) $\sim\sim CONH \sim\sim \longrightarrow \sim\sim COOH + H_2N \sim\sim \xrightarrow{R^1CHO,\ R^3NC} \sim\sim CON(CHR^2\text{-}CO\text{-}NHR^3) \sim\sim$

if $R_3NC = CN\text{-}(CH_2)_n\text{-}NC \quad \cdots\cdots\longrightarrow \sim\sim CON(CHR^2\text{-}CO\text{-}NH(CH_2)_nNC) \sim\sim$

nedy (39) copolymerized acrylamide with 2,3-epoxyethylacrylate in the presence of crosslinking agents, thus obtaining a modified poly--acrylamide gel able to bind proteins covalently.

We ourselves have studied new monomers which,though easily prepared and able to homo- or copolymerize easily with conventional means, contained chemical functions in the side chains able to react specifically and quantitatively with free hydroxyl or amino groups giving ester or amidic bonds.

Polymers of acryloyl and methacryloyl chlorides were formely used as polyamide precursors (46). These monomers however have considerable disadvantages. They are aggressive substances, very sensitive to water, and cannot be copolymerized with many other functional monomers. Furthermore, they often yield cyclized or crosslinked products when reacted with amines.

On the other hand, it is known that O-acyl derivatives of N--hydroxysuccinimide react smoothly with primary or secondary amino groups, giving the corresponding amides. This prompted us to study the synthesis and polymerization of O-acryloyl and methacryloyl derivatives of N-hydroxysuccinimide, together with the ability of the resulting polymers to give polyacrylamides or polymethacrylamides subsequently by reaction with amines (41).

Coupling of acrylic or methacrylic acid with N-hydroxysuccinimide in the presence of dicyclohexylcarbodiimide proved to be a suitable way to prepare the monomers. The polymerization of N-acryloxy- and N-methacryloxysuccinimide was readily accomplished with radical initiators in dioxane or benzene solution (Scheme 12). The polymers proved to be able to react quantitatively with aliphatic or cycloaliphatic primary or secondary amines,giving polyacrylamides which in some cases were difficult to obtain by other means (Scheme 13). For instance, when poly-N-acryloxysuccinimide was allowed to react with allylamine, we obtained a soluble, presumably linear, polymer which gave a correct analysis as poly(N-allylacrylamide). N-allyl derivatives or acrylamides or methacrylamides are not expected to give linear, polyvinyltype polymers by radical polymerization. They may either give crosslinked polymers, or cyclopolymerize, giving soluble polymers whose structure is not basically polyvinylic (42).

In the conditions we used, the reaction of N-acryloxy- and methacryloxysuccinimide with aromatic amines does not seem to be substantially quantitative. Furthermore, the activate esters of poly--acrylic and methacrylic acids described previously do not react easily with alcohols, and consequently are of little use as polye-ster precursors.

In order to obtain ester linkages, the heterocyclic amides (a-zolides) of acrylic and methacrylic acids appear to be the most sui-table monomers. It is in fact known that the azolides of carboxylic acids yield esters quite readily on alcoholysis.

Although we have not so far succeeded in synthesizing N-acryl-oylimidazole in a pure state, we have already obtained N-methacryl-oylimidazole (Scheme 14). This monomer does not homopolymerize sat-isfactorily when radical or ionic initiators are use. It does ho-wever copolymerize readily with a number of vinyl monomers. It may be also used in order to introduce pendant polymerizable vinyliden-ic groups in polymers under very mild conditions, thus affording a way for grafting polymerizations. This is illustrated in the fol-lowing example, which we performed in order to obtain a water-solu-ble copolymer containing hydrophobic, hydrophylic and tertiary ami-no blocks, to be tested as an antimetastases agent (Scheme 15). The tertiary amino groups of this polymer were subsequently trans-formed into betaine groups by means of β-propiolactone, and the new polymer so obtained is also being studied as an antimetastases agent.

It is known that the benzotriazolides of simple carboxylic acids react smoothly with alcohols or amines to give esters or ami-des (43). This prompted us to study the synthesis and the polymeriz-ation of 1-acryloyl-benzotriazole, as well as its exchange ability with alcohols and amines (44).

The monomer was prepared in fair yields by reacting acryloyl chloride with 1-H-benzotriazole in the presence of triethylamine. 1-Acryloyl-benzotriazole polymerized easily by radical initiators

Scheme 12

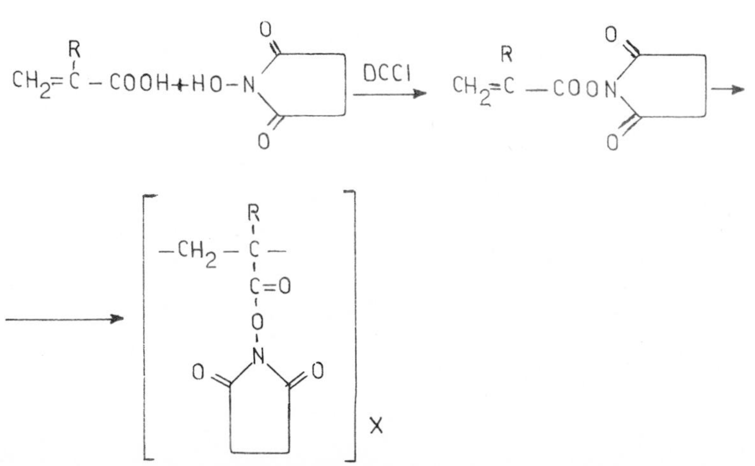

$CH_2{=}C(R){-}COOH + HO{-}N$ (succinimide) $\xrightarrow{DCCl}$ $CH_2{=}C(R){-}COON$ (succinimide) $\longrightarrow$ $[-CH_2-C(R)(C{=}O{-}O{-}N\text{(succinimide)})-]_x$

Scheme 13

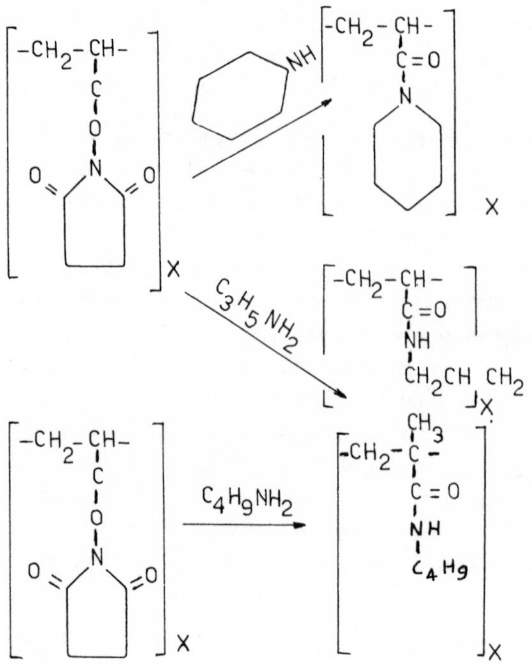

$$\left[-CH_2-\underset{\underset{\underset{\text{succinimide}}{N}}{O}}{\underset{|}{CH}}-C-\right]_x \xrightarrow{\text{piperidine (NH)}} \left[-CH_2-\underset{C=O-N(\text{piperidine})}{CH}-\right]_x$$

$$\left[-CH_2-CH(-C(=O)-O-N\text{(succinimide)})-\right]_x \xrightarrow{C_3H_5NH_2} \left[-CH_2-CH(-C=O-NH-CH_2CH\,CH_2)-\right]_x$$

$$\left[-CH_2-CH(-C(=O)-O-N\text{(succinimide)})-\right]_x \xrightarrow{C_4H_9NH_2} \left[-CH_2-C(CH_3)(-C=O-NH-C_4H_9)-\right]_x$$

Scheme 14

$$CH_2{=}\overset{CH_3}{\overset{|}{C}}-COOH + HN(\text{imidazole}) \xrightarrow{DCCI} CH_2{=}\overset{CH_3}{\overset{|}{C}}-C{=}O-N(\text{imidazole}) \xleftarrow[-IM.HCl]{} CH_2{=}\overset{CH_3}{\overset{|}{C}}-COCl + 2\,HN(\text{imidazole})$$

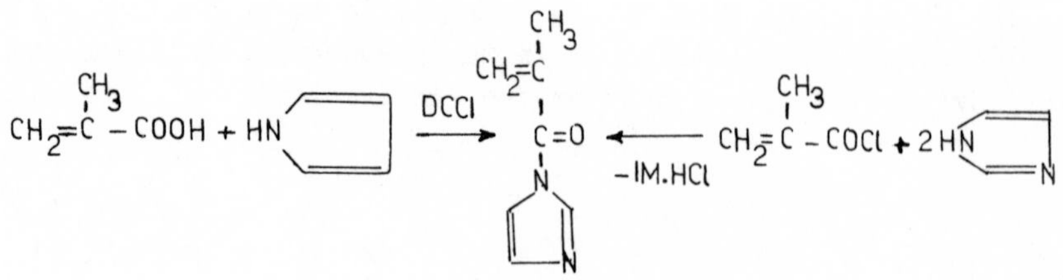

Scheme 15

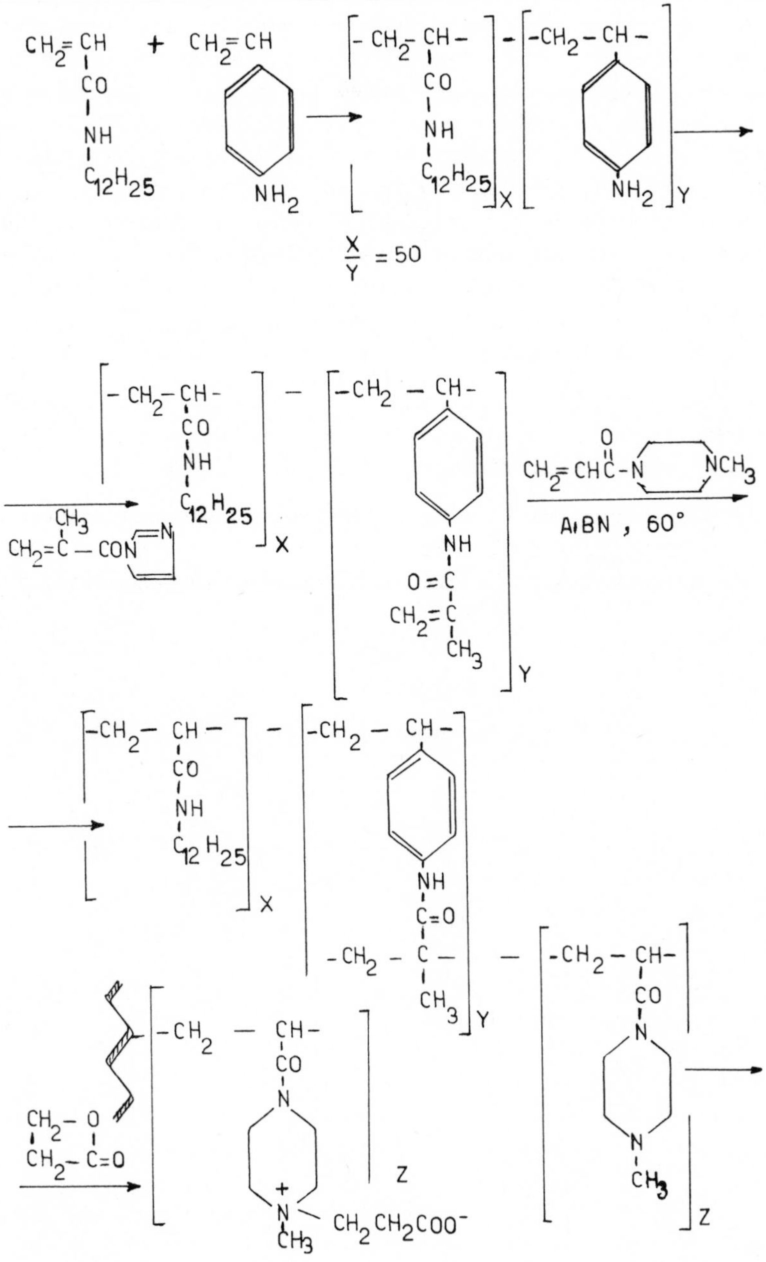

$CH_2{=}CH{-}CO{-}NH{-}C_{12}H_{25}$ + $CH_2{=}CH{-}C_6H_4{-}NH_2$ → $[-CH_2-CH(CONHC_{12}H_{25})-]_X-[-CH_2-CH(C_6H_4NH_2)-]_Y$ →

$\frac{X}{Y} = 50$

→ ($CH_2{=}C(CH_3){-}CON$-imidazole) $[-CH_2-CH(CONHC_{12}H_{25})-]_X-[-CH_2-CH(C_6H_4NHC(=O)C(CH_3){=}CH_2)-]_Y$ → ($CH_2{=}CHC(=O){-}N(C_4H_8)NCH_3$; AIBN, 60°)

→ $[-CH_2-CH(CONHC_{12}H_{25})-]_X-[-CH_2-CH(C_6H_4NHC(=O)C(CH_3)(-CH_2-))-]_Y-[-CH_2-CH(CO{-}N(C_4H_8)N{-}CH_3)-]_Z$ →

→ (β-propiolactone, $CH_2{-}O{-}C(=O){-}CH_2$) $[-CH_2-CH(CO{-}N(C_4H_8)N^{+}(CH_3)CH_2CH_2COO^{-})-]_Z$

to a high molecular weight polymer (Scheme 16). This polymer was able to react with amines in a way very similar to the poly-acryloxy-succinimide (Scheme 17) described previously. Poly-1-acryloyl-benzo-triazole was also able to react with alcohols and aromatic amines (Scheme 17). For example, we caused it to react with methanol or cy-clohexanol. The products were found by infra-red spectroscopy to be identical with authentic samples of poly-methylacrylate and poly-cy-clo-hexylacrylate, prepared by radical polymerization of the corre-sponding monomers. We also reacted it with allyl alcohol, thus obtai-ning a soluble and presumably linear poly-allylacrylate. Allylacryla-te does not yield a linear, poly-vinyl-type poly-allylacrylate on ra-dical polymerization, but it behaves in a way similar to the previous-ly cited allylacrylamide.

The exchange ability of poly-acryloylbenzotriazole with aromatic amines was tested with aniline. It was found that pure poly-N-phenyl-acrylamide could be obtained.

Finally, when poly-1-acryloyl-benzotriazole was reacted with compounds containing both amino and hydroxyl groups, only the former reacted, as illustrated by the case reported in Scheme 18.

Scheme 16

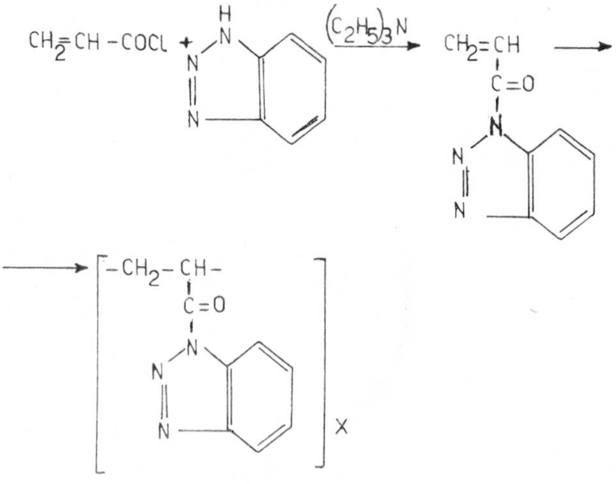

$CH_2{=}CH-COCl$ + benzotriazole $\xrightarrow{(C_2H_5)_3N}$ $CH_2{=}CH-C(=O)-N$(benzotriazolyl) $\longrightarrow$ $[-CH_2-CH(-C(=O)-N\text{(benzotriazolyl)})-]_x$

Scheme 17

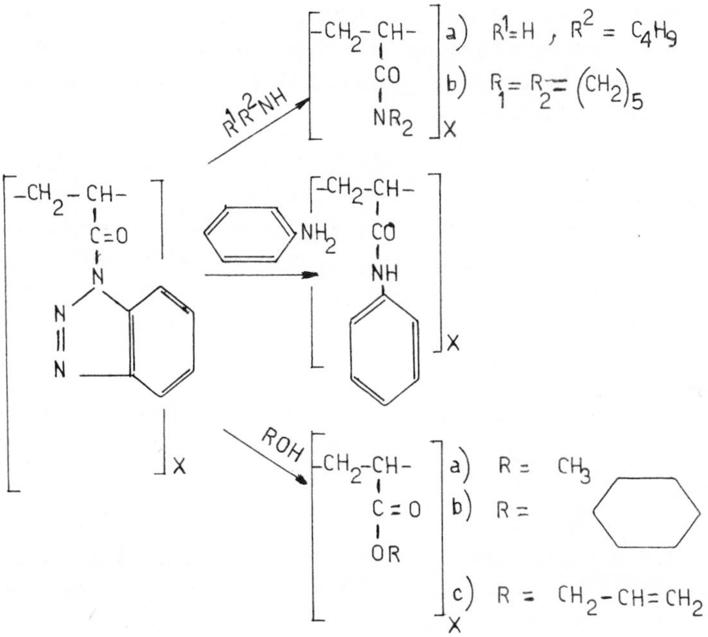

$$\left[-CH_2-CH(C{=}O)-N\text{(benzotriazol-1-yl)}\right]_x \xrightarrow{R^1R^2NH} \left[-CH_2-CH(CO-NR_2)-\right]_x$$

a) $R^1 = H$, $R^2 = C_4H_9$
b) $R_1 = R_2 = (CH_2)_5$

$$\left[-CH_2-CH(C{=}O)-N\text{(benzotriazol-1-yl)}\right]_x \xrightarrow{C_6H_5NH_2} \left[-CH_2-CH(CO-NH-C_6H_5)-\right]_x$$

$$\left[-CH_2-CH(C{=}O)-N\text{(benzotriazol-1-yl)}\right]_x \xrightarrow{ROH} \left[-CH_2-CH(C{=}O-OR)-\right]_x$$

a) $R = CH_3$
b) $R =$ cyclohexyl
c) $R = CH_2-CH=CH_2$

Scheme 18

$$\left[-CH_2-CH(CO-N\text{(benzotriazol-1-yl)})-\right]_x \xrightarrow{H_2N\,CH_2CH_2OH} \left[-CH_2-CH(CO-NH-CH_2CH_2OH)-\right]_x$$

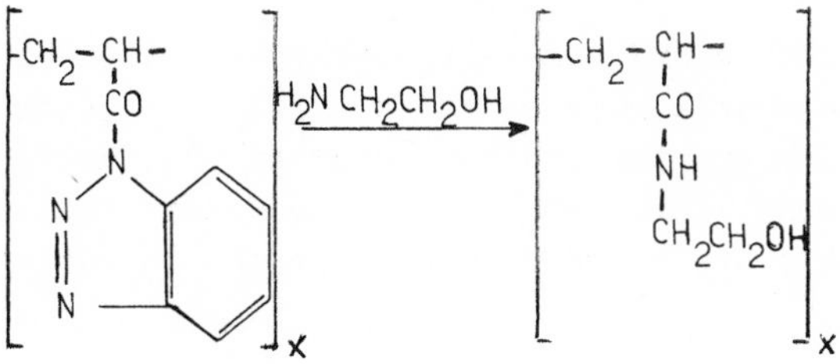

Conclusions

We are now able to draw some conclusions about the functionalization of polymers by coupling reactions of suitable chemical groups present in the polymers with polyfunctional substances, either of low or high molecular weight.

This method allows, at least in principle, the introduction of any function or set of functions in the side chains. Hence, it may be considered the most generally applicable method to synthesize multifunctional polymers. Furthermore, by proper selection of precursor functions, a single "mother" macromolecule may be utilized as the starting material for the synthesis of whole families of multifunctional polymers.

The coupling groups may be introduced either by a modification of a preformed polymer, often of conventional type, or by homo- or copolymerization of purpose-tailored monomers.

The shortcomings of the first method are essentially the same as those outlined in the first section of this paper when dealing with the functionalization of conventional polymers. They may be summarized in unsatisfactory control of the amount and disposition of the coupling groups, as well as in the possibility of side reactions, thus introducing some undesired functions at this stage. These factors are of no great importance, however, when covalent binding of proteins, especially enzymes, to polymeric matrices has to be achieved for industrial purposes.

The synthesis and homo- or copolymerization of purpose-tailored monomers does enable much better control to be exerted on the amount of the coupling groups, and may suppress the danger of untoward side-reactions. Hence, this method is particularly indicated when multifunctional polymers are to be prepared for pharmaceutical or biomedical applications.

References

1. G.D. Jones, in "Chemical Reactions on Polymers" edited by E.M. Fettes, Interscience, New York (1964), pag. 250-255

2. G.D. Jones, in "Chemical Reactions on Polymers" pag. 248

3. T.Alfrey, Jr., in "Chemical Reactions on Polymers" pag. 2

4. R.M.Wheaton and M.J.Hatch in "Ion Exchange", edited by J.A. Marinsky, M.Dekker, N.York (1969), vol. 2, pag. 191; F.Helffrech, "Ion Exchange", Mc Graw-Hill Book Co., Inc., N.York (1962); I.M.Abrams and L.Benezza, "Encyclopedia of Polymer Science and Technology" edited by H.F.Mark, B.G. Gaylord and N.M.Bikales, Interscience, New York (1967), Vol. 2, pag. 692.

5. K.Fries and G.Fickewirth,Berichte, 41 367 (1908); H.Kunz-Krause and P.Manicke, Arch. Pharm. 267 555 (1929); A.R. Bader, J.Am.Chem.Soc. 77 4155 (1955); W.J.Dale and H.E.Hennis, J.Am.Chem.Soc. 80 3645 (1958).

6. B.J.F. Hudson and R.Robinson, J.Chem.Soc. 715 (1941); W.J.Dale and H.E.Hennis, J.Am.Chem.Soc. 80 3645 (1958)

7. W.J.Dale and H.E.Hennis, J.Am.Chem.Soc. 80 3645 (1958)

8. B.Carson, W.J.Heintzelmann, L.H.Schwartzmann, H.E.Tiefental, R.J.Lokken, J.E.Nickels, G.R.Atwood, and F.J.Pavlik, J.Org. Chem. 23 544 (1958)

9. C.G.Overberger, J.C.Salamone, and S.Yaroslawsky, J.Am.Chem. Soc. 89 6231 (1967)

10. R.C.Sovish, J.Org.Chem. 24 1345 (1959); F.Danusso, P.Ferruti and C.Gazzaniga Marabelli, Chim. Ind. (Milan) 47 55 (1965); M.Kato, J.Polym.Sci. A-1, 7 2175 (1969)

11. M.Kato, J.Polym. Sc. A-1, 7 2405 (1969)

12. H.G.Cassidy, M.Ezrin, and I.H.Updegraff, J.Am.Chem.Soc. 75 1615 (1953); M.Ezrin, I.H.Updegraff and H.G.Cassidy, J.Am. Chem.Soc. 75 1610 (1953)

13. H.G.Cassidy and K.A.Kun, "Oxidation-Reduction Polymers", Interscience, N.York (1965)

14. C.G.Overberger, J.C.Salamone and S.Yaroslawsky, J.Am.Chem. Soc. 89 6231 (1967

15. F.Danusso, P.Ferruti and C.Gazzaniga Marabelli, Chim. Ind. (Milan) 47 493 (1965)

16. P.Ferruti, Chim. Ind. (Milan) 47 496 (1965)

17. G.Moraglio, P.Ferruti and A.Feré, Chim. Ind. (Milan) 50 742 (1968)

18. J.D.Reynolds, J.A. Cathcart and J.L.R.Williams, J.Org.Chem. 18 1709 (1953); K.A. Kun and H.G.Cassidy, J.Polym.Sci. 56 83 (1962); G.Manecke and G.Bourwieg, Chem.Ber. 92 2958 (1959); G.Manecke and H.J.Forster, Makromolek. Chem. 52 147 (1962)

19. P.Ferruti and A.Feré, J.Polym.Sci.C 9 3671 (1971)

20. H.W.Schlipköter and A.Brockaus, Klin. Wschr. 39 315 (1961) H.W.Schlipköter, R. Dolguer and A.Brochaus, Dtsch.Med.Wschr. 39 1895 (1963); G.Natta, E.C.Vigliani, F.Danusso, B.Pernis, P.Ferruti and M.A.Marchisio, Rend.Accad.Naz.Lincei 40 11 (1966); E.C.Vigliani, B.Pernis, M.A.Marchisio, P.Ferruti and E.Parazzi, 15th Congrés International du Médicine du Travail, Wien, Spt. 19-24, 2 665 (1966)

21. M.A.Marchisio, C.Sbertoli, G.Farina and P.Ferruti, Europ. J. Pharmacol. 12 236 (1970); M.A.Marchisio, T.Longo, P.Ferruti and F.Danusso, Europ.Surgical Res. 3 240 (1971); M.A. Marchisio, P.Ferruti and T.Longo, Europ.Surgical Res. 4 312 (1972); M.A.Marchisio, T.Longo and P.Ferruti, Experientia 29 93 (1973)

22. F.Danusso and P.Ferruti, Polymer 11 88 (1970)

23. F.Danusso, P.Ferruti and G.Peruzzo, Atti Accad. Naz. Lincei 39 498 (1965)

24. P.Ferruti, D.Pocar and G.Bianchetti, Gazz. Chim. Ital. 97 109 (1967)

25. F.Danusso, P.Ferruti and A.Feré, Europ. Polym. J. 6 1261 (1970)

26. J.Petit and B. Houel, Compt. Rend. 246 1427 (1958); B.Houel, Compt. Rend. 246 2488 (1958); H.L.Cohen and L.M.Minsk, J. Org. Chem. 24 1404 (1959); H.L.Cohen, D.G. Borden and

L.M.Minsk, J.Org.Chem. 26 1274 (1961)

27. A.Veno and C.Schnerch, J.Polym.Sc. (B) 3 3 (1965)

28. P.Ferruti and A.Bettelli, Polymer 13 184 (1972)

29. P.Ferruti, F.Danusso, G.Franchi, N.Polentarutti and S.Garattini, J. Med. Chem. 16 496 (1973)

30. C.H.H.Neufeld and C.S.Marvel, J.Polym. Sc. A-1, 5 537 (1967)

31. H.L.Panzik and J.E.Mulvaney, J.Polym.Sci. Polym. Chem. Ed. 10 3469 (1972)

32. P.Ferruti, G.Audisio, G.Cottica and A.Feré, in preparation

33. P.J.Flory, J.Am.Chem.Soc. 61 1518 (1939)

34. H.C.Isliker, Ann. N.Y.Acad.Sci. 57 225 (1953)

35. N.Grubhofer and L.Schleith, Z.Physiol. Chem. Hoppe-Seyler's, 297 108 (1954)

36. H. Brandemberger, Angew. Chem. 67 661 (1955); J.Polym.Sci. 20 215 (1956); Helv.Chim.Acta 40 61 (1957)

37. G.Manecke, Pure and Appl.Chem. 4 507 (1962)

38. L.Goldstein, Communication at the International Symposium on Insolubilized Enzymes, Milan, 15-16 June, 1973

39. J.F.Kennedy, Communication at the International Symposium on Insolubilized Enzymes, Milan, 15-16 June, 1973

40. R.C.Schultz, P.Elzer and W.Kern, Chimia 13 235 (1959); R.C.Schultz, P.Elzer and W.Kern, Makromol.Chem. 42 197 (1961); S.Boyer and A.Rondeau, Bull.Soc.Chim.France 240 (1958); P.E.Platz, J.Polym.Sci. 58 755 (1962)

41. P.Ferruti, A.Bettelli and A.Feré, Polymer 13 462 (1972)

42. L.Trossarelli, M.Guaita, A.Priola and G.Saini, Chim. Ind. (Milan) 46 1173 (1964); W.Kawai, J.Polym.Sci. A-1, 4 1191 (1966)

43. H.A.Staab, Chem. Berichte 90 1320 (1957)

44. P.Ferruti, A.Feré and G.Cottica, submitted to the "Journal of Polymer Science"

DISCUSSION SESSION

Discussion Leader - G. Challa

Harwood: How did you obtain copolymerization of N-vinyl morpholine and acrylonitrile starting from the addition product, 1,1 dimorphlolinoethane?

Ferruti: BY mixing 1,1 dimorpholinoethane(the enamines of acetaldehyde are in equilibrium with the amine and the vinylamine)with acrylonitrile in the presence of AIBN and heating at 60-80°. The copolymers precipitate in a swollen state during the reaction and their composition varies with the monomer feed. Up to about 40 mole % morpholine monomer could be incorporated in the copolymers. Since the dissociation of 1,1 dimorpholinoethane gives equimolar amounts of N-vinyl morpholine and morpholine, one molecule of β cyanoethyl morpholine is produced for each molecule of N-vinyl morpholine which is incorporated into the copolymer. The radical copolymerization of enamines with acrylonitrile is a general reaction (F. Danusso, P. Ferruti and A. Fere, Europ. Polymer J. 6 1261 (1970)).

Daly: Is it possible to copolymerize N vinyl morpholine with monomers other than acrylonitrile?

Ferruti: Some of our unpublished work indicates that copolymers with methyl methacrylate can also be obtained.

Salamone: Has anyone tried to homopolymerize Schiff bases?

Ferruti: The Schiff bases we used did not homopolymerize and I would not expect such polymers to be stable.

Smets: Could I have further information about copolymerization of Schiff bases with acrylonitrile?

Ferruti: Even hindered Schiff bases, such as those derived from ketones, form copolymers with acrylonitrile. I would also like to point out that the copolymers may have two types of backbones:

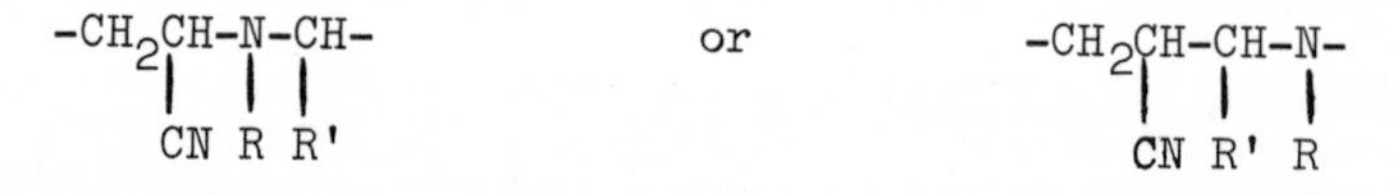

The $-CH_2-\dot{C}H-CN$ radical is electrophillic and should attack a C=N group at the nitrogen atom giving a carbon radical end unit. On the other hand, since a nitrogen radical is more stable than a carbon radical, it might be formed preferentially. Consequently, both types of structure are to be expected, a priori. A study to elucidate this point is in progress.

Bamford: I should expect that the copolymerization of Schiff bases is confined to monomers with electron deficient double bonds. Is it also possible to copolymerize with maleimide?

Ferruti: I agree. I have not yet tried such monomers other than acrylonitrile and methyl methacrylate.

Daly: What degrees of conversion were obtained for the polymer modification reactions which you reported?

Ferruti: They were all quantitative.

Smets: What were the experimental conditions for the reaction of amines or alcohols with polyacryloxy succinimide or polyacryloyl-benzotriazole?

Ferruti: The reaction of primary or secondary amines with both polymers took place after mixing the polymers dissolved in DMF or $CHCl_3$ with an excess of amine. The reaction was complete in 10-15 min. The reaction of alcohols with poly(acryloylbenzotriazole) was brought to completion by mixing a solution of the polymer in $CHCl_3$ or dioxane with a 1.5 fold excess of the alcohol and of tri-ethyl amine, and heating the mixture at 60-80° for 12-24 hr.

Goethals: I would like to mention our synthesis of polymers from trimethyl azetidine (H. Schacht and E. Goethals, Makromol. Chem. <u>167</u>, 155 (1973)):

$$(H_3C)_2C\langle CH_2, CH_2\rangle N\text{-}CH_3 \xrightarrow{Et_3O^+BF_4^-} \text{living polymer}$$

The growing end group is a quaternary ammonium group, as could be shown by NMR .

Smets: Are the polymers branched?

Goethals: Yes. We have measured 6-10% of tertiary amine groups.

Ferruti: Is it possible to prepare block copolymers with different azetidines?

Goethals: I did not prepare them, but I think it should be possible.

A KINETIC STUDY OF THE QUATERNIZATION OF TERTIARY AMINE FUNCTIONAL GROUPS IN THE POLY(VINYLPYRIDINE) SERIES WITH ALKYL BROMIDES

Claude Loucheux

Universite des Sciences et Techniques de Lille
Boite postale 36
59650 Villeneuve D' ASCQ, France

INTRODUCTION

The aim of this paper was to study the reactivity of a functional group in a small molecule in comparison to that group bound to a macromolecular backbone. The main parameters which determine the value of the reaction velocity constant of such a functional group are:

- Steric hindrance
- That one or both of the two neighbors of a considered reactive group may have already reacted, in the case of polymer bound groups.

Polyvinylpyridine was chosen to investigate the influence of these parameters.

Copolymers of low vinylpyridine content and styrene were prepared to measure the reactivity of a pyridine group isolated on the macromolecular backbone. In addition homopoly(vinylpyridines) were examined.

The chosen reaction was the quaternization reaction. It is easily carried out in solvents of high dielectric constants such as tetramethylene sulfone (TMS). All of the types of quaternizing reagents may be studied but, alkyl bromides (especially butyl and octyl) have been used.

Poly(2 vinylpyridine), poly(2 methyl 5 vinylpyridine) and poly(4 vinylpyridine) have been prepared by classical radical polymerization.

The quaternization reaction may be written:

$$\text{-CH-CH}_2\text{- (4-methylpyridine ring)} + C_4H_9Br \rightarrow \text{-CH-CH}_2\text{- (N}^+\text{-}C_4H_9\text{ pyridinium ring, 4-}CH_3\text{)}\ Br^-$$

Potentiometric titration of Br^- ion gives the extent of the reaction at any time. Kinetic data have been interpreted according to a second order rate law:

$$\frac{dx}{dt} = K(a-x)(b-x) \tag{1}$$

a = initial concentration of butyl bromide
b = initial concentration of pyridine groups
x = Br^- concentration at time t
K = velocity constant.

Introducing $\tau = \frac{x}{b}$ and integrating (1) yields (2)

$$\frac{1}{a-b} \ln \frac{1 - \frac{b}{a}\tau}{1 - \tau} = F(\tau) = Kt \tag{2}$$

The function $F(\tau)$ should be linear versus time if a second order law is obeyed.

All the experimental results are given for small model compounds, the copolymer and the homopolymer corresponding to the same repeat unit.

RESULTS AND DISCUSSION

$F(\tau)$ functions have been plotted for low molecular weight model compounds:

2 Ethylpyridine (2EP) = model for 2 vinylpyridine unit

2 Ethyl 5 picoline (5E2P) = model for 2 methyl 5 vinylpyridine unit

and for copolymers:

ST/2VP copolymer 17% 2 vinylpyridine units

ST/4VP copolymer 20% 4 vinylpyridine units

ST/2M5VP copolymer 20% 2 methyl 5 vinylpyridine units.

For ST/4VP, ST/2M5VP copolymers and for the two model small molecules studied, the F(τ) function plotted versus time is linear up to extents of reaction of 100% (see Figures 1 and 2). Values of K may then be calculated from the slopes of the curves. Table I collects these values for the quaternization with butyl bromide at 70°C.

Table I

Material	a (mole ℓ^{-1})	b (mole ℓ^{-1})	K_1 (mole^{-1} mn^{-1})
4 Ethylpyridine*	0.23	0.09	28.4×10^{-3}
ST/4VP Copolymer	0.23	0.07	25.3×10^{-3}
5E2P	0.24	0.08	35.8×10^{-4}
ST/2M5VP Copolymer	0.23	0.06	39×10^{-4}
2 EP	0.23	0.08	9×10^{-4}

*Results from Fuoss and coworkers[1]

In the limit of experimental error, the values of K may be considered the same for the model compounds and for groups situated between two unreactive groups on the polymer chain, i.e., in the cases of 4VP and 2M5VP units.

The situation is quite different for the ST/2VP copolymer. At the beginning of the reaction a very small part of the curve F(τ) versus time may be considered as linear. After 5% of reaction, the plot exhibits a decided downward curvature (see Fig. 3). From the "linear" part of the curve, a value for K may be deduced. This value is 2.9×10^{-1} mole ℓ^{-1} and is three times less than the corresponding value for 2EP given in Table I. This behavior may only be explained by the very poor accessibility of the nitrogen atom situated so close to the macromolecular backbone.

The behavior of homopolymers is represented in the curves of

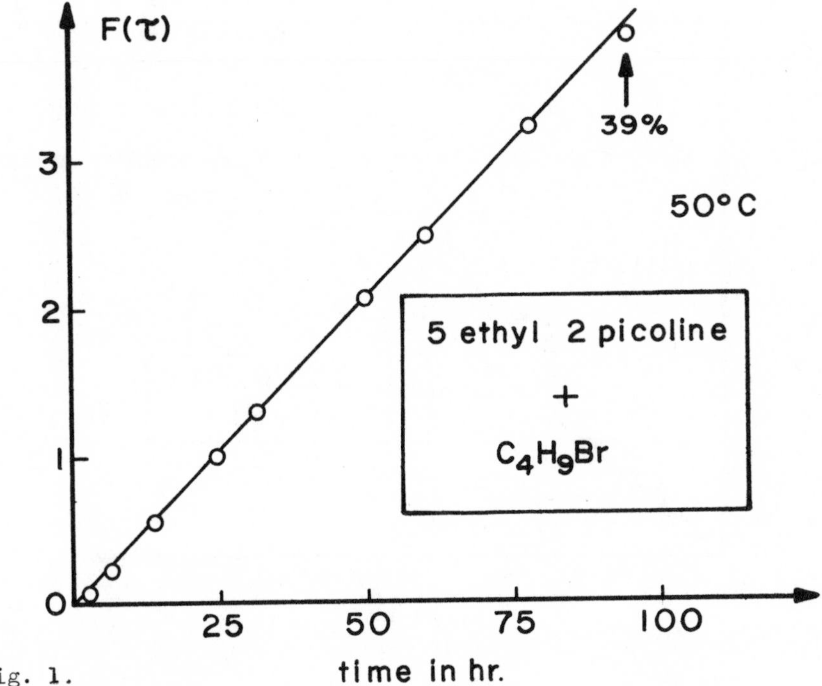
F(τ)
39%
50°C
5 ethyl 2 picoline
+
C4H9Br
0
1
2
3
25
50
75
100
time in hr.

Fig. 1.

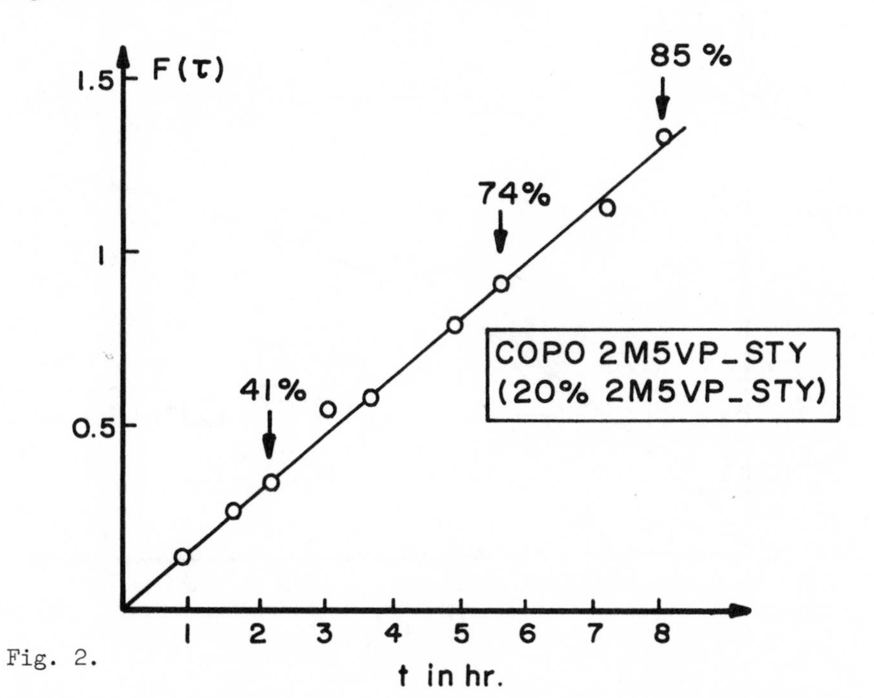
F(τ)
85 %
74%
41%
COPO 2M5VP_STY
(20% 2M5VP_STY)
1.5
1
0.5
1
2
3
4
5
6
7
8
t in hr.

Fig. 2.

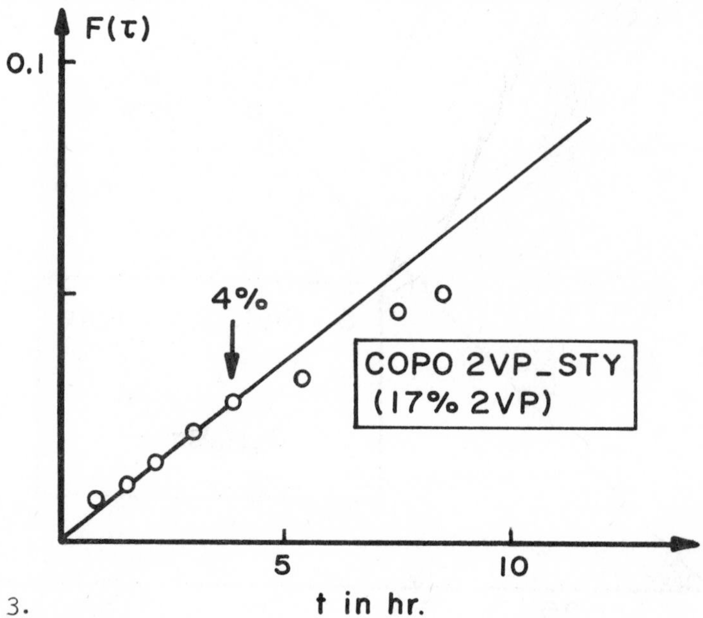
F(τ)
0.1
4%
COPO 2VP_STY
(17% 2VP)
5
10
t in hr.

Fig. 3.

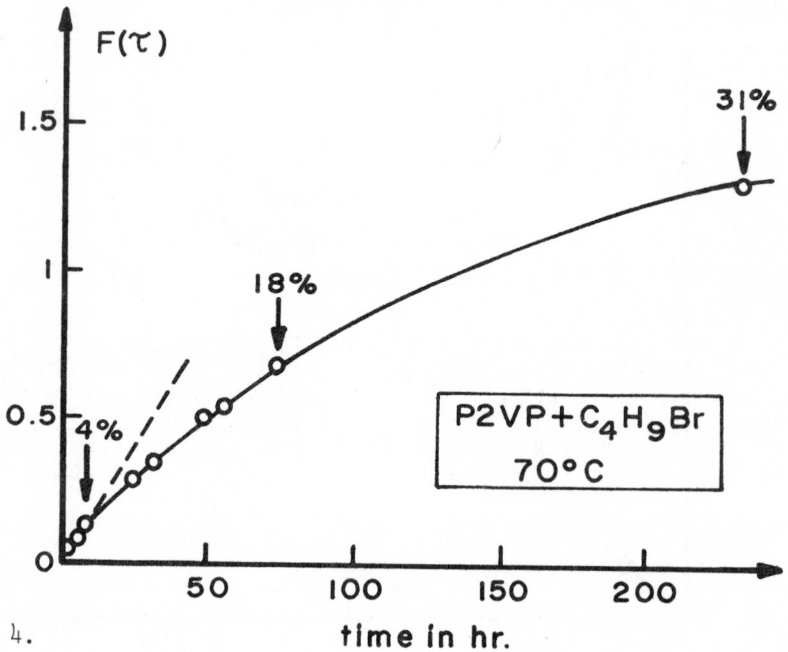
F(τ)
1.5
1
0.5
0
4%
18%
31%
P2VP+C4H9Br
70°C
50
100
150
200
time in hr.

Fig. 4.

Figures 4, 5 and 6. Each of these curves exhibits a linear part until a limiting value which depends stronly on the nature of the repeat unit.

τ_{limit}	50%	for Poly(4 vinylpryidine) (P4VP)
τ_{limit}	30-35%	for Poly(2 methyl 5 vinylpyridine) (P2M5VP)
τ_{limit}	4%	for Poly(2 vinylpyridine) (P2VP) (if a linear part of curve exists at all)

From these linear portions, initial values of the velocity constant K may be calculated. They are given in Table II for quaternization by butyl bromide at 70°C.

Table II

Material	a mole ℓ^{-1}	b mole ℓ^{-1}	K mole^{-1} mn^{-1}
P4VP	0.23	0.06	17.4×10^{-3}
P2M5VP	0.54	0.34	27.65×10^{-4}
P2VP	0.46	0.15	2.03×10^{-4}

For P4VP it is necessary that both neighbors of a considered group have reacted to change its reactivity. The initial velocity constant has decreased from 28.4×10^{-3} to 17.4×10^{-3} if comparison with the model compounds is made. This difference may be a measurement of the steric hindrance due to the macromolecular backbone.

For P2M5VP, one quaternized neighbor only is sufficient to perturb the reactivity of a considered group. Here the comparison with 5E2P ($K = 35.8 \times 10^{-4}$) also gives a measurement of the steric hindrance of the backbone.

For P2VP the behavior observed is very close to that of the ST/2VP copolymer. However, the initial value of K is four times less than the corresponding value for 2EP.

The change of kinetic order with τ may be discussed, adopting a point of view similar to that of Fuoss and coworkers. It is possible to write:

$$K_t = k_0 e^{-\alpha t} + K_2(1-e^{-\alpha t})$$

where

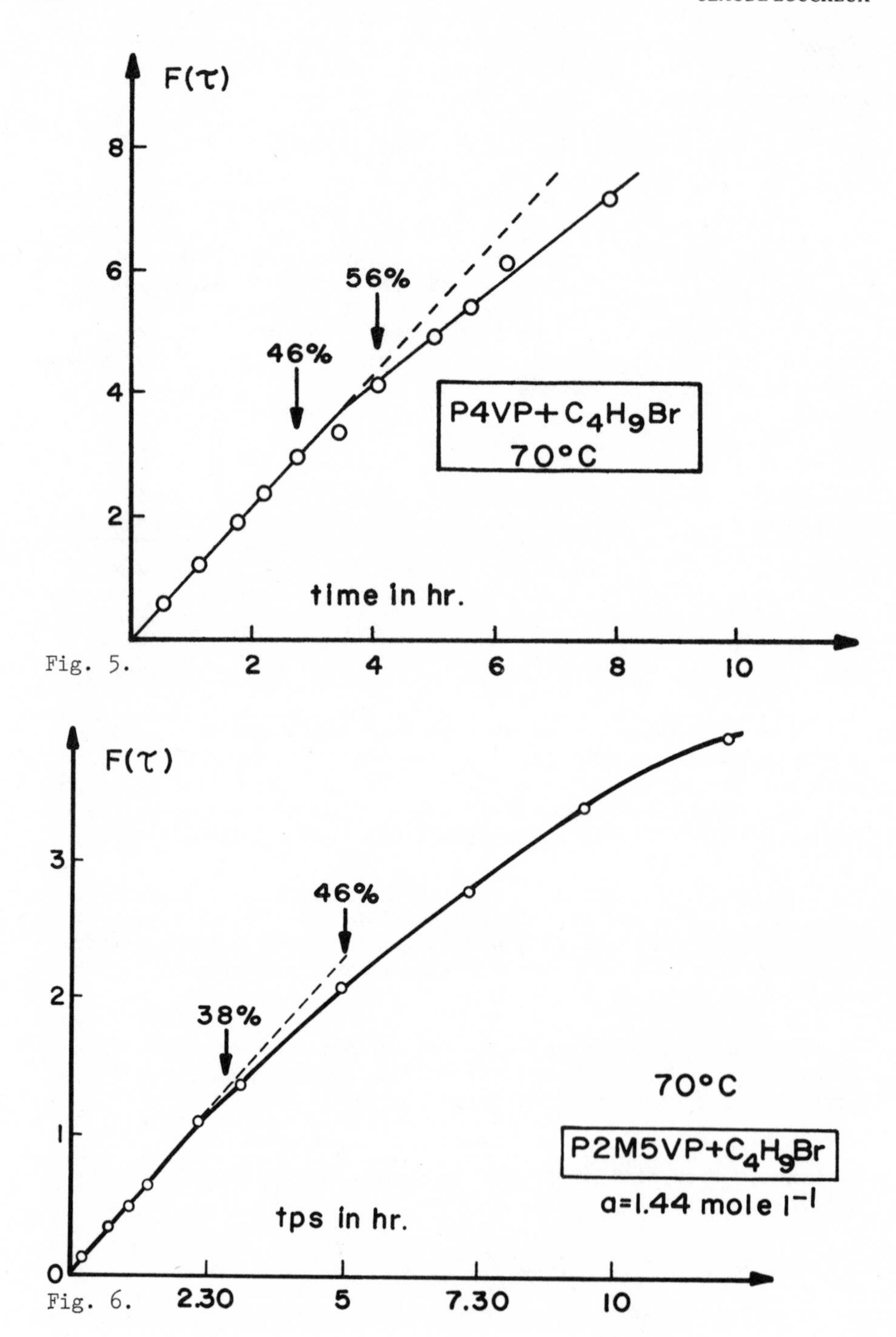
F(τ)
8
6
4
2
56%
46%
P4VP+C4H9Br
70°C
time in hr.
2
4
6
8
10
F(τ)
3
2
1
0
46%
38%
70°C
P2M5VP+C4H9Br
a=1.44 mole l⁻¹
tps in hr.
2.30
5
7.30
10

Fig. 5.

Fig. 6.

K_t = velocity constant at time t

K_0 = initial velocity constant

K_2 = final velocity constant

α = adjustable constant

Then

$$\frac{dx}{dt} = K_0 e^{-\alpha t} + K_2(1-e^{-\alpha t})(a-x)(b-x)$$

After integration:

$$\frac{F(\tau)}{t} = K_2 + (K_0-K_2)\ f(\alpha t)$$

with $$f(\alpha t) = \frac{1 - e^{-\alpha t}}{\alpha t}$$

Practically, K_0 and K_2 are obtained when $\frac{F(\tau)}{t}$ is linearized with a suitable value of α.

This calculation has been made for P2M5VP. The values of velocity constants are:

$K_0 = 24.6 \times 10^{-4}$ mole^{-1} mn^{-1}

$K_2 = 2.6 \times 10^{-4}$ mole^{-1} mn^{-1}

$$\frac{K_0}{K_2} \simeq 10$$

It would probably be more convenient to describe the kinetic results for P2M5VP by introducing three velocity constants instead of two, but that has not yet been done.

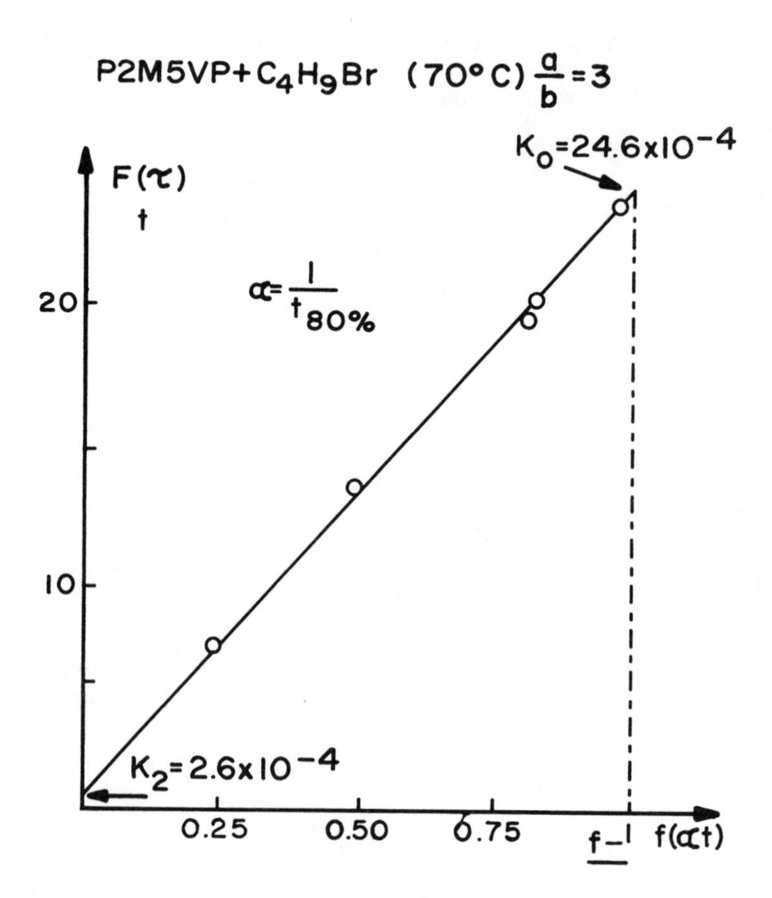

Fig. 7.

REFERENCES

1. R. M. Fuoss and Coleman, J. Amer. Chem. Soc., 77, 5472 (1955)

INSOLUBLE HALOGENATING AGENTS

Carmela Yaroslavsky

The Weizmann Institute of Science
Rehovot, Israel

Insoluble polymeric reagents are being used extensively in various fields. Due to their insolubility, they are easily separated from the reaction mixture and then available for reuse. The widespread use of such reagents is mainly a result of this technical advantage.

Insoluble polymeric reagents are being employed as catalysts, for exchange reactions and as insoluble protecting groups. In addition, specific affinities of functional groups bound to the insoluble polymeric backbone can be used in order to selectively isolate the desired compound from a mixture, or to obtain high concentrations of a compound out of dilute solutions. Multifunctional reagents, immobilized on polymeric carriers, will undergo intramolecular, rather than intermolecular reactions. In all of these examples the polymeric backbone serves as a carrier. The only prerequisites are that the carrier remains stable and undamaged during reaction or recovery, and that it possesses appropriate physical properties for the reaction involved. The influence of the polymeric backbone on the reactivity of the attached functional groups has been investigated in soluble as well as insoluble polymeric catalysts. The electrostatic charge of the polymer, apolar interactions of substrates with apolar pendent groups on the polymeric catalyst, as well as the neighbouring group effect may facilitate or retard the chemical reaction catalyzed by the polymeric catalyst.

To study the effect of the polymeric backbone on the chemical reactivity of the attached functional group we had to choose a monomeric reagent that will fulfil the following requirements:

A. Sensitivity to the reaction medium, i.e. different products can be obtained due to media of different polarity.

B. Capability of being incorporated into a polymeric backbone.

C. Wide application in organic chemistry.

All of these requirements are fulfilled by N-haloamido-, imido- and sulfonamido- compounds (1).

We started our study with N-bromopolymaleimide (PNBS) cross-linked with divinylbenzene, as an insoluble polymeric analog of the widely used low molecular weight reagent, N-bromosuccinimide (NBS) (2). Different reaction products were obtained when olefins and alkylaromatic compounds were reacted with PNBS and NBS in boiling carbontetrachloride and in the presence of free radical initiators.

Thus, the reaction of cumene with PNBS yielded α, β, β'-tribromo-cumene. Neither α-bromocumene nor α, β-dibromocumene were formed. The last two compounds, however, are obtained when N-bromosuccinimide reacts with cumene. Upon reaction of the polymeric brominating agent with ethylbenzene, α-bromo- and α, β-dibromoethylbenzene are obtained, while the analogous reaction of NBS with ethylbenzene yields α-bromoethyl-benzene. Cyclohexene reacts with PNBS to yield 1,2-dibromocyclohexane while the monomeric reagent reacts with cyclohexene to give 3-bromo-cyclohexene.

The unusual reactions which occur in the presence of PNBS may be due to its polymeric structure, by which neighbouring groups effect the chemical properties of the active functional group of the polymer. Indeed, reaction of NBS with benzylic and olefinic compounds in polar media gave similar results to those obtained with PNBS in carbontetrachloride.

The above reactions were performed using free radical initiators. In the absence of free radical initiators and solvent, different reaction products are obtained upon reaction of PNBS and alkylaromatic compounds. For example, reaction of PNBS with toluene yields three different products: bromotoluene, benzylbromide and phenyltolylmethane. The distribution of these products depends on the reaction temperature and the weight ratio of the polymeric reagent to the alkylaromatic substrate.

The following Table illustrates the distribution of products obtained on reaction of PNBS with toluene:

TABLE I

DISTRIBUTION OF PRODUCTS OBTAINED BY REACTION OF PNBS WITH TOLUENE [a]

Polymer/Toluene		Reaction	Product distribution (%)		
Weight ratio	Molar ratio	temp. (°C)	Bromo-toluene	Benzyl-bromide	Phenyltolyl-methane
2.7	0.88	80	> 99	traces	-
2.4	0.78	108	93	7	-
2.4	0.78	145	59	24	17
1	0.3	120	44	30	26
0.5	0.19	130	31	50	19
0.2	0.06	120	traces	>99	traces

[a] Reaction time was 16 hr.

The analogous reaction of NBS and toluene yielded benzyl bromide.

N-Chlorosuccinimide (NCS) is used extensively as an oxidizing agent (3). Its use for chlorination is limited since the chlorinations carried by NCS are not as specific as the brominations obtained by NBS. However, N-chloropolymaleimide (PNCS), the polymeric analog of NCS, proved to be a specific, and therefore useful chlorinating agent (4). In the absence of free radical initiators and solvent, at temperatures of 100-140° C, for 16-20 hr, alkylaromatic compounds reacted with PNCS specifically to yield the corresponding aryl-chlorosubstituted products. In contrast, reaction of the low molecular weight analog, NCS, with alkylaromatic compounds under the same conditions yields amixture of products in which α-chlorinated compounds are prevalent.

We prepared several N-chloro-amido, imido and sulfonamido-polymers. Table II describes the distribution of products obtained by reacting polymeric and low molecular weight reagents with the same alkyl-aromatic compounds.

TABLE II

DISTRIBUTION OF PRODUCTS OBTAINED BY REACTION OF ALKYLAROMATIC COMPOUNDS WITH HIGH AND LOW MOLECULAR WEIGHT CHLORINATING AGENTS

Substrate	Chlorinating agent	Reaction temp.	Product distribution %				
			Aromatic substitution	Side chain substitution			
				α	β	α, α	α, β
t-Butylbenzene	N-chloropolymaleimide	140	100				
t-Butylbenzene	N-chlorosuccinimide	140	86		14		
Cumene	N-chloropolymaleimide	138	100				
Cumene	N-chlorosuccinimide	138		mixture of products			
Ethylbenzene	N-chloropolymaleimide	120	100	-			
Ethylbenzene	N-chlorosuccinimide	120	traces	66	17		17
Toluene	N-chloropolymaleimide	110	100		-		
Toluene	N-chlorosuccinimide	110	16	66		17	
Toluene	N-chloropolyureaformaldehyde	90	>99	< 1			
Toluene	N-chloropolystyrenesulfonamide	120	98	2			
Toluene	N-chlorobenzenesulfonamide	120	50	50			
Toluene	N-chloropolymethacrylamide	110	>99	< 1			

The difference in specificity of the polymeric reagent and the low molecular weight analog is most probably due to the influence of the polymeric backbone on the chemical reactivity of its functional groups. For example, in PNCS each N-chlorosuccinimide residue is situated on the polymer in a polar environment provided by adjacent succinimide residues. As was expected, marked changes occurred in the distribution of products when alkyl aromatic compounds reacted with NCS in the presence of succinimide. The predominant product, under these conditions, was the aryl-chloroalkylaromatic compound, rather than the side chain substituted product prevailing in the absence of succinimide.

REFERENCES

1. HORNER, L. & WINKELMANN, E.H. Methods of Prep. Org. Chem. 3:151, 1964.
2. YAROSLAVSKY, C., PATCHORNIK, A. & KATCHALSKI, E. Tetrahedron Letters, 3629, 1970.
3. FILLER, R. Chem. Rev. 63:21, 1963.
4. YAROSLAVSKY, C. & KATCHALSKI, E. Tetrahedron Letters, 5173, 1972.

MODIFICATION OF POLYMERS USING IODINE ISOCYANATE. II. IMPROVED METHODS FOR IODINE ISOCYANATE ADDITION TO POLYISOPRENE AND SOME BIOMEDICAL IMPLICATIONS.*

Charles G. Gebelein and Audrey Baytos

Department of Chemistry,
Youngstown State University,
Youngstown, Ohio, 44503

SUMMARY

Polyisoprene can be reacted with iodine isocyanate by adding preformed solutions of iodine isocyanate in tetrahydrofuran to methylene chloride solutions of polyisoprene. In this manner it is possible to react up to 100% of the double bonds in the polymer with iodine isocyanate. These iodoisocyanate derivatives can be converted to methyl iodocarbamate or iodourea derivatives of polyisoprene by reaction with methanol or ammonia, respectively. The synthesis of a N'-(4-sulfonamidophenyl) iodourea derivative (a polyisoprene with pendant sulfanilamide groups) is described.

INTRODUCTION

In an earlier publication we noted that iodine isocyanate could be added to polyisoprene and that the resulting iodoisocyanate could be converted into methyl iodocarbamate and iodourea derivatives (1). We further noted that this reaction could potentially lead to the introduction of such polar functional groups as amines, aziridines, oxazolidones and oxazoles onto the polymer backbone but that these reactions were not always as simple to run as with a low molecular weight alkene. In this present study we wish to report some improved experimental methods for adding iodine isocyanate to polyisoprene and to discuss some biomedical implications of these reaction products.

The addition of iodine isocyanate to alkenes has been studied extensively by several groups (2-6). The reaction appears to occur via an electrophilic attack of positive iodine on the

* This article has been reproduced from a photocopy, as the original typed sheets were not available.

alkene to form an iodonium ion. The attack of the nucleophilic isocyanate group occurs from the backside resulting in an overall trans-addition and this attack normally occurs at the more highly substituted carbon atom. The reaction has been shown to lead to a wide variety of polar nitrogen-containing functional groups.

There are two basic experimental techniques for adding iodine isocyanate to alkenes. The first method involves a two-step process in which a solution of iodine isocyanate is generated by the reaction of iodine on silver cyanate in some appropriate solvent, such as ether or tetrahydrofuran, and this is then added to the alkene (2,6). The second method involves the *in situ* formation of INCO by reacting a mixture of alkene, iodine and silver cyanate in a suitable solvent. While most studies have used solvents which form a brown iodine solution, such as ether or tetrahydrofuran, non-complexing solvents, such as methylene chloride or pentane, can also be used in the *in situ* reaction (7). While most synthetic work reported in the literature used the *in situ* generation method, this technique is not well suited for the addition to unsaturated polymers since the separation of the polymer and the insoluble silver salts would be extremely difficult. Altho the separate generation of INCO (followed by the separation of the silver salts prior to reaction with alkene) does avoid this problem, it introduces other difficulties since the solvents most suitable for INCO generation (ether, tetrahydrofuran and dimethoxyethane) are not particularily good polyisoprene solvents and some good polyisoprprene solvents (methylene chloride, benzene and pentane) are poor solvents for the generation of INCO. Experimentally, it is necessary to generate the INCO as a concentrated solution, in a solvent such as tetrahydrofuran, and add this solution to a solution of the polymer under conditions that avert coagulation.

EXPERIMENTAL

Materials

The polyisoprene sample used in these studies was obtained from Shell Chemical (Shell Isoprene Rubber 309) and had a cis-1,4 content of 91.5% (balance was 1,2 structure) with a molecular weight of about two million. A resublimed grade of iodine was used and the silver cyanate was prepared as described in the literature (4). All other chemicals and solvents were reagent grade or better.

Addition of INCO To Polyisoprene

Several combinations of INCO generation solvent and polyisoprene solvent were examined and the best combination appears to be the use of tetrahydrofuran as the INCO generation solvent and methylene chloride as the polyisoprene solvent. Some typical experimental conditions are summarized in Table I. In all cases the INCO was generated following the procedure of Rosen and Swern (6) and then added to the polyisoprene solution after removal of the silver salts. The INCO generation was run for 1-2 hrs. at about -20°C using a 2:1 mole ratio of $AgOCN:I_2$. The reaction with the polyisoprene was allowed to proceed 2-3 hrs. at room temperature. In all cases, these solutions were homogeneous. The resulting iodoisocyanate modified polymers were converted to other derivatives by three different work-up procedures as described below. The entire procedure was run with the reaction vessels wrapped in aluminum foil to exclude light.

Procedure A: The solutions were divided into two parts. The first part was mixed with an equal volume of methanol to convert the polymer into a methyl iodocarbamate. The sample was allowed to stand overnight or longer. This procedure invariably resulted in coagulating the polymer but the conversion to carbamate was always complete. The second part was treated with an equal volume of a saturated solution of ammonia in ether and allowed to stand overnight or longer. This procedure also lead to coagulation but the iodoisocyanate polymers were always converted to the iodourea polymers. In both cases, the polymers were rinsed thoroughly with methanol and vacuum dried.

Procedure B: In each case 100 ml. methanol was added with stirring and allowed to stand overnight or longer. No coagulation occurred. The solution was concentrated on a flash evaporator and the polymer was isolated by pouring this solution into a large excess of methanol. These derivatives were iodocarbamates.

Procedure C: A solution of 12.9 g. (0.075 M.) sulfanilamide in 70 mls. acetone was added slowly with stirring and stirred overnight. Stirring is essential since the sample has marginal solubility. The polymer was isolated by pouring into a large excess of methanol. This polymer is a substituted iodourea derivative in which a sulfa drug is affixed to the polymer chain at one end of the urea groups.

Table I

Experimental Conditions For The Addition Of Iodine Isocyanate To Polyisoprene.

Run	g. Poly-isoprene	ml. CH_2Cl_2	moles INCO	ml. THF	theory % reaction	type work-up
1	2.72	200	0.004	10	10	A
2	2.72	200	0.01	25	25	A
3	6.80	300	0.025	100	25	B
4	6.80	300	0.025	100	25	C
5	2.72	200	0.02	50	50	A
6	2.72	200	0.03	75	75	A
7	2.72	200	0.04	100	100	A
8	1.70	400	0.03	200	100	B

Polymer Properties

Infrared spectra were run for these polymer samples and the results were consistent with the proposed structures. Some typical spectra were published previously (1). Polymer densities were run on some samples of the methyl iodocarbamate derivatives. These were 1.24 and 1.64 g./cc. at 25% (Run 2) and 100% (Run 7) addition, respectively. Elemental analyses were determined on some samples to confirm the composition.

RESULTS AND DISCUSSION

The equations shown below describe the reactions summarized in Table I. These experiments clearly show that iodine isocyanate can be added to polyisoprene by the preformed INCO solution technique and that any amount of double bond addition, up to 100%, is possible. Similar experiments using the *in situ* method also result in derivative formation but the entrapped silver salts can not readily be removed from the polymer sample.

In all cases, the initial product was the iodoisocyanate (I). Where the percent reaction is less than 100, the remaining double bonds would be expected to be randomly located along the polymer backbone. These iodoisocyanates were converted to methyl iococarbamates (II), iodoureas (III) or a N'-(4-sulfonamidophenyl) iodourea (IV). The latter derivative is actually a "sulfa drug" bound to the polymer backbone. Other derivatives are possible since structure (II) could be converted to an aziridine or an oxazolidone derivative and (III) could be converted into an oxazole as noted earlier (1).

$$-\!\!(CH_2-\overset{\overset{CH_3}{|}}{C}=CH-CH_2)\!\!-_n \xrightarrow{INCO} -\!\!(CH_2-\underset{\underset{NCO}{|}}{\overset{\overset{CH_3}{|}}{C}}-\overset{\overset{I}{|}}{CH}-CH_2)\!\!-_n$$

(I)

$$(I) \xrightarrow{CH_3OH} -\!\!(CH_2-\underset{\underset{NHCOCH_3 \atop \| \atop O}{|}}{\overset{\overset{CH_3}{|}}{C}}-\overset{\overset{I}{|}}{CH}-CH_2)\!\!-_n$$

(II)

$$(I) \xrightarrow{NH_3} -\!\!(CH_2-\underset{\underset{NHCONH_2}{|}}{\overset{\overset{CH_3}{|}}{C}}-\overset{\overset{I}{|}}{CH}-CH_2)\!\!-_n$$

(III)

$$(I) \xrightarrow{NH_2-C_6H_4-SO_2NH_2} -\!\!(CH_2-\underset{\underset{NH-C(=O)-NH-C_6H_4-SO_2NH_2}{|}}{\overset{\overset{CH_3}{|}}{C}}-\overset{\overset{I}{|}}{CH}-CH_2)\!\!-_n$$

(IV)

All these type derivatives are known to possess biological activity in low molecular weight compounds. The question naturally arises- do these polymeric derivatives also possess biological activity? While this question can not be answered completely at this time for these materials, there are several examples in the literature of synthetic polymeric drugs. Sulfonamide copolymers with formaldehyde (8) or dimethylolurea (9) have been shown to possess antibacterial activity and polymers containing phenethylamine groups have been shown to effect the blood pressure in cats (10). Many other examples of biologically active polymers are in the literature (11-13) and, of course, proteins and nucleic acids are polymers. Specific catalytic effects similar to enzymatic activity have been observed for some synthetic copolymers (14). It is beyond the scope of this article to review this field, however, the presence of biological activity in the polymers reported here would not be completely unexpected. Studies are in progress in this area.

The primary purpose of this paper is to report methods of synthesing these modified polymers and to discuss some of the problems inherent in this field. The only satisfactory way to prepare the iodoisocyanate derivatives (I) is to add the preformed INCO solution to a solution of the polymer. While many possible solvent combinations exist, the systems described in the experimental section appear to be the best for several reasons. Tetrahydrofuran is the best INCO generation solvent since it permits the rapid formation of fairly concentrated solutions and the INCO has fair stability in this solvent. The solvent volatility is also low enough to permit easy handling of the INCO solutions. For these reasons, this solvent is better than either ether or dimethoxyethane (6). The maximum concentration of INCO that can be generated in methylene chloride is low (less than 0.06M) even though this is an excellent solvent for *in situ* reactions of alkene, iodine and silver cyanate. The extent of INCO formation in pentane is nil (15). These factors limit the choice of generation solvent severely since polyisoprene is not especially soluble in the solvents that are best for INCO generation.

The choice of polyisoprene solvent is crucial also since it would be undesirable for the polymer to coagulate when the INCO solution is added. Methylene chloride has been found to be very satisfactory here since it permits the use of relatively high polyisoprene concentrations without coagulation occurring when the tetrahydrofuran solution of INCO is added. In this solvent it is even possible to convert the iodoisocyanate to the methyl iodocarbamate (II) under homogeneous reaction conditions (see runs 3 and 8 in Table I). While this conversion could occur under heterogeneous reaction conditions, the rate would be slower. It is worth noting that this addition of INCO is an electrophilic reaction and would proceed faster in the polar methylene chloride solutions than in nonpolar hydrocarbon solutions. The reaction can be run in benzene solutions but the tolerance for tetrahydrofuran is much lower and even small amounts of methanol tend to coagulate the polymer. These difficulties seldom ever arise with simple monomeric alkenes.

The preparation of the N'-(4-sulfonamidophenyl)-iodourea (IV) illustrates the difference between the unsaturated polymers and the simple alkenes quite well. While an iodoisocyanate derivative of a simple alkene (such as cyclohexene) can be prepared in many solvents (e.g. ether, tetrahydrofuran, methylene chloride and pentane) the conversion to the N'-(4-sulfonamidophenyl)-iodourea requires solubility of the sulfanilamide in the media. This effectively restricts the reaction solvent to tetrahydrofuran. The reaction can be run under *in situ* conditions with the simple alkene and the modified sulfa drug can be isolated readily. When an unsaturated polymer is used, the reaction is restricted to the preformed INCO solution route. Sulfanilamide has low solubility in methylene chloride and a small amount of acetone was used to achieve borderline solubility. The isolation of the product is, however, simplier in the case of the polymer since these can be precipitated by pouring into an excess of methanol. The simple alkene derivatives are usually soluble in excess methanol.

In general, the polymeric derivatives are more difficult to prepare than the corresponding product from a monomeric alkene because of these solubility problems but the final products are usually easier to isolate. Studies are continuing to determine more effective techniques for preparing the sulfa drug derivatives and other potentially useful polymers.

This work was partially supported by a grant from the Youngstown State University Research Council.

REFERENCES

1) C. G. Gebelein, *J. Macromol. Sci.-Chem.*, A5, 433 (1971).
2) L. Birckenbach and M. Linhard, *Ber.*, 63, 2544 (1930); 64, 961, 1076 (1931).
3) G. Drefahl and K. Ponsold, *Chem. Ber.*, 93. 519 (1960); G. Drefahl, K. Ponsold and G. Kollner, *J. Prakt. Chem.*, 23, (1964).
4) A. Hassner, M. E. Lorber and C. Heathcock, *J. Org. Chem.*, 32, 540 (1967); A. Hassner, R. P. Hoblitt, C. Heathcock, J. E. Kropp and M. Lorber, *J. Amer. Chem. Soc.*, 92, 1326 (1970) and references cited therein.
5) C. G. Gebelein and D. Swern, *J. Org. Chem.*, 33, 2758 (1968); C. G. Gebelein, S. Rosen and D. Swern, *J. Org. Chem.*, 34, 1677 (1969) and references cited therein.
6) S. Rosen and D. Swern, *Anal. Chem.*, 38, 1392 (1966).
7) C. G. Gebelein, *Chem. Ind.*, 23, 136 (1970).
8) L. G. Donaruma and J. Razzano, *J. Medicinal Chem.*, 14. 244 (1971).
9) J. R. Dombroski and L. G. Donaruma, *J. Medicinal Chem.*, 14, 460 (1971).
10) B. -Z. Weiner, M. Tahan and A. Zilkha, *J. Medicinal Chem.*, 15, 410 (1972).
11) M. K. Vogel, R. A. Cross, H. Bixler and R. J. Guzman, *J. Macromol. Sci. -Chem.*, A4, 675 (1970).
12) T. F. Yen, M. Davar and A. Rembaum, *J. Macromol. Sci. -Chem.*, A4, 693 (1970).
13) A. Rembaum, S. P. S. Yen, R. F. Landel and M. Shen, *J. Macromol. Sci. -Chem.*, A4, 715 (1970).
14) C. G. Overberger and J. C. Salamone, *Acc. Chem. Res.*, 2, 217 (1969).
15) C. G. Gebelein, 158th. Nat. Meeting, Amer. Chem. Soc., New York, N. Y., Sept. 1969, abstract ORGN 93.

DISCUSSION SESSION

Discussion Leader - W. H. Daly

Ferruti: What are the solubility properties of the polyisoprene molecules with oxazolidone or oxazole groups? Are they water soluble?

Gebelein: They are not water-soluble. We have not studied the solubility characteristics fully but I would expect some solubility in methylene chloride or benzene.

Goethals: Have you actually made the derivatives containing the aziridine ring on the chain and, if so, have you tried to polymerize these aziridine groups to obtain a crosslinked polymer?

Gebelein: We have only had limited success in converting the polymeric iodocarbamates into the aziridines, mainly because of solubility problems. The ring closure requires hydroxide ions which are not readily soluble in the solvents which dissolve the iodocarbamate polymers. We have not yet explored crosslinking reactions based on the aziridine groups.

Salamone: Have you attempted to crosslink the isocyanate derivatives by reaction with long chain diols or diamines?

Gebelein: No. We have been working only with soluble polymers.

Carraher: Have you examined block copolymers containing isoprene in your reaction system?

Gebelein: Most of our studies have concentrated on polyisoprene because this yields a single mode of addition (NCO group at the α methyl group). We have examined some copolymers, such as butadiene-styrene copolymers.

Fox: You stated that INCO could not be prepared well in methylene chloride, yet the in situ reaction worked well with iodine and silver cyanate, in this solvent. What is the function of the methylene chloride?

Gebelein: In methylene chloride only dilute solutions (0.60M) of INCO can be prepared. Concentrated INCO solutions are best prepared in oxygen-containing, aprotic solvents (e.g. ether, THF). In the in situ process, the main reaction appears to be between an iodine-alkene complex and the insoluble silver cyanate. This seems to work just as well on the polymers but separation of the modified polymer and the residual silver salts is very difficult.

Bamford: Are any of these INCO additions slow?

Gebelein: The reaction with 1 alkenes is fairly slow.

Bamford: Could you do a photolytic or radical catalyzed addition in these slow reactions? This should result in the attachment of the iodine atom at the methyl-bearing carbon atom.

Gebelein: We have always attempted to exclude other reaction pathways, e.g. by excluding light from the reaction vessels.

Smets: The INCO molecule may be too polar for photolytic cleavage to occur.

Rivin: What conditions were used for the cyclization reaction?

Gebelein: The cyclization to oxazolidone and oxazole rings are done thermally above 125° for the methyl iodocarbamate or iodourea derivatives of polyisoprene. The formation of the aziridine ring requires strong base, such as hydroxide ions.

Daly: Conversion of the addition products to _tert_- butyl carbamates would yield a system which should be more soluble and should also thermolyze readily , perhaps to the aziridine-containing polymer.

Gebelein: Thermal ring closure of the alkyl carbamates leads to the formation of an oxazolidone ring.

Daly: I think elimination of isobutylene might be competitive and you would also have the alternative of using acid hydrolysis on that particular blocking group. Have you tried to use oranic-soluble bases, e.g. Triton-B or sodium amyloxide,on the methyl carbamates?

Gebelein: No, but I do plan to use some of these and also crown compounds in this reaction.

DeLaMare: Do you have any relative rate data on the reaction of INCO with the _cis_- and _trans_-isomers of polydienes?

Gebelein: No.

DeLaMare: Since INCO is an electrophillic reagent, would you expect aromatic substitution on a block copolymer containing styrene?

Gebelein: Not under our reaction conditions. The reaction of INCO with styrene monomer, while slow,occurs solely at the double bond.

Lewin: Have you considered the possibility of using the reaction as an analytical tool? Have you considered running this reaction solely at a solid surface?

Gebelein: We have avoided surface reactions.

Ferruti: Did your sulfonamido polymers show any biological activity?

Gebelein: They have not been evaluated as yet.

Moore: Have you attempted to react the INCO with other funcional groups on a polymer, such as hydroxyl, rather than the double bond?

Gabelein: No.

Moore: Such iodine-rich polymers might be useful for X-ray diagnostic work.

Lewin: It might be possible to use this material or something similar to introduce antibacterial properties into textiles,but iodine might be too expensive for this purpose. Can the bromo compound be used similarly?

Gebelein: Only INCO functions well in this reaction, Cl NCO and Br NCO tend to dimerize too rapidly.

MODIFICATION OF POLYMERS CONTAINING LEWIS BASE FUNCTIONAL GROUPS VIA CONDENSATION WITH LEWIS ACIDS

Charles E. Carraher, Jr.

University of South Dakota
Department of Chemistry
Vermillion, South Dakota 57069

INTRODUCTION

Polymer modification can be effected via many routes. Our work currently involves the condensation of polymers containing Lewis Bases (nucleophiles) with mono and dihalo acid chlorides containing metals (1-14) as illustrated below.

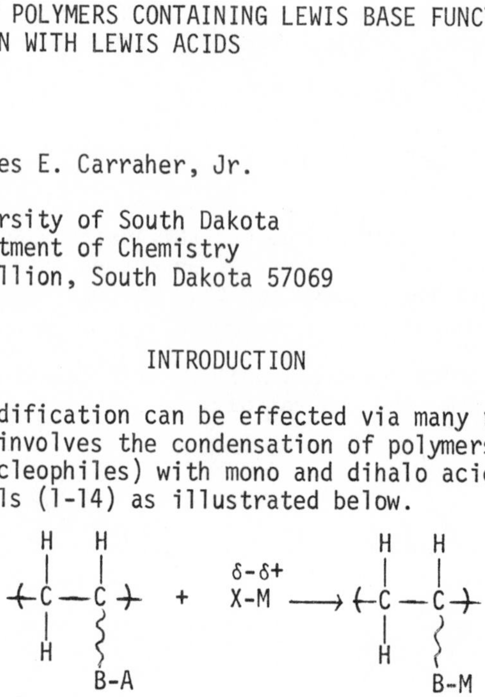

$$\underset{\text{I}}{-\!\!\left(\begin{array}{c}\text{H}\\|\\\text{C}\\|\\\text{H}\end{array}-\begin{array}{c}\text{H}\\|\\\text{C}\\\wr\\\text{B-A}\end{array}\right)\!\!-} + \overset{\delta-\,\delta+}{\text{X-M}} \longrightarrow \underset{\text{II}}{-\!\!\left(\begin{array}{c}\text{H}\\|\\\text{C}\\|\\\text{H}\end{array}-\begin{array}{c}\text{H}\\|\\\text{C}\\\wr\\\text{B-M}\end{array}\right)\!\!-}$$

Analogous condensations with nonmetallic acid halides have been accomplished by others (for instance 15-38).

Through studying such systems it should be possible to adjust conditions allowing from near zero inclusion to maximum inclusion of the modifying agent into the polymer. Also it should be possible to evaluate at least some reaction parameters associated with the reactivity of functional groups contained on a polymer.

DISCUSSION

The interfacial and solution (where applicable) techniques were utilized as the synthetic techniques for several reasons including the following: 1) Both allow the use of simple, inexpensive equipment; 2) both allow the synthesis of products near room temperature minimizing thermally dependent competing reactions and permitting the use of, or synthesis of, products unstable to elevated temperatures; and 3) the interfacial technique allows the synthesis of products under nonequilibrium conditions thus eliminating the undesirable dependence on entropy.

We have worked largely with the modification of poly(acrylic acid), poly(vinyl alcohol) and the amideoxime from polyacrylonitrile (1-14). Some of our modifications are indicated below. The analogous bifunctional reagents to those shown are effective crosslinking agents. These polymers were chosen for several reasons: first, because of their ease of availability and relatively low cost; second, because of the similarity in pendent functionality to reactants already employed by us in other studies; third, because of the similarity in pendent functionality to other interesting potential reactants - for instance, the pendent functionality for both poly(vinyl alcohol) and carbohydrates such as starch and cellulose in the hydroxyl group. The latter two points are based on the existence of a similarity between the reactivity of functional groups attached to a polymer chain to that of functional groups contained within a "smaller" molecule. (This similarity has not always been exploited to the best advantage in polymer modifications.)

$$-\!\!(CH_2\text{-}\underset{\displaystyle CO_2^-}{\underset{|}{C}H})\!\!- \xrightarrow[M=Sn,Ge,Si]{R_3MCl} -\!\!(CH_2\text{-}\underset{\displaystyle CO_2\text{-}MR_3}{\underset{|}{C}H})\!\!-$$

$$-\!\!(CH_2\text{-}\underset{\displaystyle C(=N\text{-}OH)NH_2}{\underset{|}{C}H})\!\!- \xrightarrow{R_2P(O)Cl} -\!\!(CH_2\text{-}\underset{\displaystyle C(=N\text{-}OP(O)R_2)NH_2}{\underset{|}{C}H})\!\!-$$

$$-\!\!(CH_2\text{-}\underset{\displaystyle C(=N\text{-}OH)NH_2}{\underset{|}{C}H})\!\!- \xrightarrow{R\text{-}SO_2CL} -\!\!(CH_2\text{-}\underset{\displaystyle C(=N\text{-}O_3SR)NH_2}{\underset{|}{C}H})\!\!-$$

$$-\!\!(CH_2\text{-}\underset{\displaystyle C(=N\text{-}OH)NH_2}{\underset{|}{C}H})\!\!- \xrightarrow{R_3MCl} -\!\!(CH_2\text{-}\underset{\displaystyle C(=N\text{-}OMR_3)NH_2}{\underset{|}{C}H})\!\!-$$

$$-CH_2-CH(OH)- \xrightarrow{R_2P(O)Cl} -(CH_2-CH(O-P(O)R_2))- ; \quad \xrightarrow{RSO_2CL} -(CH_2-CH(OSO_2R))- ; \quad \xrightarrow{R_3MCl} -(CH_2-CH(O-MR_3))-$$

As expected, products from the condensation with monohalo acid chlorides are linear. They are generally soluble in dipolar aprotic solvents but almost always exhibit a solubility range much less than that of either reactant. Products from dihalo acid chlorides are crosslinked and insoluble in all solvents.

The products often retain physical properties characteristic of the original polymer, but further exhibit properties derived from the included acid chloride-moiety. For instance, the products from the condensation of poly(sodium acrylate) with Group IV B compounds of the form Cp_2MCl_2 all exhibit the typical stability plateaus and fair high temperatures stability (3). Also, such products generally exhibit bulk resistivities intermediate between semiconductors and nonconductors (10^6 - 10^{13} ohm-cm) (39).

Differences in reaction dependencies between functional groups contained within a polymer and those contained in a "small" molecule are just becoming understood. Two examples associated with our work will be presented. The first concerns a difference in the steric requirements of polymers modified as described by us and analogous "linear" polymers. Bloomstrom (40) and Donaruma (41) reported the synthesis of poly-O-acrylamideoximes of form III by reaction of amideoximes with acid chlorides.

$$HO-N=C(NH_2)-R-C(NH_2)=N-OH + Cl-\overset{O}{\overset{\|}{C}}-R-\overset{O}{\overset{\|}{C}}-Cl \rightarrow -(O-N=C(NH_2)-R-C(NH_2)=N-O-\overset{O}{C}-R-\overset{O}{C})-$$

III

This led us to conclude that similar reactions should occur with the amideoxime of polyacrylonitrile. This proved to be correct (Fig. 1). There is a question, though, as to the true structure of the products since the amideoxime group contains both the N-OH and NH_2 groups. Reaction of acid chlorides with amines under analogous conditions is well known (for instance, 42-48). Evenso we argued that reaction occurred largely at the N-OH. This was based on work noted above and on that reported by Eloy and coworkers (49) who reported the formation of both acrylated products in the condensation of acid chlorides with formamideoxime. Under one set of reaction conditions the ratio of IV/V was about 3/1. Also, while an increase in the amount of acid chloride used increased the yield of the modified poly-O-acylamideoxime from polyacrylonitrile it never went above 100% based on products described in Fig. 1 (1,2,4,9,11). This indicates that reaction occurs probably at only one-half of the potential reaction sites (probably predominantly at the NOH group).

$$\text{HO-N=CH-NH}_2 + \text{Br-}\phi\text{-}\overset{\text{O}}{\overset{\|}{\text{C}}}\text{-Cl} \rightarrow \text{Br-}\phi\text{-}\overset{\text{O}}{\overset{\|}{\text{C}}}\text{-O-N=CH-NH}_2 + \text{Br-}\phi\text{-}\overset{\text{O}}{\overset{\|}{\text{C}}}\text{-N=CH-NH-}\overset{\text{O}}{\overset{\|}{\text{C}}}\text{-}\phi\text{-Br}$$

Recently we studied the condensation of cyclopentadienyl titanium dichloride (Cp_2TiCl_2) with diamideoximes (50). The products contained higher than predicted amounts of the Cp_2Ti (T) moiety. In fact products with an excess of T (51 to 80 mole-%) often contained an amount of T consistent with reaction occurring at all of the reaction sites (form VI).

$$\begin{matrix}\text{HO-N} & & \text{NOH} \\ & \diagdown\!\!\!\diagdown\ \text{C-R-C}\ /\!\!\!/ & \\ \text{H}_2\text{N}\ / & & \diagdown\ \text{NH}_2\end{matrix} + \text{TCl}_2 \rightarrow \begin{matrix}\text{-(-O-N} & & \text{N-O-T-)-} \\ & \diagdown\!\!\!\diagdown\ \text{C-R-C}\ /\!\!\!/ & \\ \text{HN}\ / & & \diagdown\ \text{NH} \\ | & & | \\ \text{T-OH} & & \text{T-OH}\end{matrix}$$

VI

Space-filling models of VI, VII and VIII clearly illustrate that the steric hindrance present in VII is much greater than that present in VI and VIII. It is currently believed that this difference in the steric requirements of the two types of polymers is responsible for the difference in the reactivity of the amideoxime group.

The reactivity of pendent functional groups contained in a polymer are generally equal to or less than that of functional groups contained on "smaller" molecules. Inclusion of Group IV A moieties from R_3MCl and $R\ MCl_2$ into poly(vinyl alcohol) via the

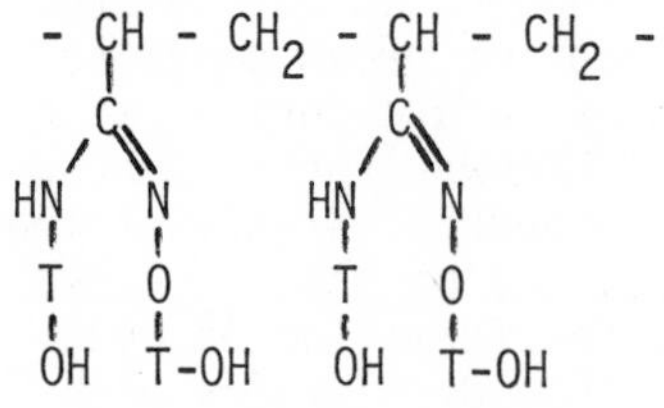

```
- CH - CH2 - CH - CH2 -              -( CH2 - CH )-
  |           |                             |
  C           C                             C
 / \\        / \\                          / \\
HN   N      HN   N                      H2N   N-O-T-OH
|    |      |    |
T    O      T    O
|    |      |    |
OH   T-OH   OH   T-OH
```

VII 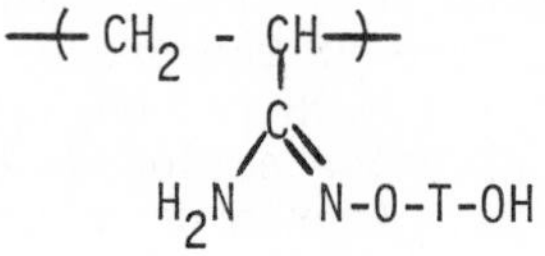 VIII

interfacial technique has been effected for only certain tin reactants (Table 1 and Ref. 13).

In comparison the condensation of R_2SnCl_2 with diols utilizing analogous interfacial systems appears to be general for a wide range of tin reactants (39). The failure of either system to incorporate similar compounds containing Si or Ge is probably

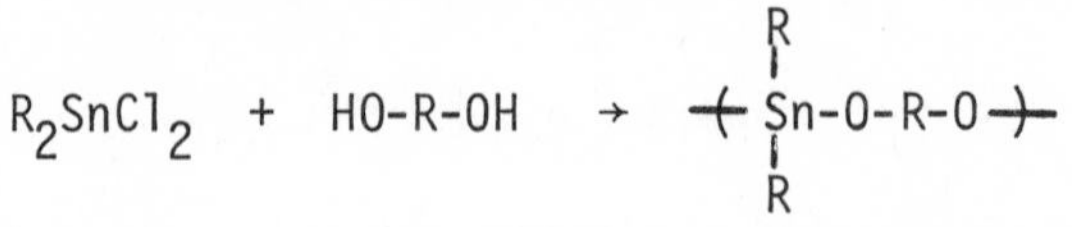

$$R_2SnCl_2 + HO\text{-}R\text{-}OH \rightarrow \left(\begin{array}{c} R \\ | \\ Sn \\ | \\ R \end{array} \text{-O-R-O} \right)$$

IX

due to the preferential hydrolysis of such reactants to form hydroxides and oligomeric oxides of form X instead of the desired inclusion product (5). Hydrolysis is in the order Si>Ge>Sn (5).

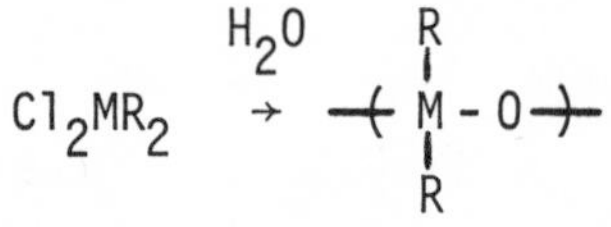

$$Cl_2MR_2 \xrightarrow{H_2O} \left(\begin{array}{c} R \\ | \\ M \\ | \\ R \end{array} \text{- O} \right)$$

X

The only successful polyester formation utilizing Ge and Si reactants has been accomplished using nonaqueous interfacial systems (51-54).

Another example concerns the failure of poly(sodium acrylate) to condense with a number of Group IV A reactants (1, 10) whereas the synthesis of Group IV A polyesters is general for condensation utilizing reactants as shown below (for instance 5, 55-58). The lesser reactivity of pendent groups is probably due to the large size of the polymer (i.e., greater steric requirements) and not to a difference in the intrinsic reactivity of the particular functional group.

Table 1. Products from PVA via condensation with Group IV A reactants.[a,d]

Organometallic	Yield[b] (%)	Mole-% Organo-metallic Reactant Included in Product	Texture and Appearance[c]
Di-t-butyltin dichloride	36	15	Powdery, brittle
Tri-n-butyltin chloride	21	7	Flexible, tough, stretchable films
Diphenyltin dichloride	20	8	Flexible, tough, stretchable
Triphenyltin chloride	26	12	Flaky, soft, flexible

[a]Reaction conditions: PVA (0.003 mole) with NaOH (0.003 mole for monohalo organometallics and 0.006 mole for the dihalo reactants) in 50 ml of water added to the organometallic reactant (0.003 mole) in 50 ml ether, at 25°C with a stirring rate of 17500 rpm (no load) for 30 seconds.
[b]Yields varied no more than ±2% for stirring times of 30, 60, and 180 seconds.
[c]All the products were white and solid.
[d]Those not successfully included into PVA include tri-n-propyltin chloride, dimethylgermanium dichloride, diethyltin dichloride, dimethyltin dichloride, dioctyltin dichloride, diphenylsilicon dichloride, triphenylsilicon chloride, dimethylsilicon dichloride, and dibutyltin dibromide.

$$R_2MCl_2 + {}^{\ominus}O\text{-}\overset{O}{\overset{\|}{C}}\text{-}R\text{-}\overset{O}{\overset{\|}{C}}\text{-}O^{\ominus} \rightarrow \left(\underset{R}{\overset{R}{M}}\text{-}O\text{-}\overset{O}{\overset{\|}{C}}\text{-}R\text{-}\overset{O}{\overset{\|}{C}}\text{-}O \right)$$

XI

In theory any polymer containing a Lewis Acid (or Base) site is modifiable via condensation with a Lewis Base (or Acid). In actuality many "good-looking" possibilities have thus far failed to yield the desired condensation products. For instance we have been unable to effect the inclusion of mono- and dichlorophosphates and phosphonates into poly(sodium acrylate) utilizing interfacial systems (13,59). The reaction site in poly(sodium acrylate) systems is believed to be very near the interface of the two solvents or in the aqueous layer because of the insolubility of the acid salt in the organic phase. Only hydrolyzed phosphorus and unreacted poly(sodium acrylate) have been recovered from their attempted condensation. Probably hydrolysis of the P-Cl group has occurred in preference to reaction with poly(sodium acrylate).

As previously noted hydrolysis is also probably responsible for the inability to modify poly(sodium acrylate) via reaction with silicon, germanium and certain tin organometallic halides. Thus hydrolysis appears to be a major stumbling block in the modification of polymers via Schotten-Baumann type reactions. We are currently experimenting with interfacial systems utilizing a third "carrier" solvent such as methyl ethyl ketone in which the unmodified polymer is soluble and which is miscible with the other liquids utilized (60). It is hoped that such systems will act to minimize hydrolysis. Tsude (38) and others (for instance 30, 61) have used similar systems to effect the modification of poly(vinyl alcohol) via condensation with organic acid chlorides. Additionally, such systems may permit a higher inclusion ratio if the "carrier" solvent is chosen to permit longer solubility of the modified polymer. Such systems may permit the use of easily hydrolizable materials in polymer modifications.

The use of such modifications, modified products and information gained from such studies is potentially great. The use of polymers as templates is just beginning and deserves special comment since it offers such futuristic possibilities as synthesis of protein for food, tailored nucleic acid syntheses, etc. Information derived from studies such as ours may be applicable in designing conditions and reactants which are used as templates or used during the replicating process. For biological systems in particular, such "anatural" templates often offer the following desirable characteristics (over biological templates): 1) better hydrolytic stability, 2) better thermal stability, and 3) less rejection (etc.) by living systems.

Less "glamorous," but still quite useful uses of modified polymers already synthesized by us include the following: 1) additives in fabrics, paints and plastics to resist "mildew" and "rot" (in particular those containing tin); 2) flame retardants (such as the phosphates and phosphonates); 3) membranes for special applications; 4) thermally stable products; and 5) materials where some electrical conductivity is required.

REFERENCES

1. C. Carraher, J. Piersma and L. Wang, Organic Coatings and Plastics Chemistry, 31, 254 (1971).
2. C. Carraher and L. Wang, Makromolekulare Chemie, 152, 43 (1972).
3. C. Carraher and J. Piersma, Makromolekulare Chemie, 152, 49 (1972).
4. C. Carraher and L. Wang, J. Polymer Sci., A-1, 9, 2893 (1971).
5. C. Carraher, Inorganic Macromolecules Reviews, 1, 271 (1972).
6. C. Carraher and L. Torre, Angew. Makromolekulare Chemie, 21, 207 (1972).
7. C. Carraher and L. Torre, J. Polymer Sci., A1, 9, 975 (1971).
8. C. Carraher, L. Torre and B. Scott, unpublished results.
9. C. Carraher and L. Wang, J. Macromol. Sci.-Chem., A7, 513 (1973).
10. C. Carraher and J. Piersma, J. Applied Polymer Sci., 16, 1851 (1972).
11. C. Carraher and L. Wang, Angew. Makromolekulare Chemie, 25, 121 (1972).
12. C. Carraher and J. Piersma, Angew. Makromolekulare Chemie, 28, 153 (1973).
13. C. Carraher and J. Piersma, J. Macromolecular Sci., in press.
14. F. Millich and C. Carraher, Ed., "Interfacial Synthesis," Marcel Dekker, Inc., N.Y., 1973.
15. E. Fettes, Ed., "Chemical Reactions of Polymers," Interscience Pubs., N.Y., 1964.
17. Jap. Pat. 17582 (1968), Inv.: M. Shiraishi.
18. G. Smirnov, I. Okhrimenko, and L. Mashlyakorski, Zh. Prikl. Khim. (Leningrad) 41, 2340 (1968); C.A. 70:30115.
19. Brit. Pat. 976392 (1964); Inv.: A. Decat, J. Lemmerling, and A. Van Paesschen.
20. E. Korneva, O. Smirnov, and V. Urarova, Zh. Prikl. Khim. (Leningrad) 39, 1876 (1966).
21. D. Reynolds and W. Kenyon, J. Amer. Chem. Soc. 72, 1584 (1950).
22. O. Klimova and V. Datsenko, Zh. Prikl. Khim. (Leningrad) 33, 2582 (1960).
23. N. Kulkkova, M. Androsova, and N. Orlov, Zh. Prikl. Khim. (Leningrad) 40, 2318 (1967).
24. N. Volgina, I. Seregina, L. Vol'f, Y. Kirilenko, S. Borisor, E. Nifant'er, N. Sviridovaard, and K. Tammik, Zh. Prikl. Khim. (Leningrad) 41, 2563 (1968) C.A. 70:48534.
25. U.S. Pat. 2733229 (1956), Inv.: N. Brace.
26. Jap. Pat. 6970 (1957), Inv.: Y. Motozato, R. Tamura, and H. Egawa.
27. Y. Motozato, H. Egawa, H. Maegaki, and K. Kunitate, Kogyo-Kagaky Zasshi 59, 479 (1966).
28. U.S.S.R. Pat. 194310 (1967), Inv.: V. Lavrishehev, Y. Bokov, and N. Vikulina.

29. Jap. Pat. 20188 (1965), Inv.: Y. Tsuda.
30. T. Shermergon and Y. Kamardin, Vysokomol. Soedin. 7, 2156 (1965).
31. Y. Kamardin, I. Shermergorn, and I. Magdeer, Vysokomol. Soedin. Ser. 89, 419 (1967).
32. U.S. Pat. 3329664 (1967), Inv.: M. Tsuda.
33. U.S.S.R. Pat. 183395 (1966), Inv.: V. Laurishchev, Y. Bokov, N. Vikulina, H. Larina, and N. Karantirov.
34. Ger. Pat. 106562 (1959), Inv.: O. Wichterle.
35. M. Tsuda, Makromolekulare Chemie, 72, 174 and 184 (1964).
36. M. Tsuda, J. Polymer Sci., A-1, 7, 259 (1969).
37. A. Yabe and M. Tsuda, J. Polymer Sci., B-9, 81 (1971).
38. M. Tsuda, "Interfacial Synthesis," Marcel Dekker, Inc., N.Y., 1973 (F. Millich and C. Carraher, Eds.).
39. A. Carraher, unpublished results.
40. U.S. Pat. 3,044,994 (1952), Inv.: D. Bloomstrom.
41. L. Donaruma, J. Org. Chem., 26, 577 (1961).
42. P. Morgan, "Condensation Polymers: By Interfacial and Solution Methods," Interscience, N.Y., 1965.
43. C. Carraher, D. Winthers and F. Millich, J. Polymer Sci., 7, 2763 (1969).
44. C. Carraher, Macromolecules, 2, 306 (1969).
45. C. Carraher and J. Greene, Makromolekulare Chemie, 30, 177 (1969) and 131, 259 (1970).
46. C. Carraher, J. Polymer Sci., 8, 3051 (1970).
47. C. Carraher and D. Winter, Makromolekulare Chemie, 141, 237 and 259 (1971) and 152, 55 (1972).
48. C. Carraher and P. Lessek, Eup. Polymer J., 8, 1339 (1972).
49. F. Eloy, R. Lenaers and C. Moussebois, Helv. Chim. Acta, 45, 437 (1962).
50. C. Carraher and R. Frary, unpublished results.
51. C. Carraher and C. Petruzzello, unpublished results.
52. C. Carraher, J. Polymer Sci., A-1, 7, 2351 and 2359 (1969).
53. C. Carraher and G. Klimiuk, Makromolekulare Chemie, 133, 211 (1970).
54. C. Carraher and G. Klimiuk, J. Polymer Sci., A-1, 8, 973 (1970).
55. C. Carraher and R. Dammeier, Makromolekulare Chemie, 135, 107 (1970); 141, 245 (1971); and 141, 251 (1971).
56. C. Carraher and R. Dammeier, J. Polymer Sci., 8, 3367 (1970).
57. M. Iskenderov, K. Plekhanova and N. Adigezalora, Uch. Zap. Azerb. Gas. Univ. Ser. Khim. Nauk 4, 71 (1965).
58. M. Frankel, D. Gertner, D. Wagner and A. Zilkha, J. Applied Polymer Sci., 9, 3383 (1965).
59. C. Carraher and J. Piersma, unpublished results.
60. C. Carraher and M. Feddersen, unpublished results.
61. I. Shermergorn and U. Kamardin, Polymer Science, U.S.S.R., 7, 2364 (1965).

DISCUSSION SESSION

Discussion Leader - R. B. Fox

Stackman: Have you studied the flammability of these materials?

Carraher: No. We have simply studied thermal stability.

Challa: Are the modified polymers soluble in organic solvents?

Carraher: Yes

Challa: The organic-soluble material contains the metal?

Carraher: In interfacial systems, many materials are soluble, e.g., ϕ_2SnCl_2, are soluble in the organic phase. The product with poly(vinyl alcohol) has limited solubility in solvents such as DMF, DMSO, etc.

Challa: Then when you hydrolyze these polymers you obtain some kind of oxides?

Carraher: Yes. Hydroxide, maybe oxide.

Challa: How fast?

Carraher: Rapidly. It depends on wetting, these materials are hydrophobic.

Challa: Does the rate of hydrolysis increase as molecular weight decreases?

Carraher: I don't know.

Fox: Under what conditions did you measure the thermal stability?

Carraher: The heating rate was 30°/min; nitrogen or air flow of 0.3 m/min.

Fox: Are the stability plateaus kinetically controlled?

Carraher: Sometimes. Some of these materials when aged isothermally will gradually lose weight at elevated temperatures over 3-4 days until 20-40% of residual weight remains, others will not lose weight to any great extent.

Moore: When you say there is weight retention in some temperature range, how much is organic and how much is inorganic and how much is residual metal?

Carraher: It depends on the system. The titanium-containing materials retain about 60% of the organic portion of the polymer, at a minimum.

SEQUENCE STUDIES DURING REACTIONS ON POLYMERS*

E. Klesper, D. Strasilla, and V. Barth

Institut für makromolekulare Chemie,
Universität Freiburg, Freiburg, BRD

1. INTRODUCTION

The investigation of the kinetics of cooperative reactions on synthetic polymers has been initiated by the studies of Morawetz (1,2), Smets (3–6), and coworkers. These studies relied mainly on the kinetics of gross conversion during the reaction, since at this time no method for the quantitative evaluation of sequences of monomer units could be applied.

For the kinetic analysis of reactions on polymers it is of advantage, however, to follow the changes in sequence probabilities of the copolymers formed during the reaction. This is particularly true if the reaction proceeds by a cooperative mechanism in the sense that groups on the polymer chain influence the rate of reaction of adjacent groups, causing thereby different rates for different types of sequences. The quantitative evaluation of sequence probabilities necessitates the application of efficient analytical tools, whereby NMR is at present the method of choice among the spectroscopic techniques.

We have previously reported on the kinetics and statistics of sequences during computer simulated reactions on polymers (7) and on general aspects of the statistics of copolymers formed during reactions on polymers (8). As an example of an actual reaction, the hydrolysis of syndiotactic poly(methyl methacrylate) (PMMA) to methyl methacrylate - methacrylic acid (MMA - MAA) copolymers was investigated by NMR (9 - 11).

* This article has been reproduced from a photocopy, as the original typed sheets were not available.

The statistics of the copolymers was found to depend on the reaction conditions, indicating a change in the mechanism for the cooperativity. The latter, mainly non-kinetic studies are being supplemented now by a preliminary NMR study on the kinetics of sequences during hydrolysis.

A first part of this communication concerns the differential rate equations suitable for the determination of relative rate constants of simple cooperative reactions on polymers by a graphical method. As has been pointed out previously (7), the statistics of copolymers obtained by reactions on polymers is fully determined by the relative rate constants, the gross conversion, and the statistics of the starting polymer. In a second part, we report some of our NMR results to date on the kinetics of sequences during hydrolysis of cosyndiotactic MMA - MAA copolymers in aqueous solution with an excess of base and the determination of the corresponding relative rate constants. A third part of this communication applies to the determination of sequences by NMR during the partial and complete esterification of syndiotactic poly(methacrylic acid) (PMAA) with diazomethane, phenyl diazomethane, and diphenyl diazomethane. The partial esterification with diazomethane yields also cosyndiotactic MMA - MAA copolymers, while the partial esterification with phenyl diazomethane or diphenyl diazomethane results in cosyndiotactic benzyl methacrylate - methacrylic acid (BMA - MAA), or diphenyl methyl methacrylate - methacrylic acid (DPMMA - MAA) copolymers. The esterification with the three diazo alkanes proceeds without a cooperative effect and Bernoullian copolymers are therefore obtained. The complete esterification of cosyndiotactic MMA - MAA, BMA - MAA, and DPMMA - MAA copolymers with a second diazo alkane is briefly described.

Attention is also drawn to the possibility to prepare cosyndiotactic, multicomponent copolymers of Bernoullian monomer distribution by successive partial esterification of syndiotactic PMAA with different diazo alkanes. These multicomponent copolymers may be useful as model copolymers for further physical-chemical studies.

2. DIFFERENTIAL RATE EQUATIONS FOR THE DETERMINATION OF RELATIVE RATE CONSTANTS

A scheme has been devised for establishing the differential rate equations for sequences for cooperative reactions on polymers in which the two flanking monomer units possess a kinetic influence (8,12). If this scheme is applied to write the rate equations for NMR observable triads in a second order reaction, one obtains:

$$1)\quad \frac{d\,P(AAA)}{d\,t} = -\,k(AAA)\,[R]\,P(AAA) - 2\,k(AAA)\,[R]\,P(AAAA) - k(AAB^{+})\,[R]\,P(AAAB^{+})$$

$$2)\quad \frac{d\,P(ABA)}{d\,t} = k(AAA)\,[R]\,P(AAA) - k(AAB^{+})\,[R]\,P(AABA^{+}) - k(BAB)\,[R]\,P(ABAB^{+})$$

$$3)\quad \frac{d\,P(AAB^{+})}{d\,t} = 2\,k(AAA)\,[R]\,P(AAAA) - k(AAA)\,[R]\,P(AAAB^{+}) - 4\,k(AAB^{+})\,[R]\,P(BAAB)$$

$$4)\quad \frac{d\,P(ABB^{+})}{d\,t} = 2\,k(AAB^{+})\,[R]\,P(AABA^{+}) + k(BAB)\,[R]\,P(ABAB^{+}) - k(BAB)\,[R]\,P(BABB^{+})$$

$$5)\quad \frac{d\,P(BAB)}{d\,t} = k(AAA)\,[R]\,P(AAAB^{+}) + 2\,k(AAB^{+})\,[R]\,P(BAAB) - k(BAB)\,[R]\,P(BAB)$$

$$6)\quad \frac{d\,P(BBB)}{d\,t} = k(AAB^{+})\,[R]\,P(AABB^{+}) + k(BAB)\,[R]\,P(BAB) + k(BAB)\,[R]\,P(BABB^{+})$$

The P( ) are probabilities for sequences composed of monomer units A and/or B. The superscript + within the parentheses indicates addition of the probabilities for forward and reverse forms of sequences, e.g. $P(AAB^{+}) = P(AAB) + P(BAA)$. The rate constants k(AAA), $k(AAB^{+})$, and k(BAB) apply to the conversion of a central A into B, whereby the central A may have two A, one A and one B, or two B as flanking monomer units. Furthermore, $k(AAB^{+}) = k(AAB) = k(BAA)$. [R] is the concentration of the attacking reagent.

Because it is desired to determine the relative rate constants instead of the rate constants themselves, d t is to be replaced (8) by d P(A). Adding Eqs. 1, 3, and 5 it is obtained:

$$7)\quad \frac{d\ P(AAA) + d\ P(AAB^{+}) + d\ P(BAB)}{d\ t} = \frac{d\ P(A)}{d\ t} =$$
$$- k(AAA)\ [R]\ P(AAA) - k(AAB^{+})\ [R]\ P(AAB^{+}) - k(BAB)\ [R]\ P(BAB)$$

Division of Eqs. 1 to 6 by Eq. 7 replaces d t by d P(A). Next it is to be considered that in cosyndiotactic MMA - MAA copolymers only triads but not tetrads may be determined by NMR. Selective pairwise combination by addition or subtraction of the six equations eliminates all tetrad probabilities on the right hand sides, only triads remaining:

$$8)\quad \frac{d\{P(AAA) + 0.5\ P(AAB^{+})\}}{d\ P(A)} = \frac{-\ 2\ k(AAA)\ P(AAA) - k(AAB^{+})\ P(AAB^{+})}{N}$$

$$9)\quad \frac{d\{P(AAA) - P(BAB)\}}{d\ P(A)} = \frac{-\ 3\ k(AAA)\ P(AAA) - k(AAB^{+})\ P(AAB^{+}) + k(BAB)\ P(BAB)}{N}$$

$$10)\quad \frac{d\{P(BAB) + 0.5\ P(AAB^{+})\}}{d\ P(A)} = \frac{k(AAA)\ P(AAA) - k(BAB)\ P(BAB)}{N}$$

$$11)\quad \frac{d\{P(BBB) + 0.5\ P(ABB^{+})\}}{d\ P(A)} = \frac{k(AAB^{+})\ P(AAB^{+}) + 2\ k(BAB)\ P(BAB)}{N}$$

$$12)\quad \frac{d\{P(BBB) - P(ABA)\}}{d\ P(A)} = \frac{-\ k(AAA)\ P(AAA) + k(AAB^{+})\ P(AAB^{+}) + 3\ k(BAB)\ P(BAB)}{N}$$

$$13)\quad \frac{d\{P(ABA) + 0.5\ P(ABB^{+})\}}{d\ P(A)} = \frac{k(AAA)\ P(AAA) - k(BAB)\ P(BAB)}{N}$$

where

$$14)\quad N = -\ k(AAA)\ P(AAA) - k(AAB^{+})\ P(AAB^{+}) - k(BAB)\ P(BAB)$$

The graphical evaluation of the derivatives on the left hand sides of Eqs. 8 to 13 will allow the determination of relative rate constants. However, it can be shown that of Eqs. 8 to 13 only one equation is independent. For this reason, and for the purpose of averaging the errors made in graphically determining the derivatives, Eqs. 8 to 13 are added to yield:

$$15) \quad \frac{1}{4}\sum_{i=1}^{6} S_i = \frac{k(BAB)\ P(BAB) - k(AAA)\ P(AAA)}{N}$$

where S_1 to S_6 are the left hand sides of Eqs. 8 to 13. While the relative rate constants may be defined in different ways, we choose:

$$16) \quad k'(AAA) = \frac{k(AAA)}{k(AAA) + k(AAB^+) + k(BAB)}$$

$$17) \quad k'(AAB^+) = \frac{k(AAB^+)}{k(AAA) + k(AAB^+) + k(BAB)}$$

and

$$18) \quad k'(AAA) + k'(AAB^+) + k'(BAB) = 1$$

In consideration of Eqs. 16 to 18 the Eq. 15 is written:

$$19) \quad k'(AAA) = C\ k'(AAB^+) + D$$

where

$$20) \quad C = \frac{P(BAB)(1 + S') - P(AAB^+)\ S'}{- P(BAB)(1 + S') - P(AAA)(1 - S')}$$

$$21) \quad D = \frac{P(BAB)(1 + S')}{P(BAB)(1 + S') + P(AAA)(1 - S')}$$

$$22) \quad S' = \frac{1}{4}\sum_{i=1}^{6} S_i$$

Eq. 19 is employed for the graphical determination of $k'(AAA)$ and $k'(AAB^+)$ in the next section.

3. KINETICS OF SEQUENCES DURING HYDROLYSIS OF COSYNDIOTACTIC MMA - MAA COPOLYMERS

The hydrolysis of cosyndiotactic MMA - MAA copolymers was carried out in aqueous solution with KOH at 145°C. For

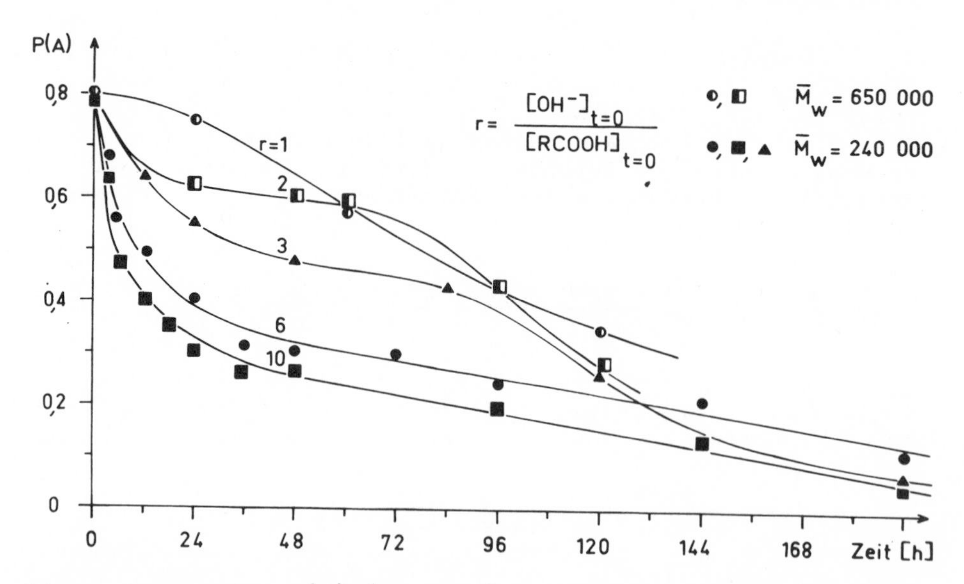

Fig.1 Conversion P(A) (A = MMA) versus hours reaction time (Zeit) at 145°C for a cosynd. MMA - MAA copolymer. The starting copolymer of P(A) = 0.79 was obtained from a PMMA of M_w of 650 000 or 240 000 .

the kinetic runs of this communication a copolymer of P(A) = 0.79 (A = MMA monomer unit) with a Bernoullian distribution of monomer units was employed as the starting material. PMMA could not be used for this purpose because of lack of solubility in the reaction medium. The course of the reaction was followed by providing runs in parallel which were stopped after different reaction times. Triad probabilities and P(A) of the copolymers could be conveniently determined from the NMR spectra recorded from pyridine solution (9,10) (see also Figs. 6 and 7 for sample spectra of cosyndiotactic MMA -MAA copolymers in pyridine).

In Fig. 1 the gross conversion during hydrolysis is plotted as P(A) versus time. The parameter r written on each of the five kinetic curves represents the ratio of the total concentration of OH^-, added as KOH, to the concentration of carboxylic acid groups which are present in the copolymer. Both concentrations apply to the beginning of the run (t=0). As indicated by the type of symbol used for the experimental points, runs have been carried out with starting copolymers which were derived from PMMA samples of two M_w . However, additional data showed that the kinetic behavior exhibited in this and the following figures is not altered when changing the molecular weight of the starting copolymer.

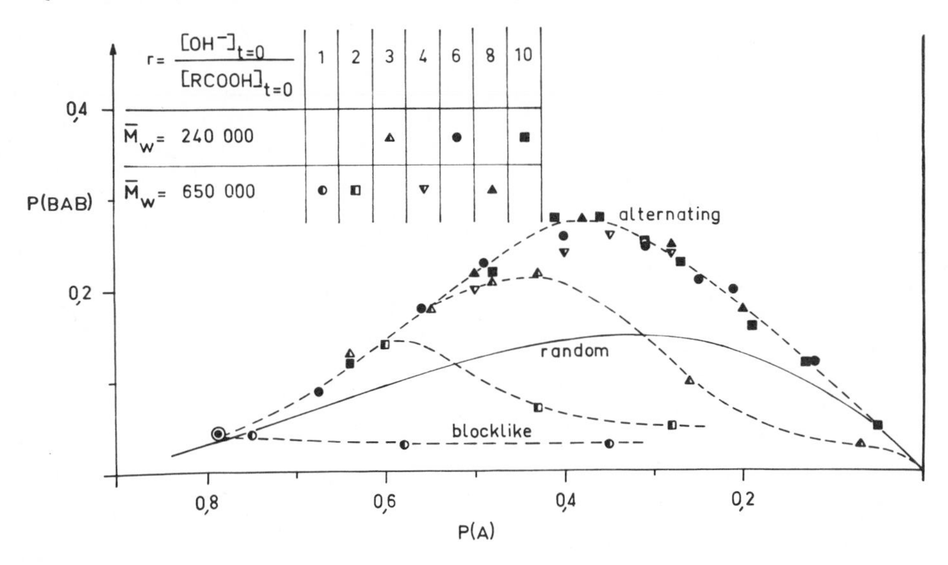

Fig. 2 P(BAB) versus conversion P(A) for the Bernoullian, cosynd. MMA - MAA copolymer of P(A) = 0.79 . Solid curve marked "random" represents calculated Bernoullian probabilities for BAB Copolymers possessing a P(BAB) above or below this curve exhibit a statistics which tends toward alternation or block character, respectively. The broken line curves are drawn to fit the experimental points.

The P(A) versus time curve of r = 2 starts with a phase of a steep decrease for P(A). After a relatively small conversion, however, the OH^- in excess of the amount required for neutralization is depleted because of the increase in carboxylic acid groups during hydrolysis. Therefore, in a second phase, the curve reaches a plateau at about P(A) = 0.6, indicating that $-COO^-$, as opposed to OH^-, does not greatly contribute to the rate of the reaction. In a third phase, the rate increases again, indicating an autocatalytic behavior connected with the increasing amount of unneutralized $-COOH$ groups on the copolymer chain. Similar phenomena are observed for r = 3 , considering that the excess of OH^- should be consumed here at about P(A) = 0.4 . Since the hydrolysis with r = 3 is carried to higher conversion than with r = 2 it can be seen that at $P(A) < 0.2$ the autocatalytic behavior yields to a decreasing rate, probably because the number of blocks of hydrolyzable A-units is rapidly decreasing in this range of conversion.

With r = 6 and r = 10 too much of an excess of OH^- is present that it might be consumed even at complete conversion. Therefore only the first phase of the hydrolysis may be

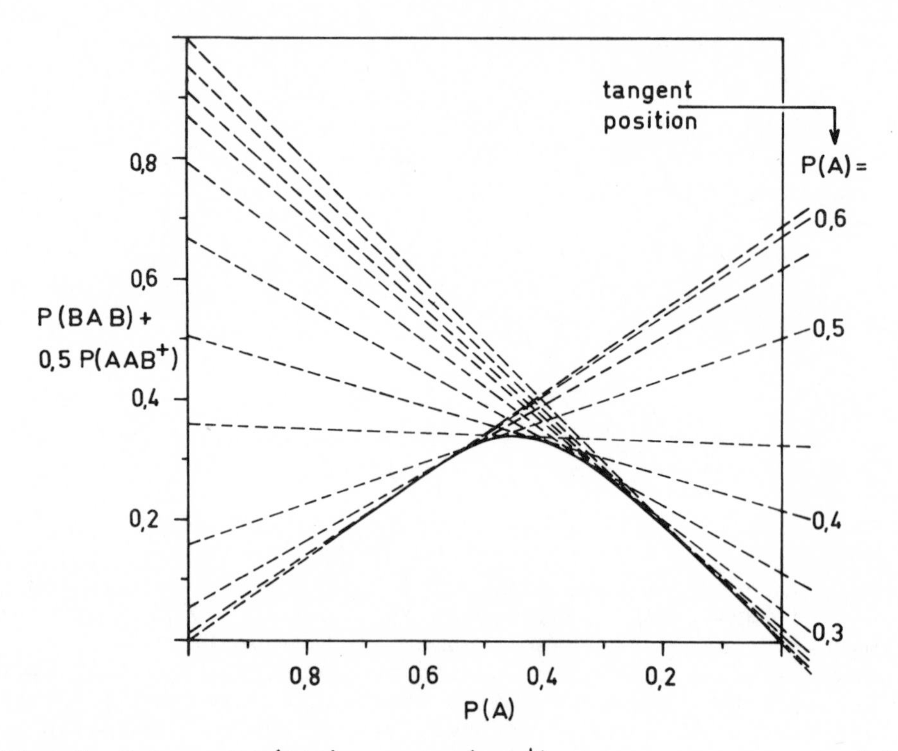

Fig. 3 Curve of $P(BAB) + 0.5\ P(AAB^+)$ versus conversion $P(A)$. Tangents are drawn to the curve at different $P(A)$.

observed. With $r = 1$ the copolymer is just neutralized at $t = 0$, no excess of OH^- being present. Thus only part of the second phase and then the third phase is seen. It should be pointed out that at all r the hydrolysis may be carried to completion by allowing sufficiently long reaction times, as was ascertained by qualitative runs (not shown).

Plotting triad probabilities versus time or versus P(A), information is obtained about the statistics of the resulting copolymers. In Fig. 2 the probability for the particularly informative triad BAB is shown in dependence of P(A). The solid curve marked "random" represents the calculated Bernoullian values for P(BAB). Experimental points above this curve indicate a statistics with a tendency toward alternation and points below the curve are an indication of a statistics having block character.

The experimental points apply to different r and to different M_w of the precursor PMMA used for the preparation of the starting copolymer. The lowermost broken line is

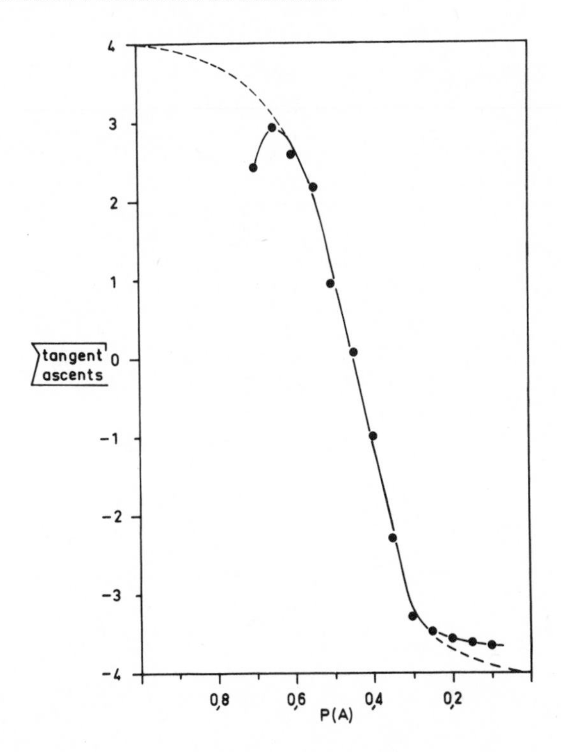

Fig. 4 Sum of tangent ascents (= 4 S' of Eq. 22) for different P(A). Broken curves are extrapolated to + 4 and - 4 .

drawn through the points of r = 1 while the broken curves next higher up apply to r = 2 and r = 3 in that order. However, the long broken curve which extends from the starting P(A) = 0.79 , and which is fully in the region for a tendency toward alternation, is formed by experimental points of all r up to r = 10 . All points of that curve have been determined at conversions where an excess of OH^- is present. Therefore, the conversion dependent kinetics of P(BAB), and in fact also of the other five NMR observable triad probabilities, does not greatly depend on the excess of OH^-, as long as there is an excess. This result is in accord with Eqs. 8 to 14 in which the factor [R] has cancelled. At about the conversion where the excess of OH^-, corresponding to a given r , has been consumed a branch is splitting off from the highermost broken curve. With r = 2 and r = 3 the branches cross the calculated Bernoullian curve indicating a change from a tendency toward alternation to block character. With r = 1 the branch nearly originates at the Bernoullian curve because of the Bernoullian character of the starting copolymer.

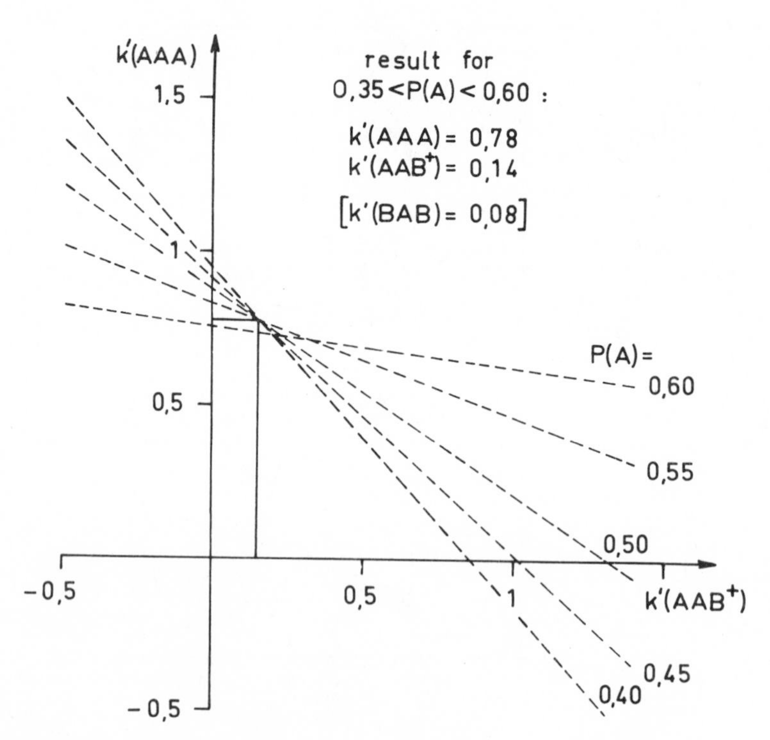

Fig. 5 Plot of k'(AAA) versus k'(AAB^+) according to Eq. 19. The data of Fig. 4 (4 S') are utilized for the range of conversion: $0.35 < P(A) < 0.60$. The value for k'(BAB) is obtained according to Eq. 18.

Thus, contrary to the runs with r = 6, 8, and 10, the runs with r = 1, 2, and 3, and possibly also 4, do not follow at higher conversions the same line, i.e. the uppermost broken curve. Because the branches differ from each other in location and shape, the conversion dependent kinetics without an excess of OH^- is greatly dependent on r.

The uppermost curve of Fig. 2 can be considered to represent the data for the "alternating" hydrolysis of the triad BAB by an excess of OH^-, the data being averaged for experimental scatter. Because analogous curves have been determined for the other five NMR observable triads, one may plot the pairs of triad probabilities written on the left hand sides of Eqs. 8 to 13 versus P(A). An example is given by Fig. 3 with the pairs of triad probabilities of Eq. 10. Drawing the tangents on this curve at different P(A) by the mirror method yields the slopes which correspond to the derivative of Eq. 10. Adding all six slopes of Eqs. 8 to 13 for a given P(A), one value for 4 S' of Eq. 22 is obtained. Repeating this addition for different P(A), Fig. 4 may be drawn where the sum of the tangent ascents is equal to 4 S'. The experimental points of Fig. 4 exhibit little scatter and trace out a smooth curve. This indicates that the sum of the graphically determined slopes is of acceptable precision. However, the experimental curve should reach + 4 at P(A) = 1 and approach - 4 when P(A) approaches 0, as is indicated by the extrapolated broken curves. This, however, is not the case. Since at the present time we have no proven understanding of the deviation of the experimental curve from the extrapolated curves, only the portion of the experimental curve between P(A) = 0.60 and P(A) = 0.35 is taken for the determination of relative rate constants.

Substitution in Eq. 19 of the values for S', taken off the curve of Fig. 4, and of the values for P(AAA), $P(AAB^+)$, and P(BAB) yields the expressions for the straight lines plotted in the k'(AAA) versus $k'(AAB^+)$ diagram of Fig. 5. The straight lines intersect in a rather narrow region giving k'(AAA) = 0.78 and $k'(AAB^+)$ = 0.14. In view of Eq. 18 this leads to k'(BAB) = 0.08. These values for the relative rate constants mean that the central A in the triad AAA is attacked 5 – 6 times more rapidly by OH^- than the central A in the triads AAB and BAA, and also that the central A of the latter triads reacts with OH^- at about two times the rate of the A in BAB. It is probable that the electrostatic repulsion between OH^- and $-COO^-$, which is smallest for AAA, greater for AAB and BAA, and most pronounced for BAB, is responsible for this finding.

4. SEQUENCES DURING ESTERIFICATION OF SYNDIOTACTIC POLY(METHACRYLIC ACID) WITH DIAZO ALKANES

4.1. Partial Esterification with Diazomethane

Syndiotactic PMAA is dissolved in a CH_3OH - H_2O mixture (2 : 1 v/v) and ethereal diazomethane solution added with stirring. Less diazomethane is used than is necessary for complete esterification. The diazomethane color disappears immediately at the point of contact of the two solutions before mixing is even nearly complete. After isolation of the resulting cosyndiotactic MMA - MAA copolymers the NMR spectra are recorded from pyridine solution. The spectrum of a copolymer of P(A) = 0.47 (A = MMA) is shown in trace I of Fig. 6, the known assignments being indicated in the figure (9). Considering the relatively small areas of the peaks for the triads ABA and BAB the copolymer appears to possess a blocklike statistics. However, fractionation of the

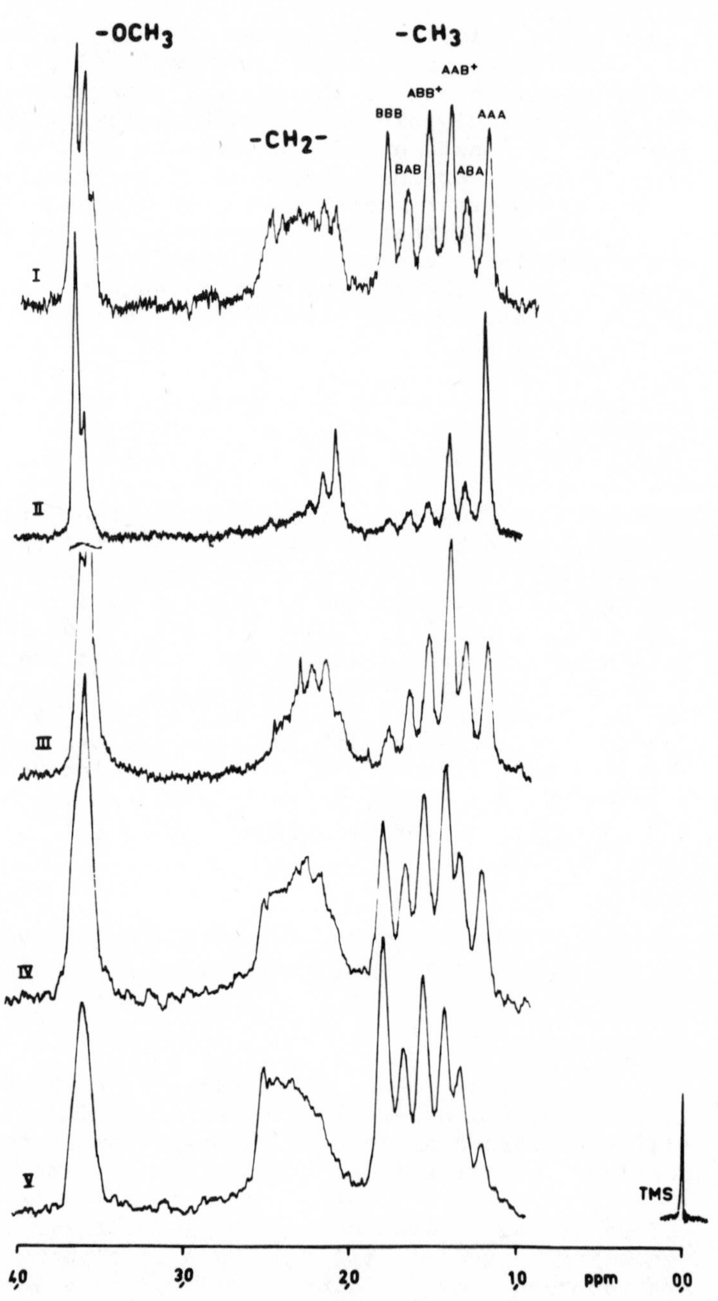

Fig. 6 NMR spectra of a cosynd. MMA - MAA copolymer prepared from unneutralized PMAA (trace I) and of fractions of this copolymer (traces II to V). Assignments as shown (A = MMA). Pyridine solutions, 100°C, 220 MHz, TMS as internal standard.

Table 1

Run numbers R compared with Bernoullian run numbers R_r of a MMA - MAA copolymer and its fractions

spectral trace Fig. 1	P(A)	R	R_r
I	0.47	39	50
II	0.73	37	39
III	0.57	51	49
IV	0.46	46	50
V	0.34	45	45

$$R = 200\ P(AB) = 100\left\{P(AB) + P(BA)\right\}$$
$$= 100\left\{P(ABA) + 0.5\ P(ABB^{+}) + P(BAB) + 0.5\ P(AAB^{+})\right\}$$
$$R_r = 200\ P(A)\ P(B)$$

copolymer by successive precipitation from aqueous solution with HCl yielded four fractions whose NMR spectra are seen as traces II to V in Fig. 6 . The fraction precipitating first corresponds to trace II, successive fractions following in order.

In Table 1 , data for P(A) and Harwood's run number (13), R, are given for the whole copolymer and its fractions. Because the four fractions differ greatly in P(A) the whole copolymer is heterogeneous with respect to conversion, i.e. individual polymer chains are esterified to a different extent. Comparing for the fractions the experimental run numbers R to the calculated Bernoullian run numbers R_r an approximately Bernoullian statistics is indicated. This is in contrast to the whole copolymer for which R is considerably smaller than R_r indicating block character. The approximately Bernoullian statistics of the fractions is corroborated by the values determined for the probabilities of the individual triads (not shown). In consequence, the whole copolymer is really Bernoullian with respect to its individual chains and the apparent block character arises by the summation of the approximately Bernoullian triad probabilities of the fractions.

The cause for the heterogeneity of conversion is the high rate of reaction for PMAA and diazomethane. The high rate precludes sufficient mixing of the reactant solutions before the diazomethane has almost completely reacted. However, when the PMAA solution is first neutralized with KOH, the rate is

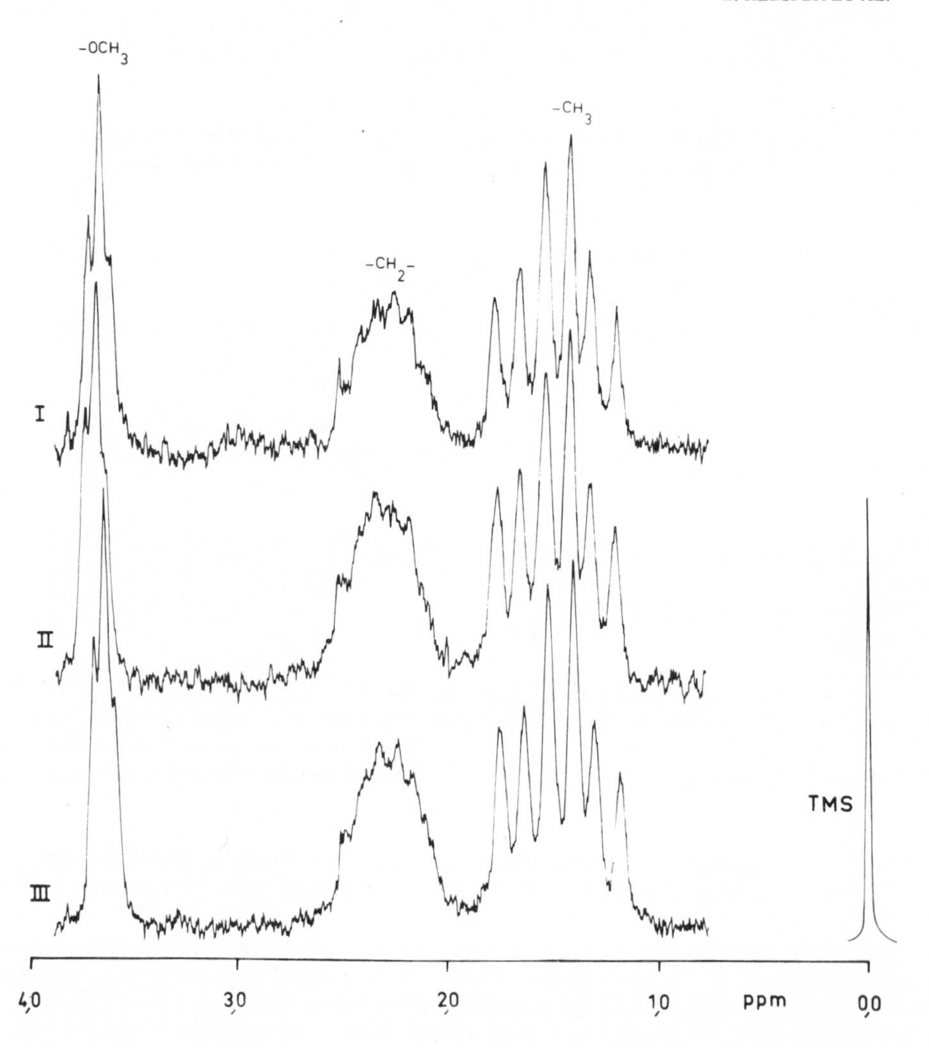

Fig. 7 NMR spectra of a cosynd. MMA - MAA copolymer prepared from neutralized PMAA (trace I) and fractions of this copolymer (traces II and III). Spectra recorded from pyridine solution, 100°C, 220 MHz, TMS as internal standard.

slowed considerably, allowing sufficient time for mixing. The decrease in rate occurs because the first and rate determining step in the reaction of acids with diazo alkanes is the protonation of the latter (14). The NMR spectrum of a cosyndiotactic MMA - MAA copolymer of P(A) = 0.48 prepared with neutralized PMAA is shown as trace I of Fig. 7 . The copolymer is fractionated as before, the spectra of the two

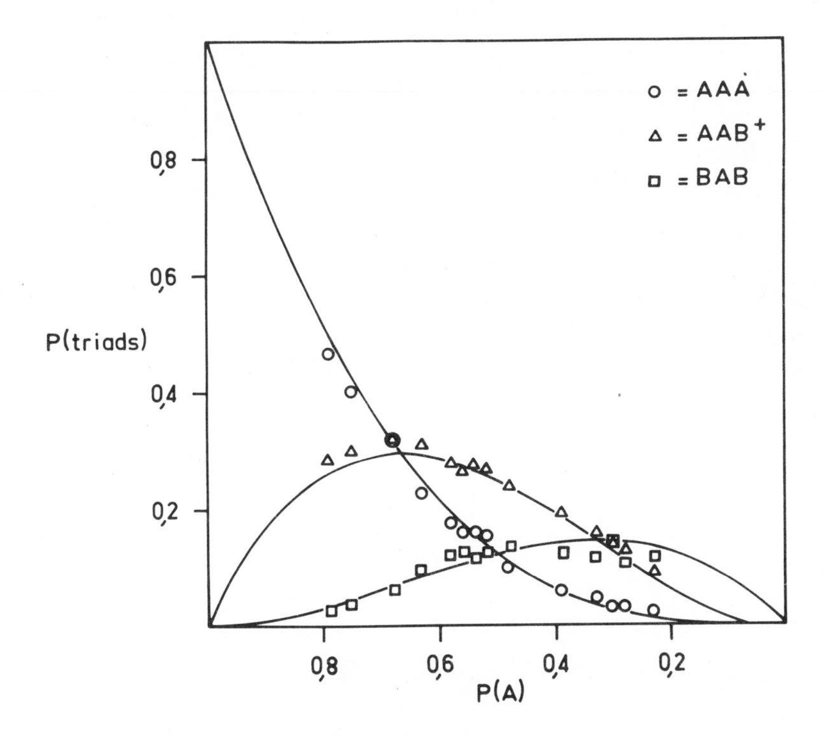

Fig. 8 Probabilities for A-centered triads in cosynd. MMA - MAA copolymers prepared with neutralized PMAA. Points are experimental triad probabilities, curves are calculated Bernoullian triad probabilities.

fractions are seen as traces II and III . Because the relative peak areas of the spectra are virtually identical, the whole copolymer is of homogeneous conversion.

Fourteen copolymers of different P(A) are obtained by esterification of neutralized syndiotactic PMAA with varying amounts of diazomethane. Their triad probabilities are determined by NMR and are plotted versus P(A) in Figs. 8 and 9. Fig. 8 shows the data for the A-centered triads and Fig. 9 for the B-centered triads. The experimental points are situated reasonably close to the curves calculated for Bernoullian statistics whereby the deviations from the curves do not exceed the experimental error. Therefore, and in view of the homogeneity of the copolymers, the individual chains must possess a nearly Bernoullian statistics over the range of P(A) investigated. In turn, the Bernoullian statistics demands the absence of cooperative kinetics, i.e. $k(AAA) = k(AAB^{+}) = k(BAB)$.

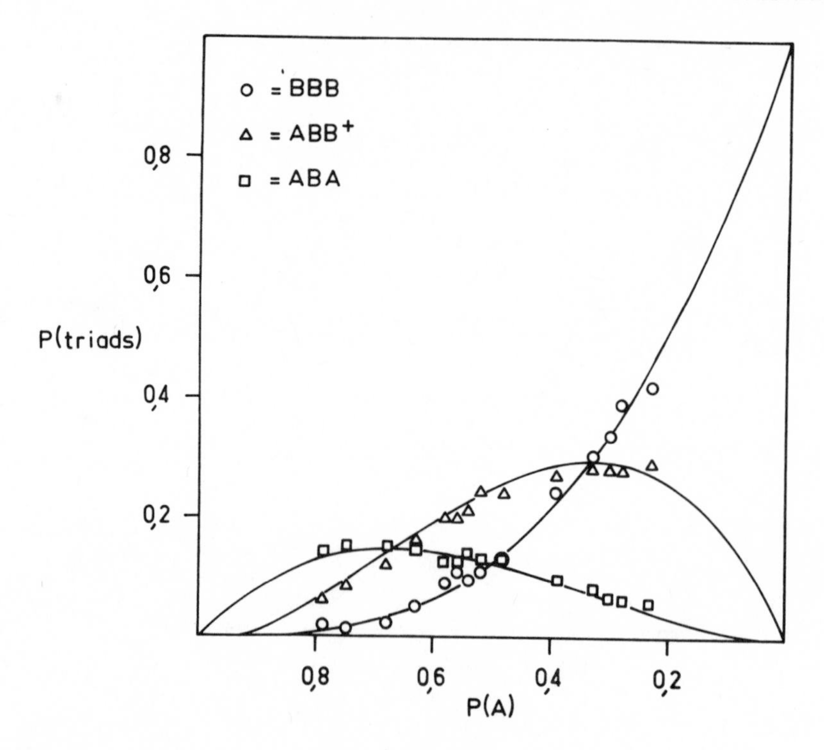

Fig. 9 Probabilities for B-centered triads in cosynd. MMA - MAA copolymers prepared with neutralized PMAA. Points are experimental triad probabilities, curves are calculated Bernoullian triad probabilities.

4.2. Partial Esterification with Phenyl Diazomethane

The esterification of syndiotactic PMAA with phenyl diazomethane in dioxane solution proceeds at a slow enough rate to allow sufficient time for mixing the reactant solutions. With partial esterification, the resulting cosyndiotactic benzyl methacrylate - methacrylic acid (BMA - MAA) copolymers should then possess homogeneous conversion. In Fig. 10 the NMR spectra from pyridine solutions of three BMA - MAA copolymers of P(A) = 0.27 , 0.46 , and 0.58 are seen as traces I to III , respectively. The α-CH_3 resonance region shows considerable sequence splitting which is due not only to triads but also to pentads of monomer units. The attainable resolution of the α-CH_3 resonance region is seen more clearly in Fig. 11 by the expanded partial spectra of the same three copolymers. The assignment of the triads and pentads is written next to trace II which shows 18 peaks in all (A = BMA). The proposed assignment has

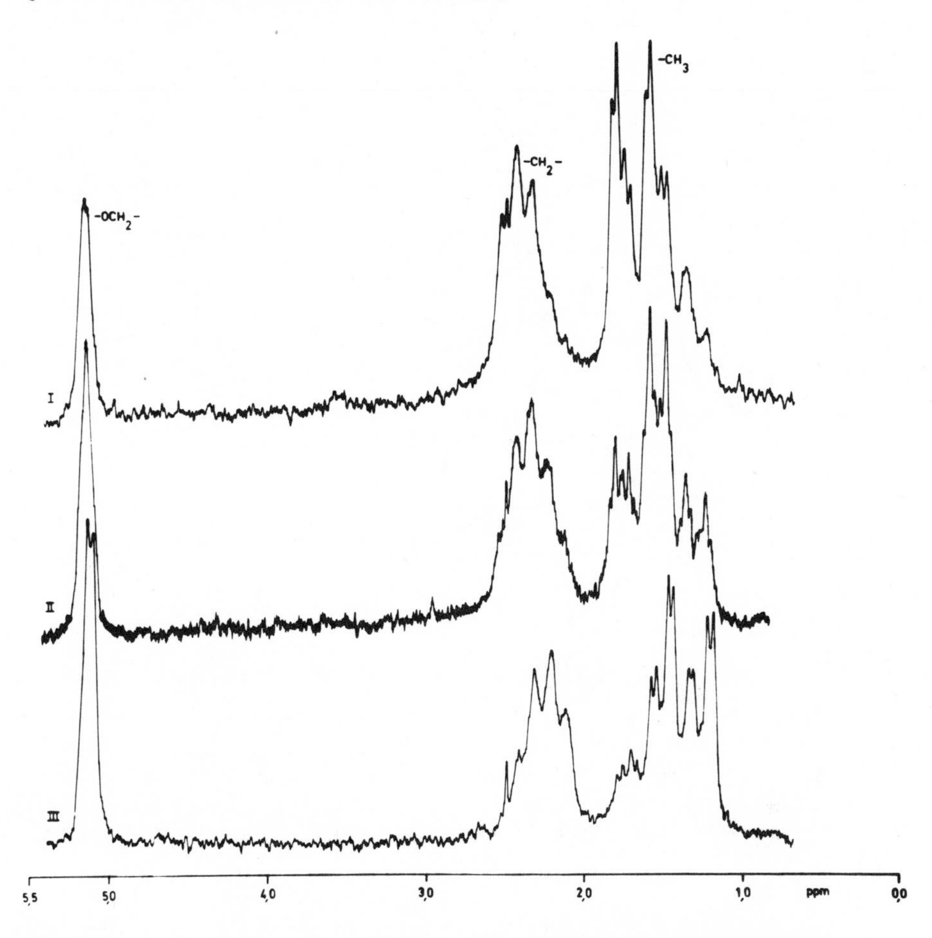

Fig. 10 NMR spectra of cosynd. BMA - MAA copolymers in pyridine at 100°C , 220 MHz , TMS as internal standard

been secured by applying generally valid statistical relations (15) . Fractionation of a copolymer of P(A) = 0.42 confirmed the homogeneity of conversion.

Recording the NMR spectra of the three copolymers of Fig. 10 from quinoline solution, the spectral traces of Fig. 12 are obtained. The α-CH_3 resonance region is overlapping strongly with the broad β-CH_2- resonance which renders the quantitative evaluation of peak areas in the α-CH_3 resonance impracticable. However, the -OCH_2- resonance is sufficiently resolved into A-centered triads to allow determination of the corresponding triad probabilities. The assignment of the A-centered triads to the three peaks of the -OCH_2- resonance

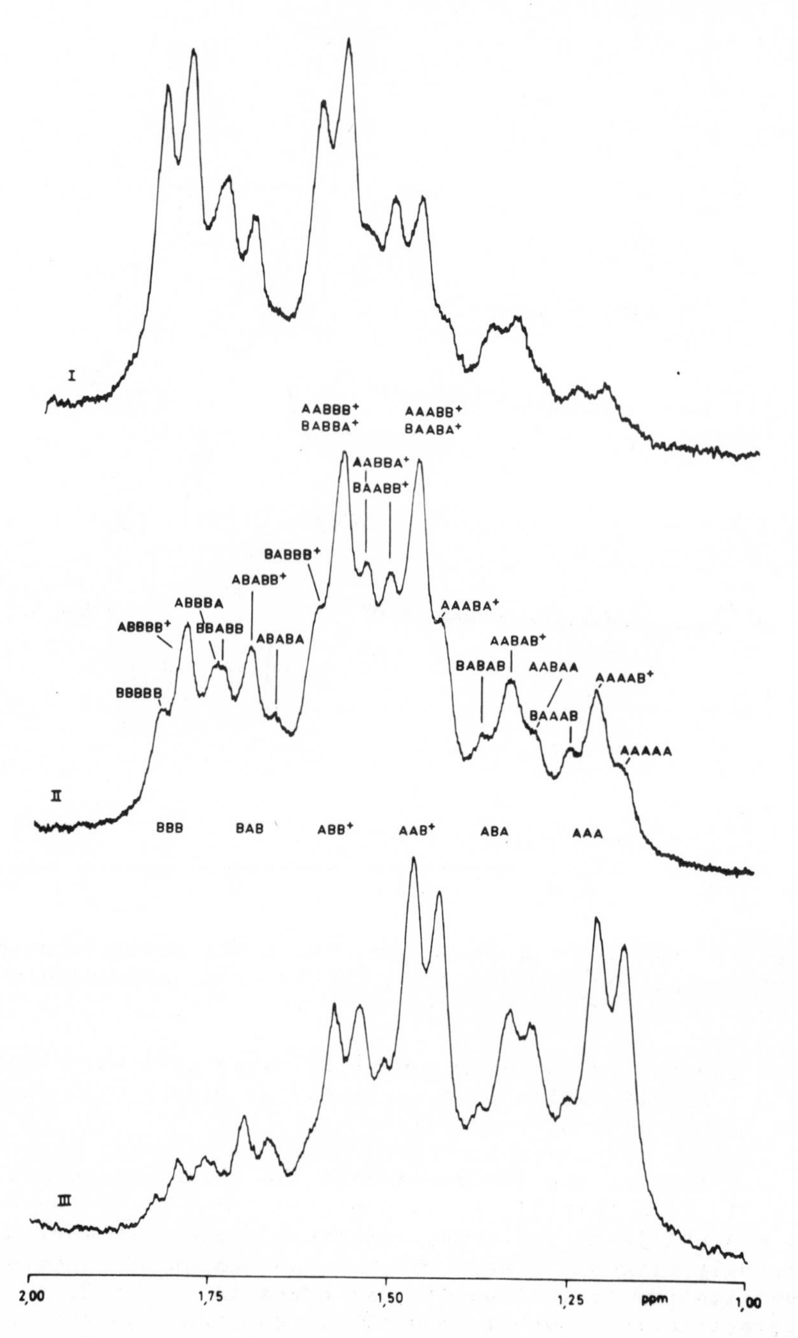

Fig. 11 Expanded partial NMR spectra of the cosynd. BMA-MAA copolymers of Fig. 10. Pyridine, 100°C, 220 MHz, TMS int. st. Assignments as shown (A = BMA).

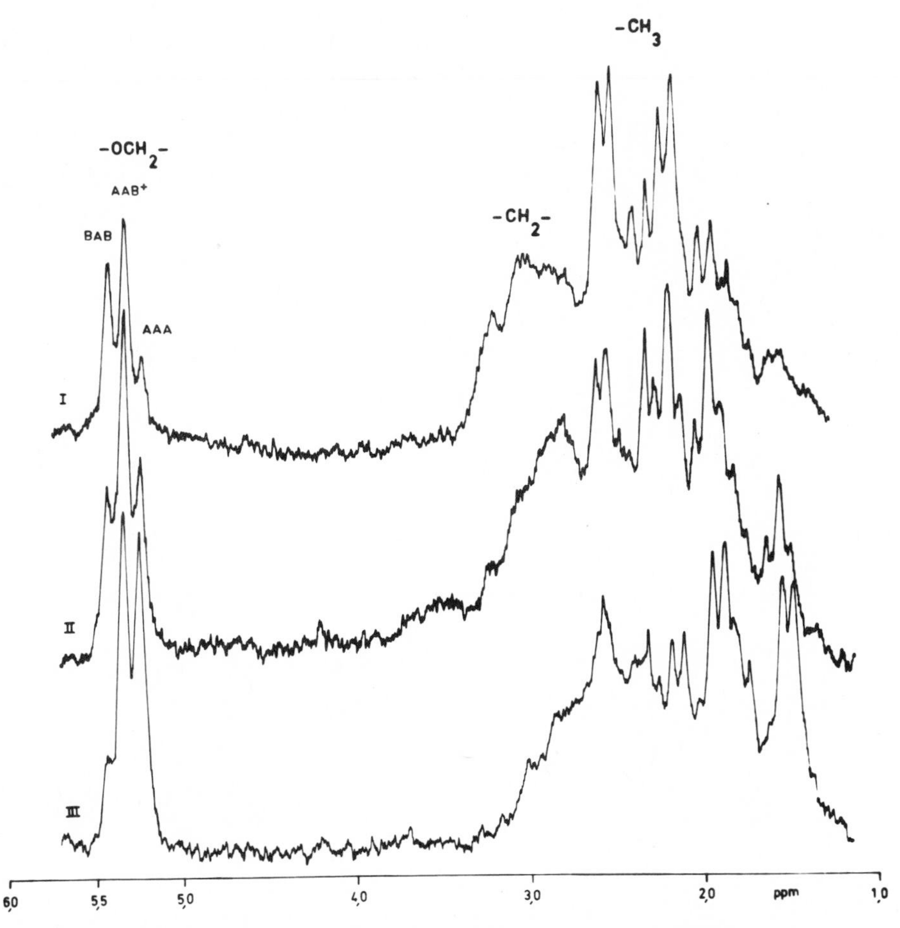

Fig. 12 NMR spectra of cosynd. BMA - MAA copolymers of Fig. 10 in quinoline solution at 100°C, 220 MHz, TMS as internal standard. Assignment of A-centered triads as shown (A = BMA).

is already rather obvious on inspection. For evaluating the statistics of the cosyndiotactic BMA - MAA copolymers the triad probabilities are determined first by the α-CH_3 resonance as recorded from pyridine solution. The experimental probabilities of the A-centered triads are plotted versus P(A) in Fig. 13 and show a nearly Bernoullian distribution. The same result is obtained for the probabilities of the B-centered triads (not shown). If, in addition, the probabilities of the A-centered triads are determined by the $-OCH_2-$ resonance from quinoline solution, the experimental points of Fig. 14 are obtained. These data, too, confirm the

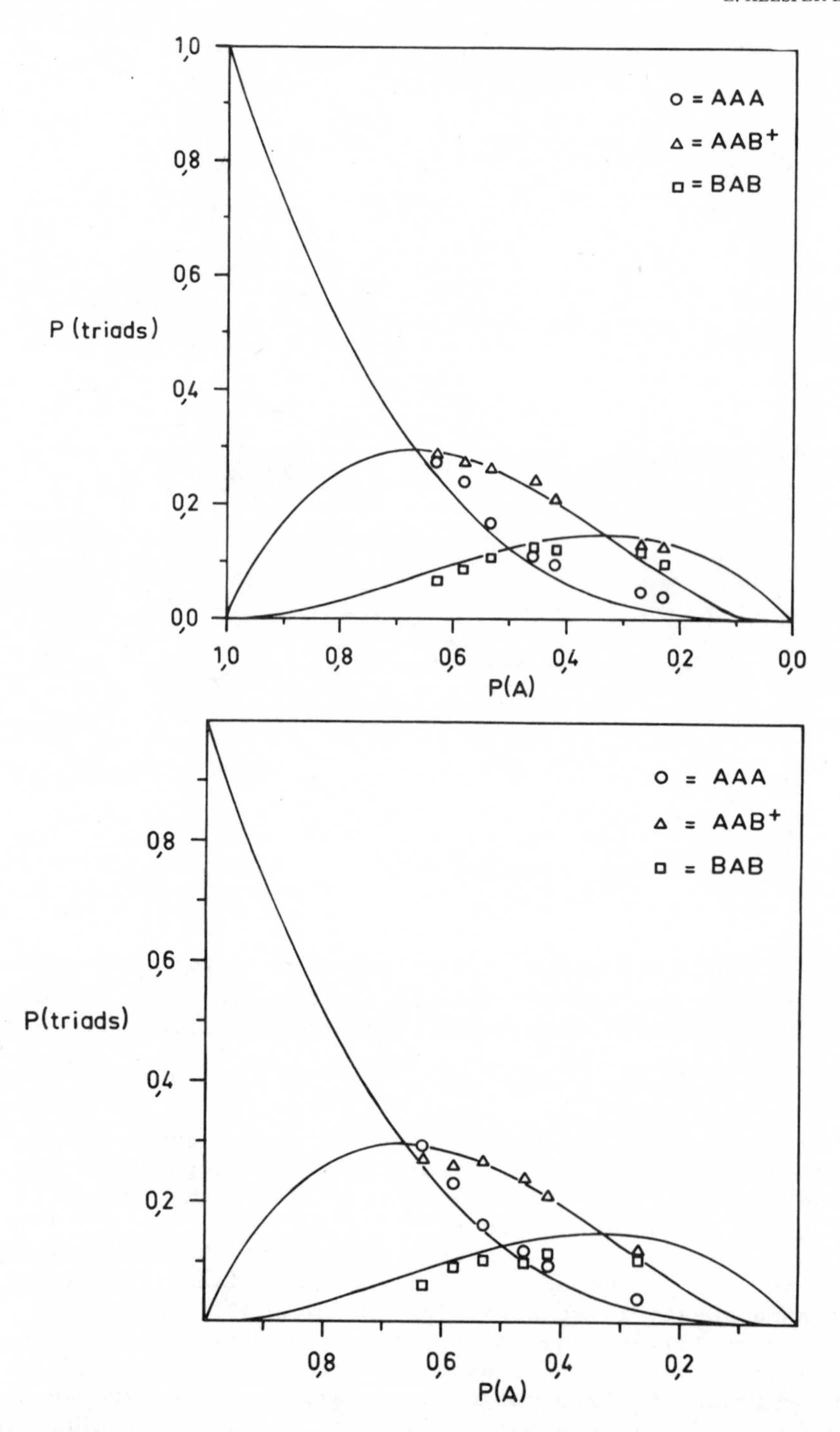

Figs. 13 (top) and 14 (bottom) Experimental probabilities (points) of A-centered triads in BMA-MAA copolymers by α-CH_3 from pyridine sol. (top) and by $-OCH_2-$ from quinoline sol. (bottom). Calculated Bernoullian probabilities: curves .

Bernoullian statistics for the BMA - MAA copolymers. The fact that nearly the same triad probabilities are found for the solutions of the copolymers in the NMR solvents pyridine and quinoline also supports the assignments made for both solvents.

An evaluation of the pentad probabilities from the α-CH_3 resonance of BMA - MAA copolymers in pyridine solution does not allow to narrow down further the Bernoullian distribution of monomer units. The pentad probabilities are found to be less accurate than the triad probabilities because of the lower resolution of the pentad peaks when compared to the resolution of the composite triad peaks.

4.3. Partial Esterification with Diphenyl Diazomethane

The esterification of syndiotactic PMAA with diphenyl diazomethane in dioxane solution proceeds still slower than with phenyl diazomethane. The partial esterification leads therefore to syndiotactic diphenyl methyl methacrylate - methacrylic acid (DPMMA - MAA) copolymers which are also of

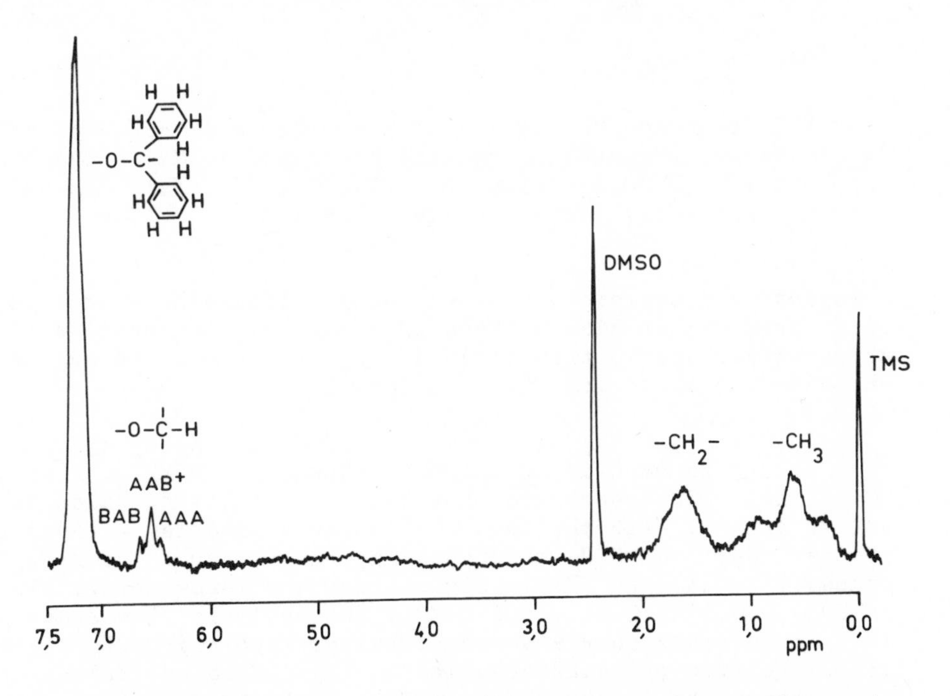

Fig. 15 NMR spectrum of a cosynd. DPMMA - MAA copolymer in DMSO.d_6 at 100°C , 220 MHz , TMS as internal standard. Assignments as shown (A = DPMMA).

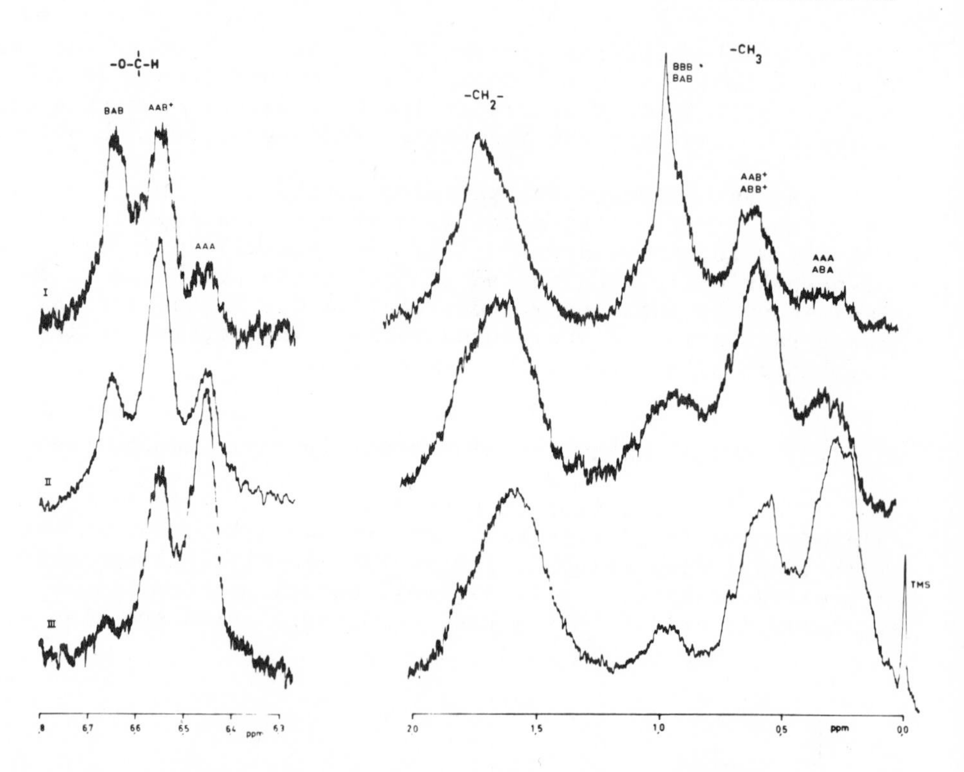

Fig. 16 Expanded NMR spectra of the α-CH_3 and β-CH_2- as well as of the -OCH= resonance regions of gosynd. DPMMA - MAA copolymers in $DMSO.d_6$ solution at 100°C , 220 MHz , TMS as internal standard. Assignments as shown (A = DPMMA).

homogeneous conversion. Since diphenyl diazomethane is more bulky than either diazomethane or phenyl diazomethane, a cooperative kinetic effect based on steric hindrance may be suspected to exist for the reaction.

In Fig. 15 a NMR spectrum of a DPMMA - MAA copolymer of P(A) = 0.51 in $DMSO.d_6$ solution is shown (A = DPMMA). Both the α-CH_3 and -OCH= resonance show sequence splitting which is due to triads. With the best resolution available, expansions of the α-CH_3 and -OCH= resonance regions appear as seen by traces I , II, and III in Fig. 16 for the three copolymers of P(A) = 0.37 , 0.51 , and 0.70 , respectively. The same three copolymers show a somewhat better resolved α-CH_3 resonance from pyridine solution as Fig. 17 demonstrates. The -OCH= resonance is not shown in Fig. 17 because it is covered by the solvent resonances.

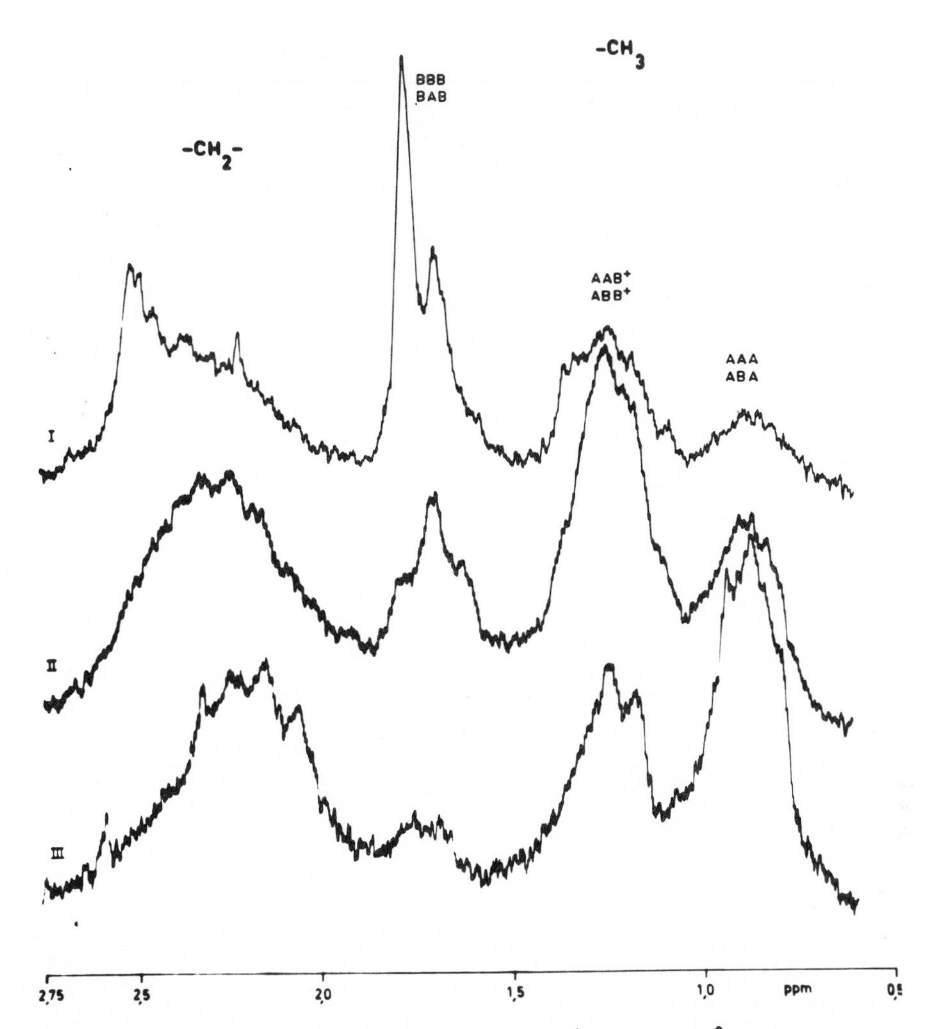

Fig. 17 Expanded NMR spectra of the α-CH_3 and β-CH_2- resonance regions of cosynd. DPMMA - MAA copolymers in pyridine solution at 100 °C , 220 MHz , TMS as internal standard. Assignments as shown (A = DPMMA).

The proposed triad assignments are seen in Figs. 16 and 17 . While the assignment of the A-centered triads in the —OCH= resonance is apparent on inspection, the assignments of the α-CH_3 resonance from both DMSO.d_6 and pyridine solution require support. According to a generally applicable statistical relation:

23) $2\,P(A) = P(AAB^+) + P(ABB^+) + 2\{P(ABA) + P(A$

Table 2

Experimental probabilities for supporting the assignments in the α-CH_3 resonance of DPMMA - MAA copolymers in solutions of $DMSO.d_6$ and pyridine

P(A)	2 P(A)	$P(AAB^+) + P(ABB^+) + 2\{P(ABA) + P(AAA)\}$ pyridine	$P(AAB^+) + P(ABB^+) + 2\{P(ABA) + P(AAA)\}$ $DMSO.d_6$
0.25	0.50	0.46	0.38
0.35	0.70	0.69	0.63
0.48	0.96	1.00	0.96
0.51	1.02	1.00	0.94
0.56	1.12	1.22	1.20
0.65	1.30	1.24	1.27
0.70	1.40	1.36	1.31

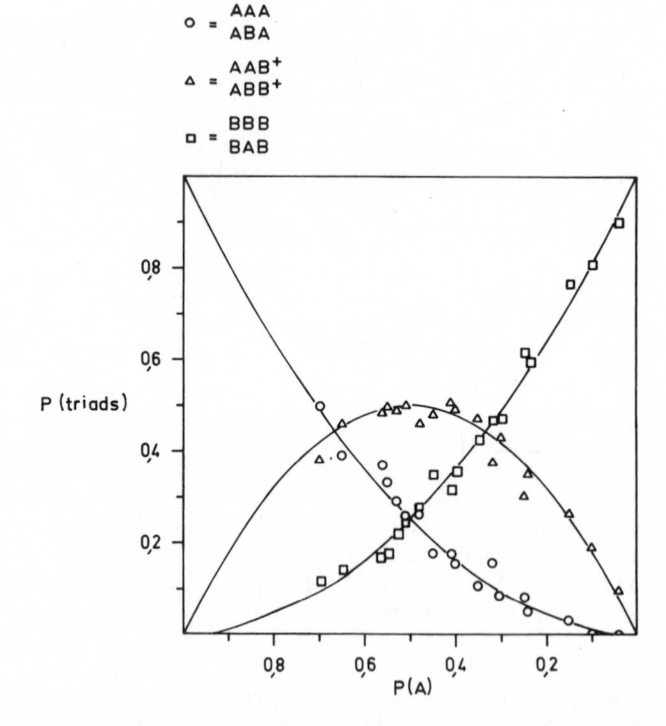

Fig. 18 Sums of probabilities for pairs of coinciding triads of cosynd. DPMMA - MAA copolymers as evaluated from α-CH_3 resonance of pyridine solutions. Points are experimental data, curves are calculated for Bernoullian statistics.

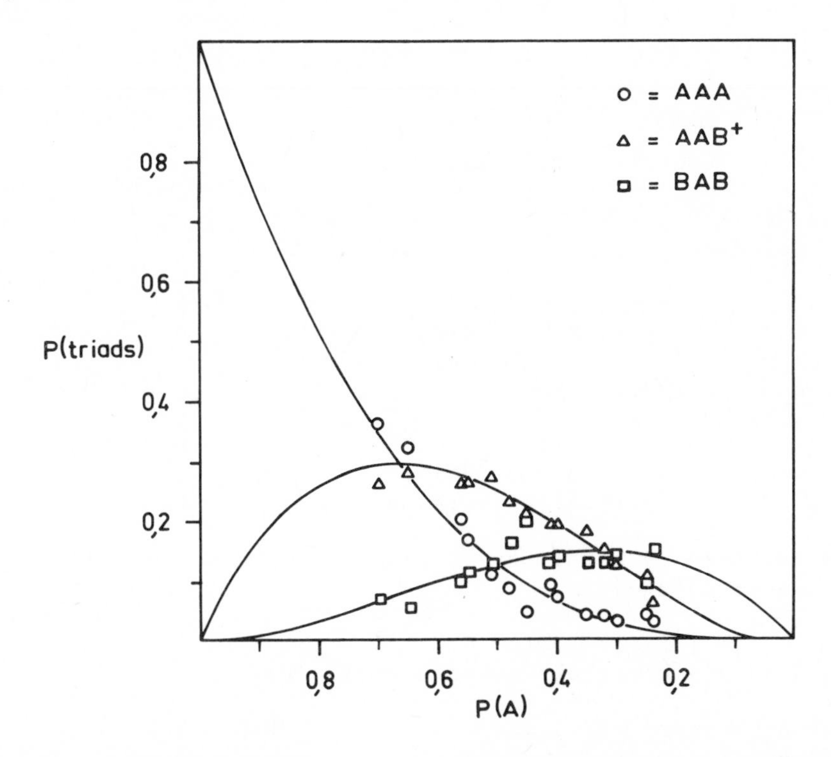

Fig. 19 Probabilities of A-centered triads in cosynd. DPMMA - MAA copolymers as evaluated by the —OCH= resonance region from DMSO.d_6 solution.

In keeping with the proposed assignments for the α—CH_3 resonance, Eq. 23 contains only directly measurable sums of triad probabilities. Experimental data corresponding to Eq. 23 are presented in Table 2 . It is seen that Eq. 23 is obeyed within experimental error, with the possible exception of the value 0.38 for DMSO.d_6 and P(A) = 0.25 .

Because of the somewhat higher resolution in the α—CH_3 resonance recorded from pyridine solution, as compared to DMSO.d_6 solution, the probabilities of the pairs of triads determined from the first solvent are shown here. Fig. 18 demonstrates that the experimental points do not scatter too much around the corresponding Bernoullian curves. The scatter is somewhat increased for the α—CH_3 data derived from DMSO.d_6 solution but there, also, can be little doubt that the statistics of the DPMMA - MAA copolymers at least approaches a Bernoullian distribution. Moreover, the probabilities for the individual A-centered triads are evaluated from the —OCH= resonance of DMSO.d_6 solutions and

Table 3

synd. PMMA (monodisperse)

↓ complete hydrolysis

synd. PMAA (polyelectrolyte) . . B B B B B B B B B B B . .

+ diazo compound I (partial conversion B ⟶ A) ↓

cosynd. binary copolymer (polyelectrolyte) . . A B B B A A B B B B B . .

+ diazo compound I I (partial conversion B ⟶ C) ↓

cosynd. ternary copolymer (polyelectrolyte) . . A B B C A A B C B B C . .

+ diazo compound I I I (complete conversion B ⟶ D) ↓

cosynd. ternary copolymer (not a polyelectrolyte) . . A D D C A A D C D D C . .

are shown in Fig. 19 . Taking into account the relatively high error in peak area determination for the weak —OCH= resonance (only one proton !), caused in part by disturbances from spinning side bands of the tenfold stronger =$C(C_6H_5)_2$ resonance, the agreement between experimental triad probabilities and Bernoullian curves of Fig. 19 may be considered satisfactory.

Summarizing the results for the esterification of syndiotactic PMAA with diazomethane, phenyl diazomethane, and diphenyl diazomethane, it is found that none of the three diazo alkanes reacts with a cooperative kinetics. Although the esterification of syndiotactic PMAA with these diazo alkanes is therefore not of great further interest from a kinetic point of view, it may be of interest for the preparation of defined model copolymers. The very fact that copolymers of Bernoullian distribution of monomer units are formed causes the statistics to be easily determined and described, which is of advantage for a model copolymer to be used for further studies.

4.4. Successive Esterification with Different Diazo Alkanes

In view of the absence of a cooperative kinetics for the esterification of syndiotactic PMAA with three diazo alkanes, the preparation of cosyndiotactic,multicomponent copolymers of Bernoullian distribution of monomer units by successive partial esterification of syndiotactic PMAA with several diazo alkanes may be considered. Toward this end one may envisage the schematic procedure of Table 3 . The starting PMAA is first partially esterified with diazo alkane I whereby a part of the B-units are converted to A-units. The resulting binary copolymer is again partially esterified with a diazo alkane II which transforms B-units into C-units yielding a ternary copolymer. All polymers thus far are polyelectrolytes. However, it might be desirable to eliminate the polyelectrolyte property by complete esterification of the ternary copolymer with a diazo alkane III to obtain a ternary copolymer of the same sequence distribution as the precursor copolymer but having no unreacted carboxyl groups. Furthermore, a monodisperse, syndiotactic PMMA can be prepared (16) which is hydrolyzed to starting PMAA used for the successive partial esterification. Thus monodisperse, highly syndiotactic, and statistically defined multicomponent copolymers may be prepared(17,18)

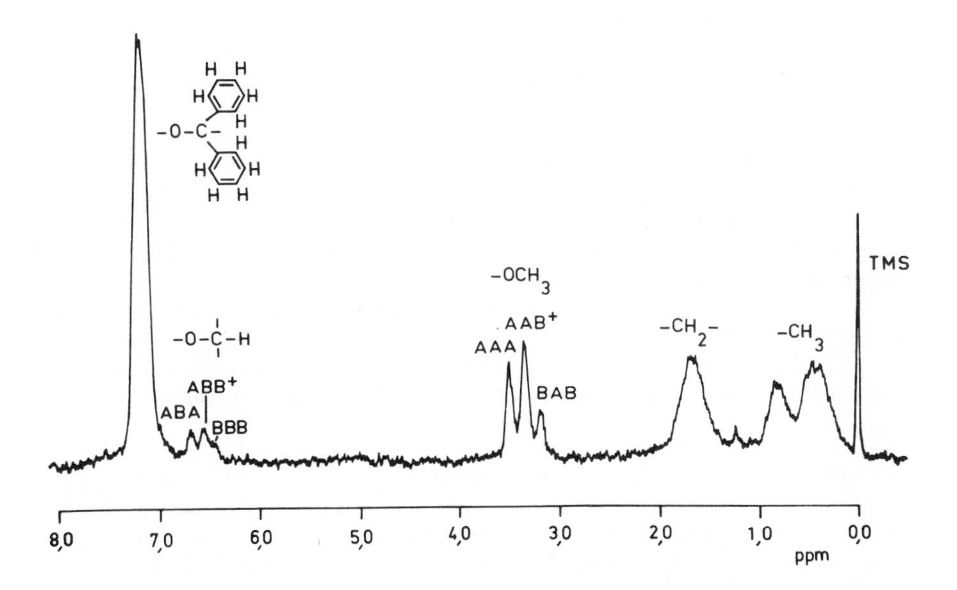

Fig. 20 NMR spectrum of a cosynd. DPMMA - MMA copolymer in $CDCl_3$ solution at 60°C, 220 MHz, TMS as internal standard. Assignments as shown (A = MMA).

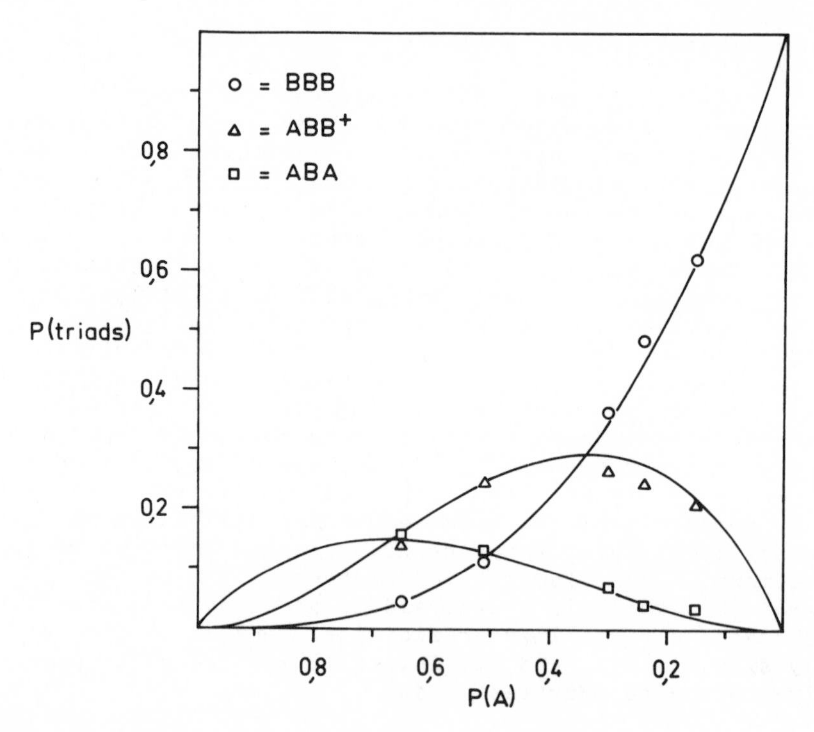

Fig. 21 Probabilities of B-centered triads (B = MMA) in cosynd. DPMMA - MMA copolymers as evaluated by the $-OCH_3$ resonance from $CDCl_3$ solution.

We have carried out complete esterifications on cosyndiotactic MMA - MAA, BMA - MAA, and DPMMA - MAA copolymers with either diazomethane, phenyl diazomethane, or diphenyl diazomethane. The triad distribution of the resulting non-electrolyte, binary copolymers was found to be the same as that of the precursor MAA copolymers. An exception was provided by the BMA - MMA copolymers in as much as no triads or other sequences could be evaluated from the NMR spectra because of lack of resolution in all NMR solvents tried.

As an example, the NMR spectrum of a diphenyl methyl methacrylate - methyl methacrylate (DPMMA - MMA) copolymer of P(A) = 0.56 , prepared by complete esterification of a DPMMA - MAA copolymer with diazomethane, is shown in Fig. 20 . The $-OCH_3$ and $-OCH=$ resonance regions show useful resolution of the A-centered and B-centered triads, respectively (A = MMA). The probabilities of the MMA-centered triads determined from the $-OCH_3$ resonance is shown in Fig. 21 (here B = MMA). The experimental points are rather close to the Bernoullian curves. In addition it is found that the P(A) of the derived DPMMA - MMA copolymers has the same value as the P(B) of the corresponding DPMMA - MAA precursor copolymers. Thus the second, complete esterification step has not altered the distribution of the B-centered triads of the precursor copolymers.

5. EXPERIMENTAL PROCEDURES

5.1. Homopolymers

The preparation of syndiotactic PMMA (19) and the total hydrolysis to syndiotactic PMAA (20) has been reported. The PMMA had 92% syndiotactic triads and approximately 8% heterotactic triads.

5.2. Hydrolysis

The cosyndiotactic MMA - MAA copolymer of Bernoullian statistics and P(A) = 0.79 , which was employed as the starting copolymer for the kinetic runs of Section 3 , was obtained by partial hydrolysis of syndiotactic PMMA. The PMMA (250 mg) was dissolved in dioxane (8 ml) in a glass vial and CH_3OH - KOH mixture (20% w/w KOH) (4 ml) was added. The air was removed by a stream of N_2 and the vial closed. On transferal of the vial to a water bath of 45°C the precipitate dissolved which had formed on addition of the CH_3OH - KOH mixture. After 19 h at 45°C with continuous, slow tumbling of the vial, the contents were dissolved in H_2O (250 ml). The copolymer was precipitated at 80°C by dropwise addition of conc. HCl , filtered while hot, and dried at 50°C in vacuo for at least 24 h . Multiple charges were also run and copolymers from different charges were combined in several cases.

For the kinetic runs, the copolymer (250 mg) was dissolved in aqueous KOH (12 ml) at 115°C for 15 min. under N_2 in a closed vial of alkali resistant glass. To ease dissolution the vial was tumbled. The solution contained the amount of KOH specified by the parameter r and sufficient KCl to raise the amount of K^+ to 5 mval for all runs. After dissolution the contents were transferred to another alkali resistant glass vial with Teflon liner, in order to prevent consumption of OH^- by the glass. After removing the air by flushing with N_2 the vial was closed and placed into holes of an aluminum block in a thermostat of 145°C . Usually several vials with identical contents were run in parallel and removed one by one after predetermined reaction times. The copolymers were isolated as before.

5.3. Esterification with Diazo Alkanes

The partial esterification of syndiotactic PMAA with diazomethane, phenyl diazomethane, or diphenyl diazomethane, as well as the complete esterification of cosyndiotactic MMA - MAA copolymers, cosyndiotactic BMA - MAA copolymers, and cosyndiotactic DPMMA - MAA copolymers with either one of the three diazo alkanes is described elsewhere (15,20).

5.4. NMR Spectra

The ^{1}H NMR spectra were recorded with 8 - 10 % w/v polymer in solution. Other conditions are specified in the figure captions. The evaluation of probabilities from the NMR spectra is described elsewhere (10,15,20).

Financial support by the Deutsche Forschungsgemeinschaft is gratefully acknowledged.

REFERENCES

1) H. Morawetz, P.E. Zimmering, J. phys. Chem., 58 753 (1954)
2) H. Morawetz, E.W. Westhead, J. Pol. Sci., 16 273 (1955)
3) W. DeLoecker, G. Smets, J. Pol. Sci., 40 203 (1959)
4) G. Smets, A.M. Hesbain, J. Pol. Sci., 40 217 (1959)
5) G. Smets, W. DeLoecker, J. Pol. Sci., 41 375 (1959)
6) G. Smets, W. DeLoecker, J. Pol. Sci., 45 461 (1960)
7) E. Klesper, W. Gronski, V. Barth, Makromol. Chem., 150 223 (1971)
8) E. Klesper, A. Johnsen, W. Gronski, Makromol. Chem., 160 167 (1972)
9) E. Klesper, J. Pol. Sci., B, 6 663 (1968)
10) E. Klesper, W. Gronski, V. Barth, Makromol. Chem., 139 1 (1970)
11) E. Klesper, V. Barth, A. Johnsen, Pure and Appl. Chem., 8 151 (1971)
12) P. Rabinowitz, A. Silberberg, R. Simha, E. Loftus, Adv. Chem. Phys., Vol. 15, 281 (1969)
13) H.J. Harwood, W.M. Ritchey, J. Pol. Sci., B, 2 601 (1964)
14) R. Huisgen, Angew. Chemie, 67 443 (1955)
15) D. Strasilla, E. Klesper, Makromol. Chem., in press
16) A. Roig, J.E. Figueruelo, E. Llano, J. Pol. Sci., C, 16 4141 (1968)
17) E. Klesper, A. Johnsen, unpublished
18) E. Klesper, D. Strasilla, unpublished
19) H. Abe, K. Imai, M. Matsumoto, J. Pol. Sci., B, 3 1053 (1965)
20) E. Klesper, D. Strasilla, W. Regel, Makromol. Chem., in press

DISCUSSION SESSION

Discussion Leader - M. Lewin

Loucheux: Do you have any results which point to hydrophobic interactions as found by Quadrifoglio and Creszenzi who observed a globule- coil transition on exceeding 25% methacrylic acid units in methyl methacrylate- methacrylic acid copolymers? This result may be connected with the influence of the α-methyl groups. Correspondingly, such a transition was not found with a methyl acrylate-acrylic acid copolymer.

Klesper: We checked the chemical shifts of the triads in the NMR spectra of our cosyndiotactic MMA-MAA copolymers dissolved in water containing as exess of KOH. There was no change in chemical shifts of the triads which function as our kinetic entities, over the range of conversion. However, one should expect a change in chemical shifts when the conformations change because of any inter- and intra-molecular interactions. Thus we conclude that there is no conformational change which might change the rate constants of our kinetic entities. However, our copolymers are fully neutralized at all degrees of conversion during hydrolysis with an excess of hydroxyl ions. In the experiments you quoted the charge on the copolymer was gradually increased during a potentiometric titration until the transition was observed.

Moore: Can 100% isotactic poly(methyl methacrylate)be prepared?

Klesper: Probably not. According to our experience, the best we have obtained is 98% isotactic triads.

Ise: Have you considered the molecular weights of your copolymers?

Klesper: We measure the molecular weights by transforming the copolymers to PMMA with diazomethane. The viscosity-molecular weight relationship for PMMA is well established. No change in molecular weight was observed on hydrolysis. However, our kinetic data are not influenced to a significant extent by the molecular weight of the starting copolymer.

Ise: Does the theory which you presented apply to low molecular weight polymers and oligomers?

Klesper: The statistical considerations presented apply to infinitely long chains of homo-and co-polymers or at least to chains which are long enough so that end effects can be neglected. By end effects we mean kinetic effects and NMR signals from end groups. The theory is not now applicable to oligomers, but suitable changes could be made which would make it useful for oligomers.

Ise: You considered the statistics of diads and triads of monomers. These sequences are certainly also present in oligomers.

Klesper: Yes, but an oligomer of, e.g. five monomer units AAAAB contains only three complete triads, which are underlined. The central A units will behave kinetically like the corresponding triads in a long chain copolymer, barring conformational effects. If we examine the terminal A group of the oligomer we see that it is not part of a complete trial and will therefore be excluded from our statistical considerations.

Lewin: Have you studied the temperature dependence of the kinetics of the hydrolysis?

Klesper: We have done single experiments at 115° and 175°. At the low temperature the reaction requires a week or more. At the higher temperature the polymer may be degrading.

Lewin: It would be of interest to determine the activation energies and see whether they differ for different sequences. This might serve as an additional tool for investigating the cooperative reaction.

Klesper: This is certainly true.

Harwood: Would you explain why your reactions accelerate after the plateau region?

Klesper: The plateau region occurs after hydroxide ion is completely consumed. Hydrolysis then occurs slowly and carboxylic acid groups are introduced into the copolymers. As the concentration of such groups increases the reaction speeds up. Apparently ester groups having acid neighbors are highly reactive.

Challa: Considering that the syndiotactic PMMA contains about 8% of heterotactic triads, one has to expect that your kinetic triad data are affected by heterotactic triads of different reactivity. Is it possible to conclude from your data whether the heterotactic triads are gathered in blocks or if they are randomly distributed?

Klesper: The 8% heterotactic triads will introduce an error in our kinetic triad measurements. We know from other work that the heterotactic triads are located under the syndiotactic triads in the proton NMR spectra in pyridine solution. So far we have not determined the probabilities of the heterotactic triads individually. We know from the NMR spectra of the PMMA from which the starting copolymer is obtained that the heterotactic triads are randomly distributed.

SOME PROBLEMS IN THE REACTIVITY OF MACROMOLECULES

N.A.Plate'
Insitute of Petrochemical Synthesis
M.V.Lomonossov State University of Moscow
Moscow, U.S.S.R.

INTRODUCTION

Consideration of the current state of the problem of chemical reactions with macromolecules as reagents should include a study of the various effects which differentiate the behavior and reactivity of units and functional groups attached to polymer chains from the behavior of low molecular weight species. These include the neighboring groups effect, the influence of microtacticity, conformational effects, the influence of electrostatic interactions and various supermolecular effects.

The neighboring group effect is one of the most important and is typical for macromolecular reactions, because of its influence on the kinetics of the reaction, on unit sequence distribution along the chain and on the composition heterogeneity of the products. The last two factors play a key role in the chemical, physical and mechanical properties of polymeric materials.

We will consider the present day theory of macromolecular reactivity in the framework of the model of the influence of reacted groups, B, on nearby initial groups, A, in a polymeric chain.

The accurate mathematical solution for the distribution sequence of units for any polymer analogous reactions will be proposed based on the multiplet method and rate constants K_0, K_1 and K_2. The results of these calculations will be compared with the various approximation methods based on the Markovian-type processes and on a Monte Carlo simulation method. The predicted, calculated data will be compared with experimental results obtained by studying the hydrolysis of several polymethaerylic esters.

Experimental methods for the determination of the individual rate constants will be discussed and the details of useage of the Monte Carlo simulation method to predict the composition heterogeneity of the products of polymer analogous reactions, as well as other simpler methods of approximation will be demonstrated. The results obtained after computer calculation will compared with the experimental data for the example of the quaternization of polyvinyl pyridine with benzyl chloride. The obtained agreement permits the use of the theoretical approach to the calculation of the unit sequence distribution and composition heterogeneity in cases where the experimental data are difficult to obtain.

All the apparatus of theoretical calculations and kinetic approach to the study of macromolecular reactions can also be used to shed light on the mechanisms of particular chemical reactions involving macromolecules and to discover other effects related to the polymeric nature of the reagent.

RESULTS AND DISCUSSION

Let us consider the case when, in quasi-isolated macromolecules, units of A are converted into units of B in an irreversible first order reaction, the reactivity of A depending only on the nature of its two immediate neighbors. We shall distinguish triads AAA, BAA (or AAB) and BAB; their fractions will be designated N_o, N_1 and N_2 and the rate constants of transformation of the central units k_o, k_1 and k_2 respectivly.

Formally a kinetic theory of such processes should describe the change in the composition of the reaction mixture with time. In the case of polymer analogous reactions one requires a description of the changes in the total concentration of A units (kinetic curve), distribution of A and B units in the macromolecule (structure of the chain) and distribution of the macromolecules by degrees of conversion (composition heterogeneity). For a particular molecule these characteristics will depend upon the ratio of the constants k_o: k_1: k_2. The patterns of the kinetic curves differ greatly when a) the neighboring group effect is absent ($k_o = k_1 = k_2$), b) acceleration ($k_o \ll k_1 \leqslant k_2$), c) retardation ($k_o \geqslant k_1 \gtrless k_2$) are strong. At $k_o = k_1 = k_2$ A and B units are distributed randomly, when acceleration is strong B blocks are formed, when retardation is strong the amount of isolated B units increases. One might also expect <u>a priori</u> that the composition heterogeneity of the product will increase, should the neighboring units produce an accelerating effect and decrease, should the effect be a decelerating one. Hence a problem arose-to describe reaction kinetics, chain structure and composition heterogeneity of the products as functions of the ratio of the constants k_o, k_1, and k_2.

Some partial solutions of this problem were put forward in 1962-1963[2,6]. Later Mc quarrie et al.[7] found an unambiguous solution for the description of the kinetics and distribution of non-reacted units by applying the multiplet method. These authors distinguish between two types of A unit sequences: j-clusters if the sequence is closed by B units and j-tuplets if the nature of the bordering units is inessential. The probabilities of cluster C_j^A and tuplets in the chain are interconnected.

$$P_j^A = \sum_{i=o}^{M=j} (i+1) C_{j+1}^A \qquad (1)$$

where M is the maximum value of j;

$$C_j^A = P_j^A - 2P_{j+1}^A + P_{j+2}^A \qquad (2)$$

The time dependence of P was found:

$$P_j^A(t) = e^{-jKot} \exp\{2(K_o - K_1)\ [\ t - (1 - e^{-Kot})/K_o]\}_{j>2} \qquad (3)$$

$$P_1^A(t) = e^{-K_2 t}[2(K_2 - K_1)e^{2(K_1 - K_o)/K_o} \int_o^t e^{(K_2 - 2K_1)t} \exp\left\{\frac{2(K_o - K_1)e^{-Kot}}{K_o}\right\} dt + (2K_1 - K_o - K_2)e^{2(K_1 - K_o)/K_o} \int_o^t e^{(K_2 - 2K_1 - K_o)t} \exp\left\{\frac{2(K_o - K_1)e^{-Kot}}{K_o}\right\} dt] \qquad (4)$$

Knowing $P_j^A(t)$ one may easily derive $C_j^A(t)$ from (2) and thus find the length distribution of A unit sequences; (4) is the direct description of the reaction kinetics. Fig. 1. shows the results of such calculations which have already been treated.[3,4]

However, the C_j^A distribution doesnot fully characterize the structure of the products. It is also of interest to elucidate the structure of the products. For example, when partially hydrolyzed polymethylmethacrlate is heated, theneighboring

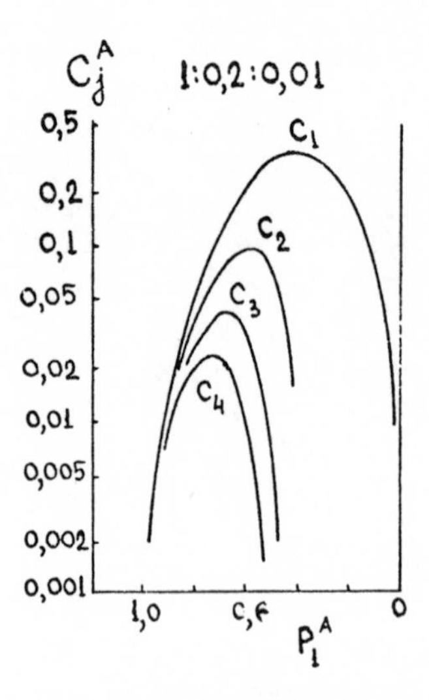

Fig. 1. Kinetcs and length distribution of A-units sequences [3,4] $k:k_1:k_2$are given.

carboxyl groups interact to form a cyclic anhydride[9]

O=C O=C | O | H   O | H $\xrightarrow{-H_2O}$ O=C C=O \ O /

We can extract equations describing the time dependence of the probability of such tuplets for any possible combination of A and B at n sites of the Xn sequence. Using these equations and McQuarrie's solution [7] one may estimate the probability of A and B units. For example, when all X=B one is dealing with the C_j^B distribution. It should be noted that an accurate solution will be rather difficult mathematically. For instance, to find the probability of the simplest tuplet, AXA, one must solve the following equations:

$$\frac{dP\{AXA\}}{d\tau} = -2K'P\{AXA\} + 2(K'-K)P\{AXA_2\} + 2(K'-K)P\{A_3\} - 2(K'-2K+1)P\{A_4\} \quad (5)$$

$$\frac{dP\{AXA_2\}}{d\tau} = -(2K+K')\,P\{AXA_2\} + (K-1)\,P\{AXA_3\} + (K'-K)\,P\{A_2XA_2\} + (K'-1)P\{A_4\} - (K'-2K+1)P\{A_5\} \quad (6)$$

$$\frac{dP\{A_2XA_2\}}{d\tau} = -4KP\{A_2XA_2\} + (K-1)P\{A_2XA_3\} + (K-1)P\{A_3XA_2\} + 2(K-1)P\{A_5\} \quad (7)$$

Here $K=K_1/K_o$, $K'=K_2/K_o$ and $\tau=k_o t$

System 5-7 may be solved as the following relationships have been proven valid:

$$P\{AX_n A_{j+1}\} = p\{AX_n A_j\}e^{-\tau} \quad \text{at } j>2 \quad (8)$$

$$P\{A_{i+1}X_n A_j\} = P\{A_i X_n A_{j+1}\} = P\{A_i X_n A_j\}e^{-\tau} \quad i,j \geq 2 \quad (9)$$

However, when n is large, the system of equations will be too cumbersome to solve. Approximations, particularly Markovian, of different orders, should be applicable. The distribution of units in the products of polymer analogous reactions, is not, strictly speaking, Markovian, since chemical processes are not unidirectional. However, since some identical parameters are present in the structure of the chains of the two kinds of copolymers it is possible to apply Markovian approximations in the case of polymer analogous reactions.

From the ultimate unit model equations one may derive the following equation for the run number R:

$$R = \frac{4P\{A\}(1-P\{A\})}{1+[1-4P\{A\}(1-P\{A\})(1-r_A r_B)]} \quad (10)$$

where $R=2P\{AB\}$, ($P\{AB\}$ is the probability of finding diad AB[11]), r_A r_B are relative reactivities. If the copolymer has a fixed composition, P {A}, one may obtain the dependence of R upon r_A r_B. For chains of the same compositon which are the products of polymer analogous reactions, the value of R at various K ($K=K_1/K_o=K_2/K_o$) may be found using the Monte Carlo simulation method,[12,13] or the McQuarrie equation [7]. In Fig.2 one may see that R values for copolymerization (the curve) and the polymer analogous reaction (the points) calculated for P {A} =0.75 and the same values of $r_A \cdot r_B$ and K. Good agreement between the curve and the points testifies to the fact that in the given polymer analogous reaction ($K_1=K_2$) a considerable element of R is the same as in the first order Markovian process of copolymerization.

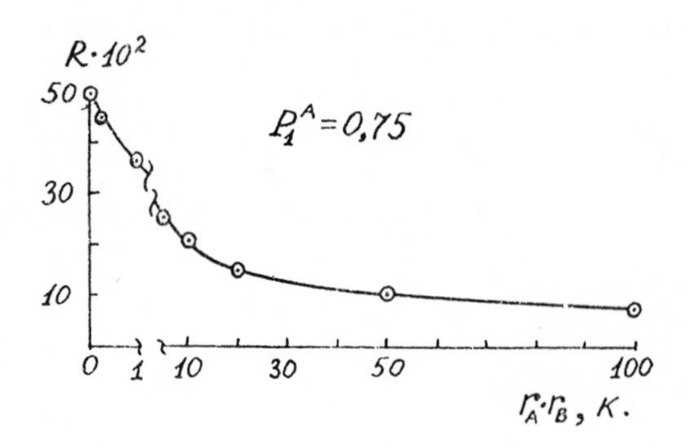

Fig. 2. Run number, R_1 as a function of $r_A \cdot r_B$ (curve) or $k=k_1/k_0=k_2/k_0$ (points) at P {A}=0.75.

Thus, there are grounds for the use of Markovian approximation chains of different order. Equations have been given[10] which describe the chain structure as first, second and third order Markovian approximations. It was thereby assumed that at any given moment of time the polymer chain is a Markovian chain of the n-th order (n=1,2,3) and the transition of the chain from one state to another in time occurs by a mechanism involving the neighboring group effect.

In the second order Markovian approximation the system has four, and in the third order approximation- eight differential equation for the respective probabilities.

Another approximation has been suggested [10], based on the kinetics of conversions of AB_jA unit sequences consisting of reacted units. Let us call this method the"B approximation". Let us designate $N_{1/2} / N_{1/2} + N_2 = \gamma$

$$\frac{dP\{AB_jA\}}{d\tau} = \delta_{j,1}N + 2(1-\delta_{j,1})P\{AB_{j-1}A\}K\gamma - 2P\{AB_jA\}[K\gamma + K'(1-\gamma)] + \frac{K'(1-\gamma)^2}{N_2}\sum_{n=1}^{j-2} P\{AB_nA\}P\{AB_{j-n-\gamma}A\} \tag{1}$$

$$\delta = \begin{cases} 1 & \text{at } j=1 \\ 0 & \text{at } j \neq 1 \end{cases}$$

To compare the above method we have estimated, using computer techniques, some distribution elements as function of time at different $K_0:K_1:K_2$ ratios in accurate equations in B-approximations and in Markovian approximation of the first, second and third order.

Some of the results are presented in Fig. 3-5 as functions of probabilities P {ABA} , P{AB_2A}and P{AB_3A}abbreviated as B_1, B_2,and B_3 respectively, versus the degree of conversion, 1-P{A}.

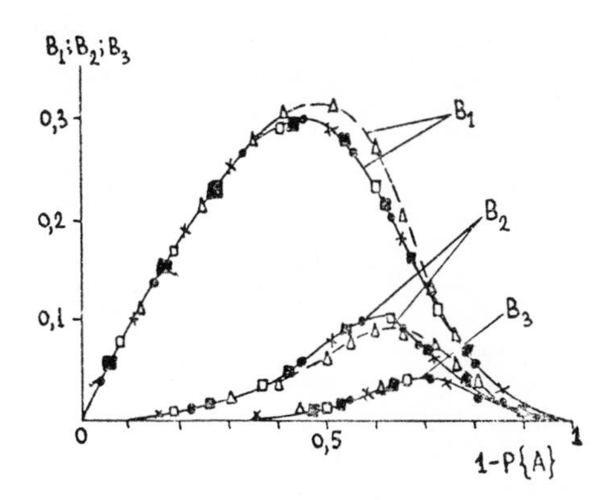

Fig. 3. Probabilities of B_1, B_2, B_3versus the degree of conversion (1-P{A}) at K=0.2, K=0.01.

- ● - accurate solution
- × - B-approximation
- △ - first order Markovian approximation
- □ - second order Markovian approximation
- ■ - third order Markovian approximation

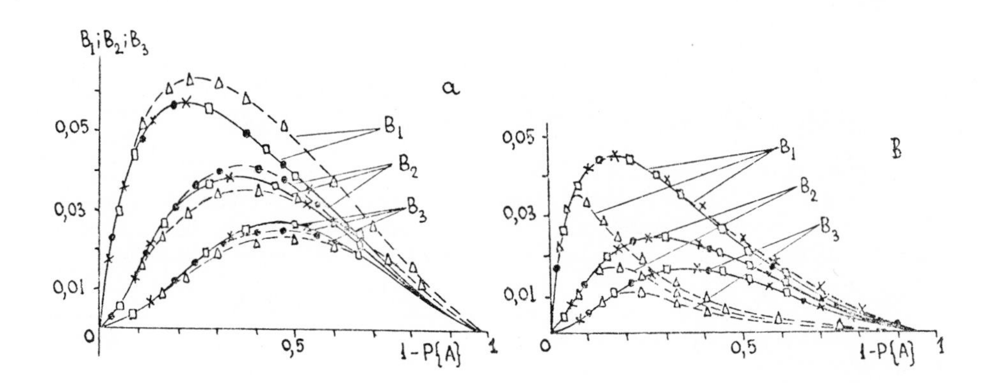

Fig. 4. Probabilities of B_1, B_2, B_3 versus the degree of conversion (1-P{A}) at K=5, K'=5 (a) and K=5, K'=100. Designations as in Fig. 3.

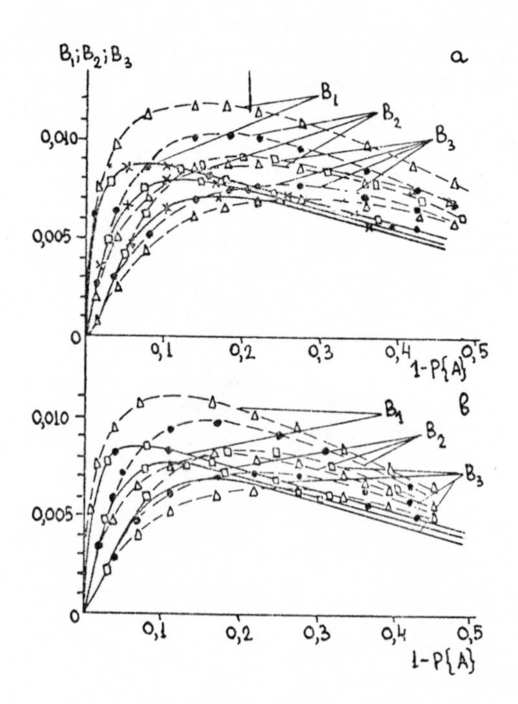

Fig. 5. Probabilities of B_1, B_2, B_3 versus the degree of conversion (1-P{A}) at K=50, K'=50 (a) and K=50, K'=99 (b). Designations as in Fig. 3.

These elements are most demonstrative, as it is in the distribution of B sequences where the maximum discrepancies between accurate calculation and approximations should be expected.

When retardation due to the neighboring group effect is at work (Fig.3) all approximate solutions are close to the results obtained by accurate methods, the only exception being the first Markovian approximation at 1-P{A}>0.4. Fig.4 shows a relatively small accelerating effect of one of the reacted neighboring units ($K_1:K_0=5$). The degree of deviation of the first Markovian approximation strongly depends upon the $K_1: K_0$ ratio (cf. 4a,4b). Other approximations give, in both cases, small deviations of almost equal value.

When acceleration is strong, $K_1: K_0=50$ (Fig.5) only B- approximations yield results close to those abtained by accurate methods for all the elements of distribution studied (in Fig.5b the points-corresponding to the B-approximation are absent because at the condition $K_0:K_1:K_2$ =1:5:99, K'=2K-1 is valid when the B-approximation equation become accurate).

The calculations made showed that B-approximations combined with McQuarrie's equations can be recommended for full description of units in the products of polymer analogous reactions.

The relationship between composition heterogeneity and the ratio of kinetic constants was studied by the Monte Carlo simulation method [12,13]. Fig.6 shows the relationship between composition distribution dispersion, σ^2, and conversion at various ratios of rate constants.

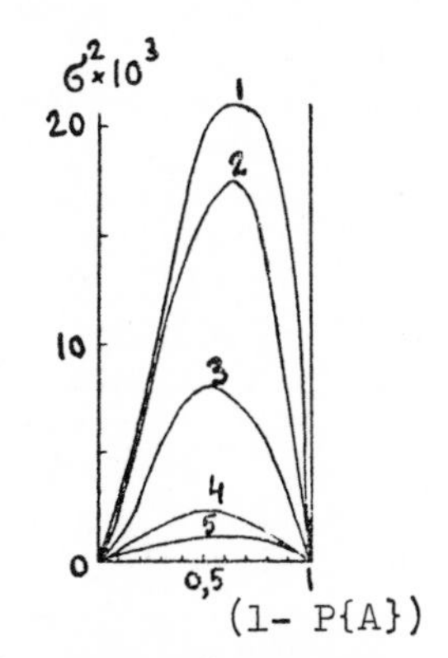

Fig. 6. Dispersion of composition heterogeneity versus degree of conversion (1-P{A}) at K_0:K_1:K_2=1:100:100 (1), 1:50:50 (2), 1:5:100 (3), 1:1:1 (4), 1:0.2:0.2 (5).

The value of σ^2 grows as the accelerating effect of the neighboring group increases. For example, the maximum value of σ^2 increases 5-fold when K_0:K_1:K_2=1:5:5 changes to K_0:K_1:K_2=1:100:100. It is interesting that in the case of strong accelerating effects the σ^2-(1-P{A}) curves are asymmetrical. At K_0:K_1:K_2=1:100:100, σ^2 at 1-P{A}=0.9, is almost five times as large as at 1-P{A}=0.1. There are grounds to believe that the results obtained by the Monte Carlo method may be the equivalent of the accurate solution and, hence, may be used to estimate the adequacy of different approximations for describing composition heterogeneity.

So we can obtain a complete description of chain structure, composition heterogeneity of products and reaction kinetics using the values of individual rate constants. Such calculations are especially valuable when the structure of the chain cannot be established experimentally while individual constants are amenable to determination.

The effect of neighboring groups in polymer analogous reactions has been described for many substances but individual rate constants have seldom been determined. Table I contains the results of investigations in which K_0,K_1 and K_2 constants are available.

Table I
Individual Rate Constants of Polymer Analogous Reactions

Reaction	Rate Constants
Quaternization of 4 vinyl pyridine [17]	$K_0=K_1=6.54x10^{-4}$ l/mole min. $K_2=2.1x10^{-4}$ $K_0:K_1:K_2=1:1:0.32$
Chlorination of polyethylene [18]	$K_0=6.1x10^{-4}$ l/mole sec $K_1=2.1x10^{-4}$; $K_2=0.48x10^{-4}$ $K_0:K_1:K_2=1:0.34:0.079$
Epoxidation of natural rubber[19]	$K_0=9.6$ l/mole min. $K_1=6.4$; $K_2=4$ $K_0:K_1:K_2=1:0.67:0.42$
Hydrolysis of polymethyl methacylate*	$K_0=0.78$ l/mole hr. $K_1=0.14$; $K_2=0.08$ $K_0:K_1K_2=1:0.18$:0.10

* See contribution of Klesper et al., in these proceedings

The meager amount of quantitative data available is a reflection of the difficulties encountered in the experimental determination of the rate constants. Let us consider the methods which may, in principle, be used for this purpose.

For the model discussed (the specificity of the process is determined only by the neighboring group effect) it is natural to use the following method. The rate of reaction may be described by equation 12.

$$-\frac{dP\{A\}}{dt} = k_0N_0+k_1N_1+ k_2N_2 \qquad (12)$$

Having determined the content of units N_0,$N_1$and N_2 in three samples isolated at different degrees of conversion and having determined from kinetic curves the rates corresponding to these conversions one may work out a system of three equations of type 12 and find K_0,K_1 and K_2. However, the values of N_0,N_1 and $N_2$are extremely difficult to determine experimentally. There are very few copolymers for which a quantitative distribution of units in the chain has been successfully determined. [20] The only reliable values of N_0,N_1 and N_2 so far determined for the product of an incomplete polymer analogous reaction are those for partially hydrolyzed polymethyl methacrylate. Klesper [21,22] used NMR at 100 and 200

MHz to determine the distribution of triads in syndio-, iso-and a-tactic copolymers of methyl methacrylate with methacrylic acid obtained both by copolymerization and by hydrolysis of poly-methyl methacrylate to various degrees of conversion. The work of Oshima and Iwai [23] gives one hope that it will be possible to find N_0,N_1 and N_2 for chlorinated polyethylene from its NMR spectrum. Unfortunately, it would be wishful thinking to say that there exists a basic method for the experimental determination of N_0,N_1 and N_2.

Another natural and general approach for the determination of constants would be to use low molecular weight compounds as models of the corresponding fragments of the macromolecular chain. For example, if the curves for a polymer and for corresponding monofunctional, low molecular weight substance have different shapes, one may assume that the neighboring group effect is occurring.

On the other hand, Khodzhaeva[19] while studying the interactions of natural rubber and squalene (the model compound) with perbenzoic acid in benzene at 25°, found the $K_0:K_1:K_2$ ratio to be similar to that in the model; 1:0.67:0.42 and 1:0.6:0.37 respectively. In this instance this approach was successful, but as long as no general principles for the selection of models have been elucidated, no *a priori* statement about the adequacy of a model is valid. This fact coupled with the arduous syntheses of many of the model compounds prevents the determination of individual rate constants by this approach.

Polymeric models, i.e. polymer samples of known N_0,N_1 and N_2 distribution may be of interest in the determination of constants. Such samples can be synthesized, e.g., by copolymerizing monomers A and B. If the relative reactivities ra and rb,are known, the N_0,N_1 and N_2 values for the copolymers obtained when the monomer ratio in the reaction mixture A/B=X can be calculated from the following formulae[11]

$$N_0=(r_A X)^2/(1+r_A X)^2 \qquad (13)$$

$$N_1= 2r_A X/(1+r_A X)^2$$

$$N_2=1/(1+r_A X)^2$$

Having determined the initial rates of the polymer analogous reaction A→B for three samples of the copolymer of different , but known composition, one may find K_0,K_1 and K_2 according to equation 12.

It should be noted that , as a rule, attempts to prepare stereoregular polymers by copolymerization have not been successful. Consequently, copolymers may be used as models (for the problem in question) only in reactions where the neighboring group effect does not depend on the stereochemical configuration of the neighboring units.

Polymer models of known distributions of N_0, N_1 and N_2 may, in some cases, be prepared in the A→B polymer analogous reaction, when the neighboring group effect is not operative, i.e. $k_0=k_1=k_2$. In that case N_0, N_1 and N_2 for a sample of composition P{A} may easily be calculated[31] :

$$N_0=(P\{A\})^2;\ N_1=2P\{A\}(1-P\{A\});\ N_2=(1-P\{A\})^2 \qquad (14)$$

The shape of the kinetic curve (the linear character of its semi-logarithmic anamorphosis) may be used as a criterion for the absence of the neighboring effect. Measuring initial rates of the A→B reaction for such samples under conditions when the effect of the neighboring units is operative, allows one to calculate the constants K_0,K_1 and K_2 from equation 12.

Polymers consisting of triads of unique types may also be used as models. For example, a homopolymer of A contains N_0-type units and in some cases K_0 may be determined from initial rates of conversion. If polymer models for all three types of triads were available, one might determine the whole set of constants. Despite the absence of this desirable situation in general, individual constants may be determined in appropriate cases.

Krentzel et al.[18] studied the photochemical chlorination of polyethylene in chlorobenzene at 50°. Polyethylene itself served as a model of N_0 units (CH_2-CH_2-CH_2) and poly vinyl chloride as a model of N_2 units (CHCl-CH_2-CHCl). Constants K_0 and K_2 were found from the initial rates of chlorination of PE and PVC, respectively. Kinetic equations[34] were used to calculate the kinetic curves for the reaction, using the determined values of K_0 and K_2 and varying K_1. At K_1 :K_0 =0.35:1, good agreement between calculated and measured data was obtained as shown in Fig.7. It should be noted that the obtained ratio of K_1:K_0 is close to the results obtained by chlorimation of low molecular weight paraffins.[35,36] A similar method was employed to determine the constants for interaction of natural rubber and perbenzoic acid.[19]

Conclusion

Thus, each of the general methods for determining individual constants discussed above may be applied to a rather limited number of objects. Further, the accuracy and reliability of these methods are questionable and must be thoroughly studied.

Thus it may be stated that the formal kinetic theory of the

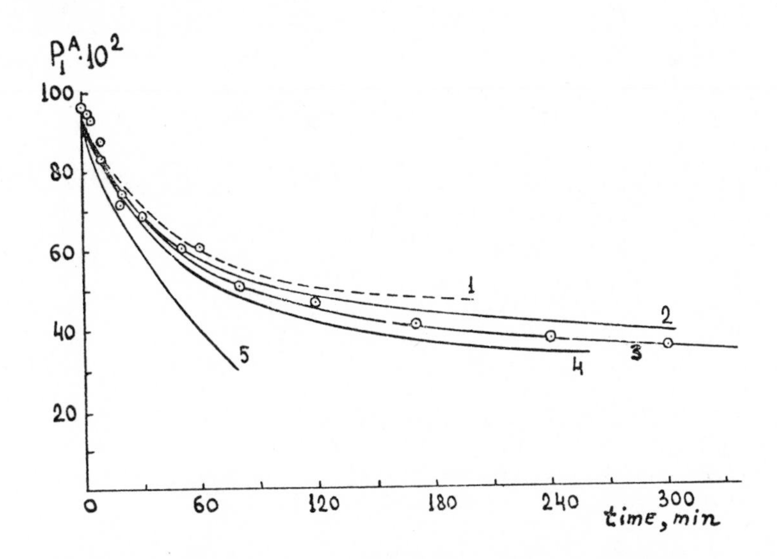

Fig.7. Calculated kinetic curves of chlorination of polyethylene at $k_0:k_1:k_2$=1:0.16: 0.08(1), 1:0.24:0.08, 1:0.35:0.08(3); 1:0.50:0.08(4); 1:1:1 (5), points-experimental data.

neighboring group effect has been considerably advanced in recent years and the first determinations of individual rate constants have been made. The main outcome of the new approach is that there is now available a tool with which to study the mechanisms of polymer analous reactions.

References

1. N.A. Plate',in the monograph, "Kinetics and Mechanism of Macromolecular Reactions", Nauka, 1968, Moscow

N.A. Plate', IUPAC International Symposium on Macromolecular Chemistry, Budapest 1969, p651, Akademiai Kiado,Budapest.

2. J.B. Keller,J. Chem. Phys. 37 2584 (1962).

3. T.Alfrey, W. C. Lloyd, ibid. 38, 318 (1963).

4. C.B. Arends, ibid. 38, 322 (1963).

5. J.B. Keller, ibid. 38, 325 (1963).

6. L. Lazare, ibid. 59, 727 (1963)

7. D.A. McQuarrie, J.P. McTague, H. Reiss, Biopolymers 3,657 (1963)

8. E.R. Cohen, H. Reiss, J. Chem. Phys. 38, 680 (1963)

9. J. Semon, J. B. Lando, Amer. Chem. Soc. Polymer Preprints 10, 1281 (1969).

10a A. D. Litmanovich, O. V. Noah, N.A. Plate', N.B. Vasijev,

10b A. L. Toom, Preprints of the International Conference on Chemical Transformations of Polymers, Bratislava, Soc. B.P 56 (1971).

11. K. Ito, Y. Yamashita, J. Polymer, Sci. A3, 2165 (1965).

12. A. D. Litmanovich, N. A. Plate', O. V. Noal, V. I. Golyakov Europ. Polymer J. Suppl. 517 (1969).

13. N. A. Plate', A. D. Litmanovich, O. V. Noah, V. I. Golyakov Visokomolekularnye Soedinania IIA, 2204 (1969).

14. O. V. Noah, A. Ya. Tjemkin, ibid. IIA, 1689 (1969).

15. A. Sippel, E. Heim, Kolloid- Z. und Z. fuer Polymere 185,97 (1962)

16. H. K. Frensdorf, O. Ekiner, J. Polymer Sci. A-2, 1157 (1967).

17. C. B. Arends, J. Chem. Phys. 39, (1963).

18. L. B. Krenzel, A. D. Litmanovich, V. A. Pastuchova (1969).

19. V. A. Agasandyan Visokomolekularnye Soedinania 1970.
I. D. Khodjaeya, Theses , Moscow

20. H. J. Harwood, Angew. Chemie 77, 405, 1124 (1965).

21. E. Klesper, J. Polymer Sci. B6, 313 663 (1968).

22. E. Klesper, W. Gronski, ibid. B7, 727 (1969).

23. T. Ohshima, T. Iwai, 18th Ann. meeting Polymer Soc. of Japan, 11-13 Nov. 1969, Part 2, p 425

24. M. L. Bender, M. C. Neveu, J. Amer. Chem. Soc. 80, 5388 (1958).

25. E. Gaetjens, H. Moravetz, ibid. 82, 5328 (1960).

26. T. C. Bruice , U. K. Pandit, ibid. 82, 5858 (1960).

27. K. Shibatani, K. Fujii, J. Polymer Sci. 8A-1, 1647 (1970).

28. G. Smets, W. DeLoecker, J. Polymer Sci. 41, 375 (1959).

29. N. W. Johnston, H. J. Harwood, ibid 22C591 (1969).

30. E. Klesper, W. Gronski, V. Barth, Makromol. Chem. 139,1(1970).

31. A. D. Litmanovich, Doklady Acad. Nauk USSR.

32. S. Mukhopadhyay, B. C. Mitra, S. R. Palit, Indian J.Chem. 7,903 (1969).

33. R. M. Fuoss, M. Watanabe, B. D. Coleman, J. Polymer Sci. 48,5 (1960).

34. L. B. Krenzel, A. D. Litmanovich, Visokomolekularnye Soedinania 9B 175 (1967)

35. D. V. Tishenko, Zhournal Obshchay Chimii 1843 (1948).

36. N. Colebourn, E. S. Stern, J. Chem. Soc. 1599 (1965).

37. A. D. Litmanovich, A. L. Isyumnikov, In "New Methods" of Investigation of Polymers"(In Russian) Mir; Moscow, 1968 page 200.

38. M. N. Gusev, Theses, Moscow (1970).

39. L. Horner, L. Schafer, Liebigs Ann Chem. 635, 31 (1960).

Current references not cited herein:

40. N. A. Plate', A. D. Litmanovich, J. Pure and Applied Chem.8 123 (1971).

41. N. A. Plate', A. D. Litmanovich, Vysokomol.Soedin. A14, 2503 (1972)

42. V. I. Kryshtol, A. D. Litmanovich and N. A. Plate', ibid. B14 326 (1972)

43. O. V. Noah, A. L. Toom, N. B. Vasilev, A.D. Litmanovich and N. A. Plate', ibid. A15,877 (1973)

DISCUSSION SESSION

Discussion Leader - R. B. Fox

Challa: Did you observe a neighboring group effect for the basic hydrolysis of co-isotactic copolymers of methacrylic acid and its diphenymethyl ester when the ester content was less than 50%?

Plate': If you have <u>syn</u>-poly (diphenyl methacrylate), there is no neighboring group effect in a weakly basic medium like pyridine-water. Under the same conditions, the isotactic form exhibits an acceleration.

Challa: At the same time, you reported an increase in viscosity for lower ester contents and considered this as supporting your hydrolysis studies?

Plate': If you start with a copolymer of relativety high ester content (~70%) and attempt to describe the kinetics with the same rate constants for different parts of the experiment curves, you will not succeed. This means that there is something which complicates the relatively simple model of hydrolysis based only upon the neighboring group effect. For lower ester contents, the description is very good.

Challa: Therefore in that region there is a constant neighboring group effect?

Plate': Yes. You are in a region where the neighboring group effect is insensitive to composition.

Challa: At high acid content (low ester content), the increased viscosity corresponds to an extension of the coil. One would expect that the neighboring group effect would then increase. This seems contradictory.

Plate': No. Viscosity is too rough a measure to detect the change. It is the micro-environment around the active site which is important. The neighboring group is on the microscale, while viscosity is on the macroscale.

Challa: Then the viscosity measurments are not strong support for your hydrolysis studies?

Plate': The changes in viscosity are an additional argument that something is going on. But they are not direct confirmation of the idea. You should consider the situation around two coils of the copolymers with different acid group contents to see what is going on at the micro level. Fortunately, the micro phenomena in this case, are probably reflected in the macroviscosity phenomena.

St. Pierre: Are there effects resulting from the juxtaposition of the various groups even when far removed from each other, as contrasted with the nearest neighbor effect?

Plate': Yes, but they are smaller than for neighboring groups. These effects are due to anchimeric assistance, as when a six-membered ring intermediate can form. We have not seen good examples of remote group effects in the polymer series. If you calculate the probability of such contacts occurring, it can be seen that less than 1/15 of the groups can take part, and therefore their influence in the normal neighboring group effect is practically negligible. (Ed. Note: Such effects seem to predominate in biological systems, such as enzymes, and may be lacking in synthetic systems because they lack the precisely controlled structure of biological macromolecules).

Rempp: Looking at the composition heterogeneity curves you have shown, one might expect that, when you have a random distribution of reacted and unreacted units ($K_0:K_1:K_2=1:1:1$) or even when you have a retardation effect ($K_1 < K_0$), you should have very little fluctuation in composition in a sample for a given average composition. This did not appear clearly in the dispersion curves you showed. How does the chain length influence the results of the calculations?

Plate': Your questions deal with modulation of the reactions in the Monte Carlo simulation when we look for the composition heterogeneity. First, we normally deal with relatively high molecular weight samples. If you put a polymer of DP=1000 into the computer as a model, the calculations require dozens of hours for even the fastest machines. Thus it is convenient to use a minimum length which has the same distributional parameters as for the infinite chain. In the case of $K_0:K_1:K_2=1:1:1$ or 1:0.1:.001, a fifty unit chain is sufficient. Only minutes of computer time are required. Further only about one hundred chains are needed to supply statistical reliability. If $K_0:K_1:K_2=1:5:100$ (acceleration) at least two hundred units are required because long blocks must be grown. Even in this case when you have a pseudo monomolecular reaction with an excess of a low molecular weight reagent, in the absence of any conformational, electrostatic or supermolecular complications, you will still have (for purely kinetic reasons) a certain degree of compositional heterogeneity, as I have shown. If you get into the acceleration effect, the compositional heterogeneity becomes essentially higher and has a maximum level at about 60% of conversion.

Rempp: But if you have a retardation effect, it seems that all sites in the molecules are equally accessible and you should not have heterogeneity in chemical composition.

Plate': Strictly speaking it is not the case. The results of computer calculations show that certain inhomogeneities in chemical composition exist. At conversions close to 0% or 100% inhomogeneity is not far from zero.

Rempp: In some cases you showed identical K_0, K_1, K_2 values and some had great differences in K_0, K_1, K_2. Isn't it speculative to deal with three adjustable parameters when the differences are not very large (i.e. slight retardation or acceleration) which is often the case? In some of your curves you compared the experimental points with calculated curves for various sets of K values. Is the experimental accuracy sufficient to ascertain the Ks in this way?

Plate': Our accuracy allows us to say whether we have the normal distribution or a deviation. Small differences among the sets of K's allow us to determine real effects such as retardation.

Salamone: You said that the sulfuric acid-catalyzed hydrolysis of both it -and syn- diphenyl methyl methacrylate proceeded at the same rate. This is surprising when you consider the case of PMMA.

Plate': I am also surprised.

Salamone: Is the hydrolysis complete?

Plate': It can easily be brought to completion in 10M H_2SO_4 .

St. Pierre: Is it possible that for the diphenyl methyl ester solvolysis is by alkyl-oxygen cleavage notinvolving the carbonyl group?

Plate': That is a possibility which we are still investigating.

Challa: Doesn't the substitution occur at those carbon atoms already bearing chlorine and wouldn't this tendency confuse your interpretation?

Plate': It has been shown that in the chlorination of polyethylene (to less than 45-50% chlorine content) no geminal dichloride groups are formed.

Harwood: Were olefin groups formed?

Plate': Under our conditions no double bonds were detected.

Harwood: In the pyridine- water solvolysis of it- and syn-poly methyl methacrylate you should be cautious in conducting strictly kinetic studies, because changes in polymer-solvent interactions can occur, obscuring the true effects.

Plate': An independent tool is needed as a control on the polymer-solvent interactions-- perhaps the end-to-end distance and the unperturbed dimensions may be useful in this regard.

CHEMICAL REACTIVITY OF COPOLYMERS

H. James Harwood

Department of Polymer Science
The University of Akron, Akron, Ohio 44325

INTRODUCTION

Chemical reactions of polymers, whether or not they yield copolymers as final products involve copolymers as intermediates. An understanding of the factors that influence copolymer reactivity is therefore basic to a general understanding of polymer reactivity. Since it is relatively easy to calculate various aspects of copolymer structure from information about variables involved in its preparation, it is of interest to evaluate the influence that neighboring substituents can have on the chemical reactivity of a monomer unit in a copolymer.

The reactivity of a functional group in a copolymer is influenced by neighboring units in many ways, including the following:

a) Steric effects. These inhibit the approach of reagents and tend to diminish reactivity. They are particularly important when the neighboring units contain quarternary carbon atoms, as in the case of methacrylates, methacrylonitrile, α-methylstyrene, etc.

b) Polar and electrostatic effects. These may hinder or enhance the approach of a charged reagent in the case of electrostatic interactions or make it difficult to introduce an electronegative substituent near another electronegative substituent, as in the case of chlorination reactions.

c) Specific interactions between neighboring groups. These may lead to cyclization reactions involving adjacent monomer units or may lead to anchimeric assisted reactions.
d) Solvation effects. These influence reactivity by influencing the concentration of reagents and solvent in the vicinity of a reactive group and can be very significant if electrostatic interactions are involved.

In some cases, the functional groups on a copolymer have vastly different reactivities, depending on the nature of their neighboring substituents, and it is then possible to interpret the reactivity of the copolymer quantitatively in terms of its structure. This paper will discuss several examples.

SPECIFICATION OF COPOLYMER STRUCTURE

If the reactivity of A-type units in a copolymer of A and B units is being considered and if only the nature of nearest neighbors determines their reactivity, then three arrangements must be considered in the initial stages of the reactions. These are shown below.

Environments of A-units in an A-B Copolymer

$$A\text{-}\overset{*}{A}\text{-}A \quad (B\text{-}\overset{*}{A}\text{-}A + A\text{-}\overset{*}{A}\text{-}B) \quad B\text{-}\overset{*}{A}\text{-}B$$

After some A units have been converted to C-units, however, other arrangements arise (e.g., $C\text{-}\overset{*}{A}\text{-}B$, $C\text{-}\overset{*}{A}\text{-}C$, etc.) and the interpretation of reactivity in terms of structure can become very complicated, especially if the A and C units differ appreciably in their influence on the reactivity of neighboring A units. This presentation will not deal with this complicating feature, but will describe methods used to calculate the proportions of A units in $A\text{-}\overset{*}{A}\text{-}A$, $(B\text{-}\overset{*}{A}\text{-}A + A\text{-}\overset{*}{A}\text{-}B)$ and $B\text{-}\overset{*}{A}\text{-}B$ triads in copolymers, as well as other structural aspects. These proportions will then be used, in conjunction with experimental results, to establish structure-reactivity correlations.

Almost all free radical initiated copolymerizations follow the terminal copolymerization model. This means that the copolymer is formed by the following propagation steps, where ∿∿A· and ∿∿B· refer to growing polymer radicals, and where A and B represent monomers.

$$\sim\!\sim\!\sim A\cdot + A \xrightarrow{k_{AA}} \sim\!\sim\!\sim A\cdot \qquad (1)$$

$$\sim\!\sim\!\sim A\cdot + B \xrightarrow{k_{AB}} \sim\!\sim\!\sim B\cdot \qquad (2)$$

$$\sim\!\sim\!\sim B\cdot + A \xrightarrow{k_{BA}} \sim\!\sim\!\sim A\cdot \qquad (3)$$

$$\sim\!\sim\!\sim B\cdot + B \xrightarrow{k_{BB}} \sim\!\sim\!\sim B. \qquad (4)$$

Methods are available [1] for evaluating ratios of the specific rate constants for Equations 1-4, i.e., $r_A = k_{BB}/k_{AA}$. These ratios, termed monomer reactivity ratios, can be used in conjunction with the ratio of monomers (A_f/B_f) present at the time a copolymer was formed to calculate various aspects of copolymer structure. Thus P(A/B), the conditional probability that an A unit follows a B unit in the copolymer chain may be calculated as shown in Equation 5

$$P(A/B) = \frac{k_{BA}(\sim\!\sim\!\sim B.)(A_f)}{k_{BA}(\sim\!\sim\!\sim B.)(A_f) + k_{BB}(\sim\!\sim\!\sim B\cdot)(B_f)} \qquad (5)$$

$$= 1/(1 + r_B\ B_f/A_f)$$

Similarly,

$$P(B/B) = 1 - P(A/B) \qquad (6)$$

$$P(B/A) = 1/(1 + r_A\ A_f/B_f) \qquad (7)$$

$$P(A/A) = 1 - P(B/A) \qquad (8)$$

It is possible to apply these conditional probabilities reversibly in the case of a copolymer formed by a terminal model copolymerization process; P(A/B) can thus also represent the probability that an A unit precedes a B unit in the chain. Thus, P(A/B) and P(B/A) can be used to calculate the probabilities that A units are in specified environments in copolymer chains. The probability that an A unit has B units on both sides, i.e., the probability, $f_{B\overset{*}{A}B}$ that it is centered in a $B\overset{*}{A}B$ triad, is the product of the separate probabilities that the A unit is preceded and followed by B units. Thus,

$$f_{B\overset{*}{A}B} = P(B/A)P(B/A) \qquad (9)$$

Similarly,

$$f_{B\overset{*}{A}A} = f_{A\overset{*}{A}B} = P(B/A)P(A/A) = P(B/A)[1-P(B/A] \qquad (10)$$

$$f_{AAA} = P(A/A)P(A/A) = [1-P(B/A)]^2 \qquad (11)$$

The quantities f^*_{BAB}, f^*_{BAA}, f^*_{AAB}, and f^*_{AAA} are termed A-centered triad fractions because they represent the fractions of A units centered in the various types of triads.

In the case that the A and B units are asymmetric, then the reactivity of A units in nine triads must be considered. The various possible configurations of A-centered triads are written below, equivalent structures being collected together in boxes. For simplicity only one-half of the possible configurations are shown.

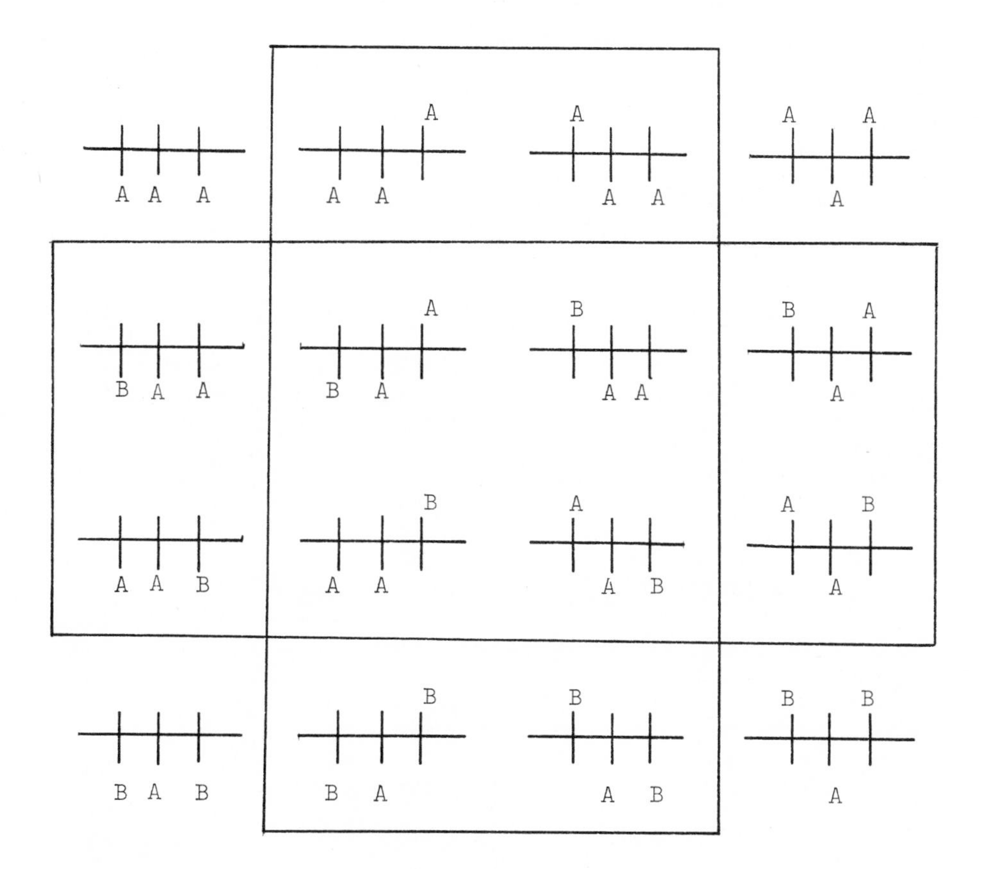

In accounting for stereochemical aspects of copolymers, parameters usually designated by σ are used to represent the probability that a given linkage has a meso (isotactic or coisotactic) configuration. These are often evaluated by nmr spectroscopy. The reader is referred to the paper by Dr. Klesper in this volume for more information on this topic. Given values of σ_{AA}, σ_{BB} and $\sigma_{AB} = \sigma_{BA}$, it is possible to calculate the fraction of A units in any of the configurations presented above. Thus,

$$f\left(\underset{A\ A\ A}{\text{+++}}\right) = (\sigma_{AA})^2 \ f_{AAA} \quad (12)$$

$$f\left(\underset{B\ A}{\overset{B}{\text{+++}}}\right) = \sigma_{AB}(1-\sigma_{AB})\ f_{BAB} \quad (13)$$

In considering the extent of cyclization obtainable in a copolymer cyclization, and for other purposes also, it is necessary to know the probability of finding a sequence of A units that contains exactly n A units. This is easily calculated by finding the probability that exactly n-1 A units follow the first A unit. Thus the probability that a sequence of A units contains exactly 4 A units is,

$$P_{(B)-A-A-A-A-B} = [P(A/A)]^3\ P(B/A) \quad (14)$$

The general expression for the probability of a sequence of n A units (Pn) is then

$$Pn = [P(A/A)]^{n-1}\ P(B/A) \quad (15)$$

It is thus possible to calculate aspects of copolymer structures from information about the reactivities and concentrations of monomers used to prepare the copolymer. In the remainder of this paper, efforts to interpret the chemical reactivity in terms of calculated aspects of their structures will be reviewed.

COPOLYMER CYCLIZATION REACTIONS

Functional groups pendant from vinyl polymers and copolymers are often 5 or 6 atoms away from other reactive groups. As a result of this, intramolecular reactions forming cyclic units are commonly encountered in the reactions of polymers. Several excellent reviews [2-5] cover the scope of such reactions.

Cyclization reactions occurring in copolymers may be classified in two ways: inter- and intra-sequence cyclization reactions. Intrasequence cyclization reactions occur within sequences of like units whereas intersequence cyclization reactions occur at sequence junctions and they involve dissimilar monomer units.

Intrasequence cyclication

∿∿BABAABAAABBBAAAABBAAAAB∿∿

Intersequence cyclization

~~BABAABABABBABABBBABABBA~~

Provided that cyclization occurs at random, the extent of cyclization obtained is not the maximum amount possible, due to limitations imposed by stoichiometric and statistical effects. Thus, in the case of intrasequence cyclization, at least one A unit in each sequence of odd numbered A units does not have a neighbor with which to react (stoichiometric limitation). In sequences of four A units, all units can cyclize only if the first cyclization reaction involves units at the end of such sequences. If the two central units cyclize first, then two A units are left without reactive neighbors. Since the two central units can be expected to cyclize first in one-third of such sequences (statistical limitation), one can expect, on an average, that 1/3 x 2 = 0.67 A units per sequence of 4 A units will not cyclize during a random cyclization process. Flory [5] has shown that the average number of unreacted A units (Sn) expected from the random cyclization of sequences of n A units is given by Equation 16, where $S_1 = 1$, $S_2 = 0$, $S_3 = 1$ and other Sn values are calculated by the equation. Table I provides a list of Sn values calculated by this equation.

$$Sn = 2(S_1 + S_2 + S_3 + \cdots + S_{n-2})/(n-1) \qquad (16)$$

The fraction of A units expected not to cyclize, fA(u), during the random cyclization of such units in a copolymer is calculated by Equation 17. The summation in the numerator of this equation is proportional to the number of uncyclizable units whereas the summation in the denominator is proportional to the total number of A units present. The ratio is thus the fraction of uncyclizable A units.

$$fA(u) = \frac{\sum_{n=1}^{n=\infty} SnPn}{\sum_{n=1}^{n=\infty} n\,Pn} = \frac{\sum_{n=1}^{n=\infty} Sn[P(A/A)]^{n-1}\,P(B/A)}{\sum_{n=1}^{n=\infty} n\,[P(A/A)]^{n-1}P(B/A)} \qquad (17)$$

Equation 17 can be used, in conjunction with Sn values provided in Table I and with conditional probabilities derived as described in the previous section, to predict fA(u) values for copolymers. However, work by Flory, Wall, and Alfrey, et al. [5-10] has led to generally applicable solutions of Equation 17. Thus, Equations 18 and 19 are applicable for intrasequence

TABLE I

Expected Number of Uncyclizable Units (Sn) in Sequences of n Units

n	Sn	n	Sn	n	Sn
1	1.000	16	2.436	31	4.466
2	0.000	17	2.571	32	4.601
3	1.000	18	2.707	33	4.737
4	0.667	19	2.842	34	4.872
5	1.000	20	2.977	35	5.007
6	1.067	21	3.113	36	5.143
7	1.222	22	3.248	37	5.278
8	1.352	23	3.383	38	5.413
9	1.489	24	3.519	39	5.549
10	1.624	25	3.654	40	5.684
11	1.759	26	3.789	41	5.819
12	1.895	27	3.925	42	5.955
13	2.030	28	4.060	43	6.090
14	2.165	29	4.195	44	6.225
15	2.301	30	4.331	45	6.361

and intersequence cyclization reactions, respectively.

$$fA(u) = e^{-2P(A/A)} \tag{18}$$

$$fA(u) = [\cosh[P(A/B)P(B/A)]^{1/2} - [P(B/A)/P(A/B)] \sinh[P(A/B)P(B/A)]^{1/2}]^2 \tag{19}$$

In addition, computer programs have been written for calculating the extent of intersequence cyclization to be expected for high conversion copolymers [11] and terpolymers [12].

Thus the extents of cyclization obtainable when copolymers are cyclized randomly are predictable from conditional probabilities that characterize the structures of the copolymers. In the following sections the extent to which such predictions agree with experimental results will be discussed.

Intrasequence Cyclization Reactions Forming Single Rings

Two interesting special cases of intrasequence cyclization are encountered when the copolymer has a random structure ($P(A/A) = P_A$, the mole fraction of A in the copolymer) and when cyclization of a homopolymer

(P(A/A = 1) is considered. In the latter case, the fraction of uncyclized units expected is e^{-2} or 0.1354. Thus, random cyclization of a homopolymer should leave 13.5 percent of the units uncyclized. Most studies of intrasequence cyclization have, in fact, involved homopolymers rather than copolymers.

Marvel, Sample and Roy (13) found that dehalogenation of poly(vinyl chloride) and poly(vinyl bromide) with zinc removes only 84-87 percent of the chlorine atoms, presumably due to the formation of cyclopropane rings. Although this result is in good accord with theoretical expectation, Russell [14] has found that unsaturated structures and crosslinks are also formed in this reaction.

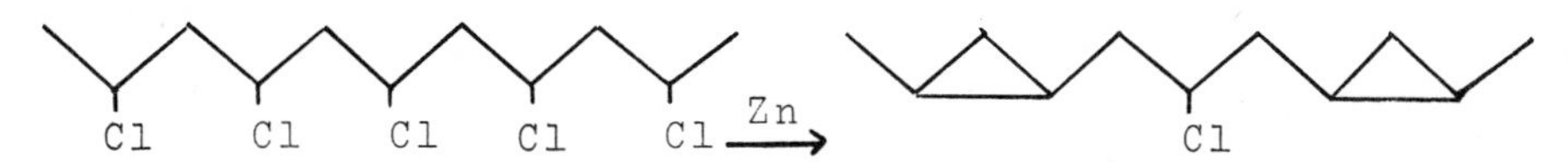

Marvel and coworkers [15] also investigated the extent of dechlorination obtained when vinyl chloride-vinyl acetate copolymers were treated with zinc. The extents of cyclization obtained seemed to agree with statistical expectation, assuming the copolymers had random structures, but it was later shown by Alfrey, _et al_.[16] that deacetylation occurs simultaneously with dehalogenation and that about 87 percent of the chlorine atoms in such copolymers are removed by zinc, regardless of copolymer composition.

Schulz and Harwood [17] reported that pyrolysis of poly(t-butyl N-vinyl carbamate) yields isobutylene, carbon dioxide, t-butanol and a polymer containing vinyl amine and cyclic urea units. The reaction was believed to involve pyrolysis of the carbamate units to yield vinyl amine units which then cyclized, where possible, with neighboring carbamate units.

NH NH → NH_2 NH → NH NH
CO CO CO CO
O-tBu O-tBu O-tBu

$+CO_2$ + C_4H_8 + t-BuOH

Since pyrolysis occurred much more slowly than cyclization, the pyrolysis-cyclization sequence could be considered as an intrasequence cyclization process.

The completely pyrolyzed polymers contained 13 percent vinyl amine units and the relative yields of CO_2, isobutene and t-butanol were consistent with those expected for a random cyclization process. The product was crosslinked due to the formation of urea crosslinks, but the amount formed was too small to interfere with statistical considerations. These workers also found that the amine contents of partially pyrolyzed polymers were in good agreement with those expected for a random cyclization process. Dever [18] investigated the pyrolysis of copolymers of styrene, N-vinyl carbazole and N-vinyl pyrrolidone with t-butyl N-vinyl carbamate and obtained fair to good agreement between calculated and observed extents of cyclization.

Other examples of intrasequence cyclization reactions, some of which result in nearly the statistically expected extents of cyclization, are shown below.

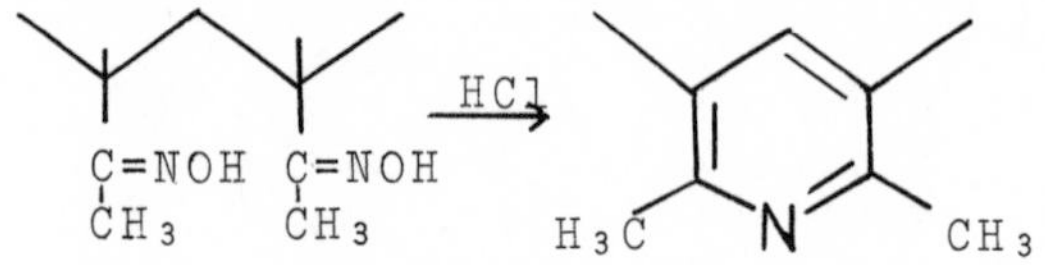

$\xrightarrow{HCl}$ Reference [19]

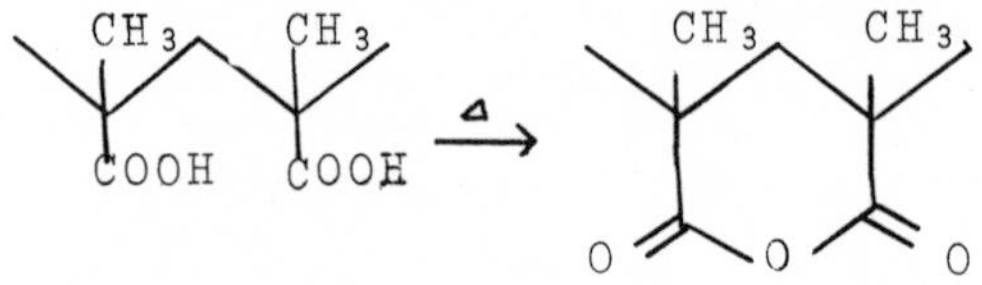

$\xrightarrow{\Delta}$ References [20-21]

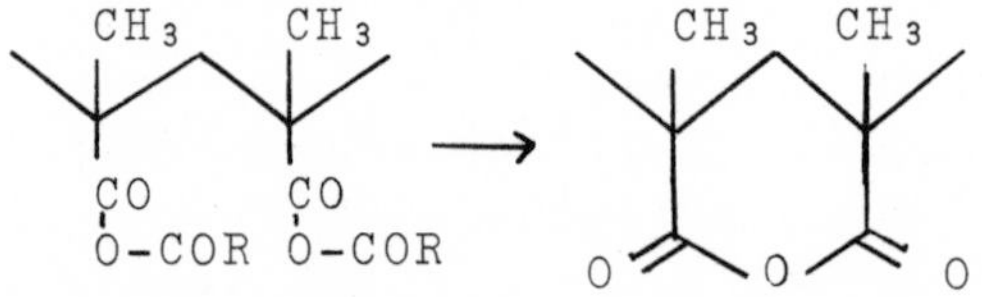

+ RCO-O-COR
Reference [22]

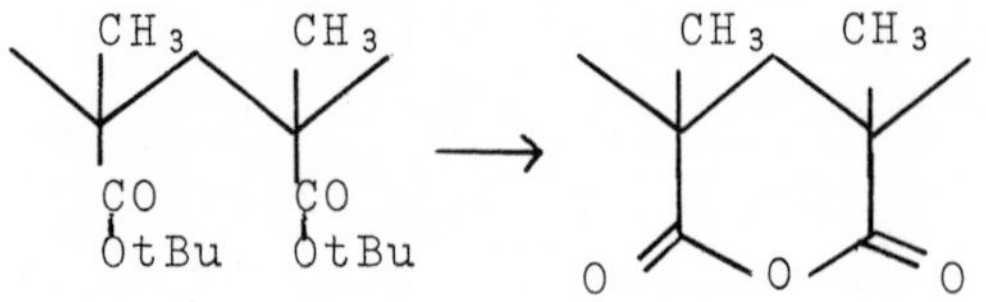

+ $CH_2=C(CH_3)_2$
Reference [23]

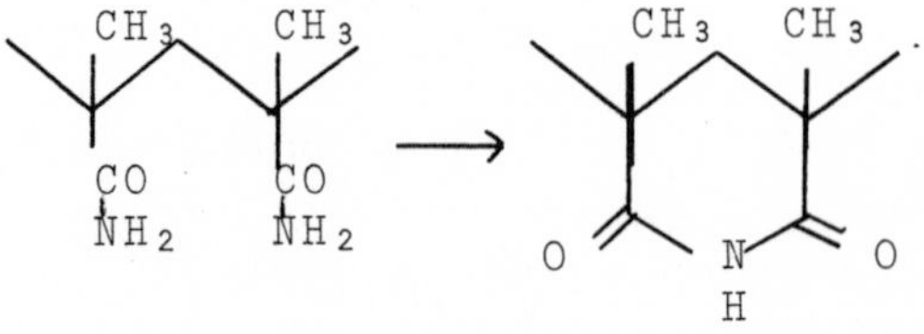

+ NH_3
Reference [24]

CO NH$_2$ CO NH$_2$ $\xrightarrow{\text{NaOCl}}$ O NH Reference [25]

Cl Cl $\xrightarrow[\text{ØCH}_3]{\text{AlCl}_3}$ CH$_3$ Reference [26]

OH OH $\xrightarrow{\text{RCHO}}$ O O CHR Reference [27]

CO Cl CO Cl $\xrightarrow{\text{NaN}_3}$ O NH Reference [28]

CHO CHO $\xrightarrow{\text{NaOH}}$ CH$_2$ OH COONa $\xrightarrow{\text{H}^+}$ CH$_2$ O O

References [29-30]

Intrasequence Cyclization Reactions Yielding Fused Rings

Some intrasequence cyclization reactions yield fused rings, bicyclic rings or spiro ring systems. Few statistical analyses of the extents of cyclization expected for such reactions have been made, although Flory [31] has shown that cyclization of homopolymers will be only 81.6 percent complete when the yield of fused rings is limited by directional effects. Thus, poly(methyl vinyl ketone) and poly(methyl isopropenyl ketone) undergo an intrasequence aldol condensation on heating and the extent of cyclization is limited to 82-86 percent [32-33].

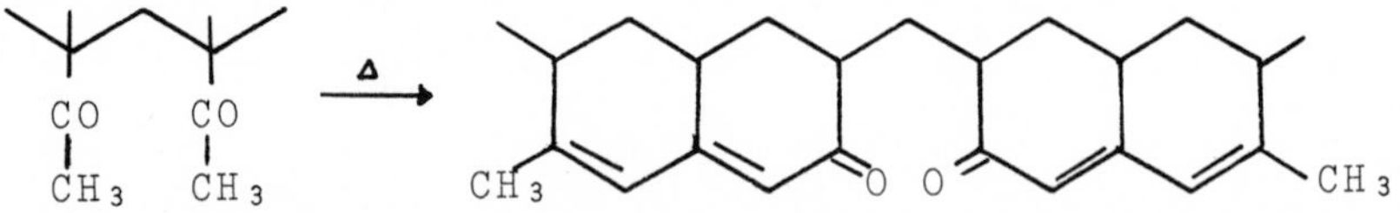

CO CO
CH_3 CH_3 $\xrightarrow{\Delta}$ CH_3 O O CH_3

In addition, Hay has measured, by ultraviolet spectroscopy, the relative amounts of single, double, triple, etc. fused rings present in partially cyclized structures and has found that they agree well with values expected for a random cyclization process.

No theory is presently available for predicting the extent of cyclization to occur in methyl vinyl ketone copolymers. Grassie and Hay [35] found that pyrolysis of methyl vinyl ketone-acrylonitrile copolymers yielded polymers with fused structures containing both acrylonitrile and methyl vinyl ketone units. Armstrong and Harwood [36] found that conventional and alternating equimolar styrene-methyl vinyl ketone copolymers differed appreciably when heated. The conventional copolymer became deep red, due to the formation of fused rings containing conjugated double bonds, and remained soluble. The alternating copolymer, in contrast, became only light yellow and it crosslinked. Apparently only intermolecular reactions, leading to crosslinking, were possible with the alternating copolymer.

Other examples of intrasequence cyclization reactions yielding fused or spiro rings are provided below.

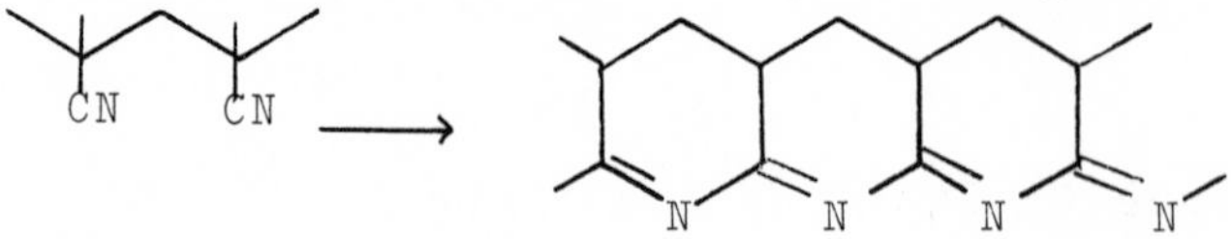

CN CN ⟶ N N N N Reference [37]

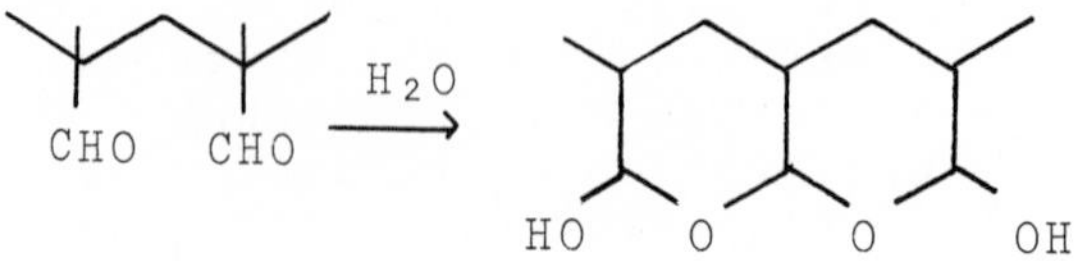

CHO CHO $\xrightarrow{H_2O}$ HO O O OH References [38-39]

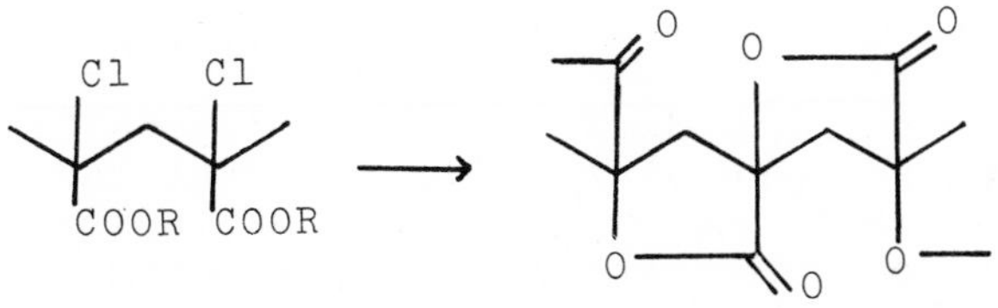

References [40-42]

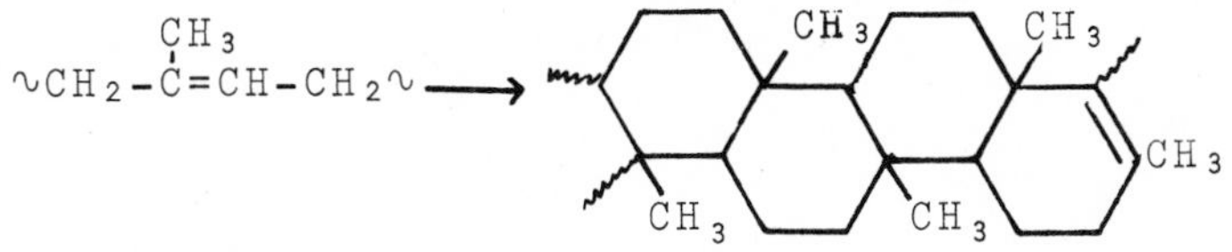

Reference [43]

Intersequence Cyclization Reactions

A large number of intersequence cyclization reactions are possible with copolymers and many of these have been studied quantitatively. Sakurada and Kawishima [44] used an ingenious conductimetric titration method to measure the extent of lactone formation obtained when hydrolyzed vinyl acetate-methyl acrylate copolymers were acidified. Copolymers prepared in various conversions from several monomer mixtures were studied and a graphical method was used to calculate the average compositions and lactone contents of the hydrolyzed copolymers. As can be seen in Table II, good agreement was obtained between theoretical and experimental results.

TABLE II

Lactone Contents of Hydrolyzed Vinyl Acetate-Methyl Acrylate Copolymers After Acidification

Mole % Methyl Acrylate			Conversion in Copolymer Preparation	Mole % Lactone Units	
Monomer Mixture	Copolymer				
	Calc'd	Found		Calc'd	Found
20	44.4	47.5	32.8	29.9	29.1
20	42.4	46.4	37.2	29.3	22.2
20	23.0	22.3	87.0	17.1	13.4
50	57.0	53.9	87.0	21.7	22.9

Similarly, Minsk, Waugh and Kenyon [45] hydrolyzed a 1:1 vinyl acetate-maleic anhydride copolymer under acidic conditions and found that 47 percent of the acid groups in the resulting polymer formed lactone units with adjacent hydroxyl groups. This result indicated that the copolymer had a highly alternating structure. In a related study, Kern [46] found that

Ac → CO OH O O O O O O

alternating copolymers of t-butyl vinyl ether with maleic anhydride or citraconic anhydride yield copolymers containing acid and lactone units when heated at 120-220°. The weight losses obtained in these studies were claimed to equal those expected. In addition, Powell and coworkers [47] found that α-hydroxymethylstyrene-MMA copolymers lactonized during extrusion, although quantitative results were not reported.

Zutty and Welch [42] reported that vinyl halide-methyl methacrylate copolymers spontaneously lactonize on heating, forming methyl halide. The reaction also occurs in low molecular weight γ-haloesters [48-49]. Quantitative studies [11, 12, 50-56] on this reaction in vinyl halide-MMA copolymers and in vinyl halide-MMA-styrene terpolymers have yielded results in excellent accord with theoretical expectation, as is shown in Table III. Johnston [57-58] reports that similar reactions occur slowly in acrylate ester-vinyl halide copolymers. Cyclization can be catalyzed by materials such as $ZnCl_2$ and dibutyltin dilaureate.

Butadiene-MMA and isoprene-MMA copolymers [59-61] undergo intersequence cyclization when treated with bromine. The reaction presumably involves attack by MMA units on neighboring bromonium ions. A similar cyclization reaction occurs during the bromination of dimethylmaleic acid [62].

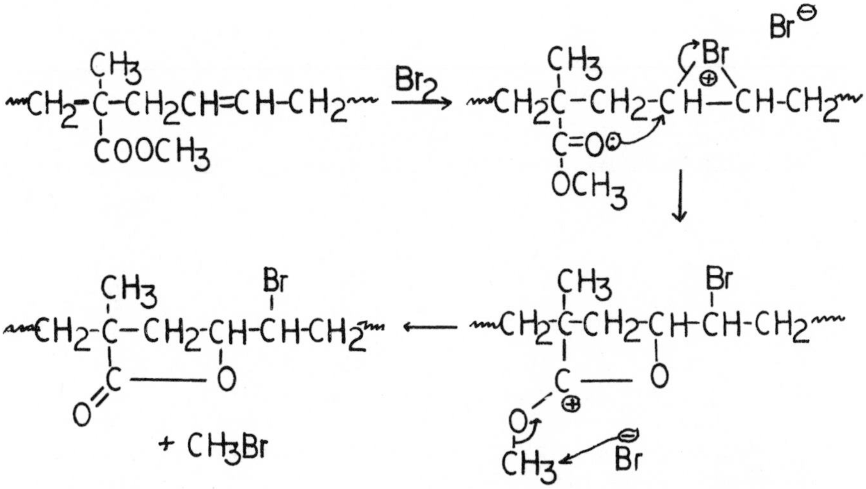

$$\sim CH_2-\underset{COOCH_3}{\overset{CH_3}{C}}-CH_2CH{=}CH-CH_2\sim \xrightarrow{Br_2} \sim CH_2\underset{\underset{OCH_3}{C=O}}{\overset{CH_3}{C}}-CH_2-\overset{\overset{Br^{\oplus}}{\frown}}{CH-CH}-CH_2\sim \quad Br^{\ominus}$$

$$\downarrow$$

$$\sim CH_2-\overset{CH_3}{C}-CH_2-CH-\overset{Br}{CH}-CH_2\sim \;(\text{lactone ring: } C(=O)-O) + CH_3Br \longleftarrow \sim CH_2-\overset{CH_3}{C}-CH_2-CH-\overset{Br}{CH}-CH_2\sim \;(\text{ring: } C^{\oplus}-O,\; O-CH_3 \leftarrow Br^{\ominus})$$

The extents of cyclization obtained when butadiene-MMA copolymers are brominated are in good agreement with theoretical expectation as can be seen in Table IV.

TABLE III

Extents of Cyclization Obtained in Copolymers and Terpolymers Containing Vinyl Chloride and Methyl Methacrylate

Polymer Composition (Mole %)			Fraction of MMA Units Cyclized ($f_{MMA(C)}$)	
VCl	MMA	Styrene	Expt'l	Calc'd
24	76	-	0.31	0.34
31	69	-	0.40	0.43
42	58	-	0.58	0.57
50*	50*	-	0.86	0.865
52	48	-	0.66	0.72
64	36	-	0.82	0.80
75	25	-	0.95	0.93
4	40	56	0.07	0.07
17	40	43	0.269	0.286
8	63	29	0.120	0.127
6	51	43	0.092	0.093
5	72	23	0.066	0.066
28	38	34	0.448	0.443

* alternating copolymer

TABLE IV

Intersequence Cyclization in Butadiene-Methyl Methacrylate Copolymers

Mole % MMA in Copolymer	Fraction of MMA Units Cyclized	
	Expt'l	Calc'd
72.5	0.33	0.28
61.0	0.50	0.50
55.0	0.66	0.70
50.0*	0.88	0.865
41.5	0.86	0.88
30.0	0.91	0.95

* alternating copolymer

Guilbault and Harwood [63] investigated the extents of cyclization obtained when phenyl N-vinyl carbamate-t-butyl N-vinyl carbamate copolymers were subjected to the following sequence of reactions.

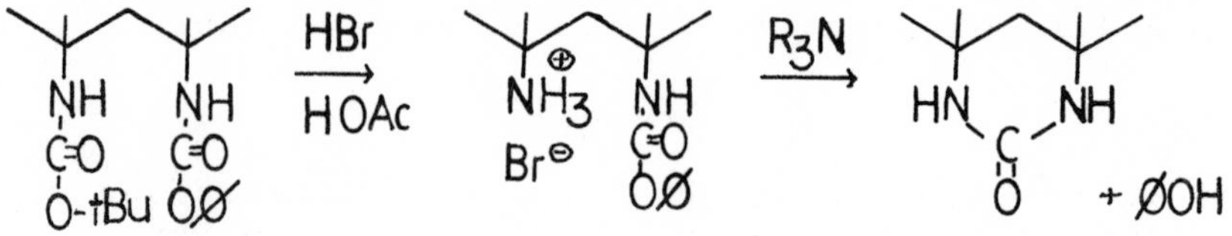

$$\text{NH-C=O-O-tBu},\ \text{NH-C=O-O}\phi \xrightarrow[\text{HOAc}]{\text{HBr}} \text{NH}_3^{\oplus}\,\text{Br}^{\ominus},\ \text{NH-C=O-O}\phi \xrightarrow{\text{R}_3\text{N}} \text{HN-C(=O)-NH} + \phi\text{OH}$$

As can be seen in Table V, the extents of cyclization obtained were in good agreement with statistical expectation, assuming the copolymers had random structures.

TABLE V

Intersequence Cyclization in Copolymers Derived from Phenyl N-Vinyl Carbamate (A)-t-Butyl N-Vinyl Carbamate (B) Copolymers

Mole % A in Copolymer	fA(u)	
	Expt'l	Calc'd
94	0.93	0.93
81	0.76	0.78
65	0.58	0.56
39	0.27	0.23
18	0.14	0.05

Other interesting examples of intersequence cyclization reactions are the following.

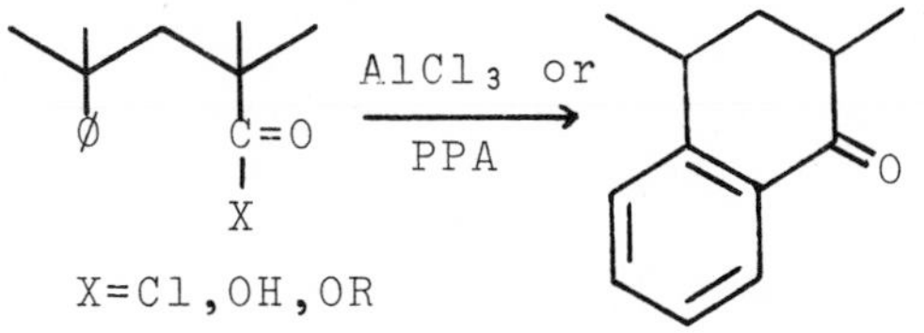

$AlCl_3$ or PPA

X=Cl,OH,OR

References [64-65]

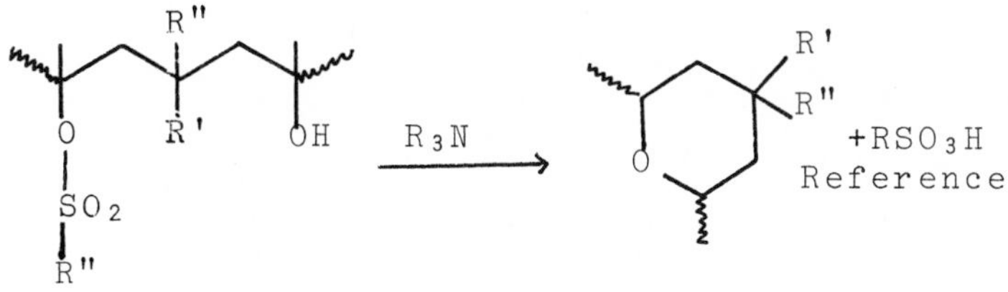

R_3N

$+RSO_3H$

Reference [66]

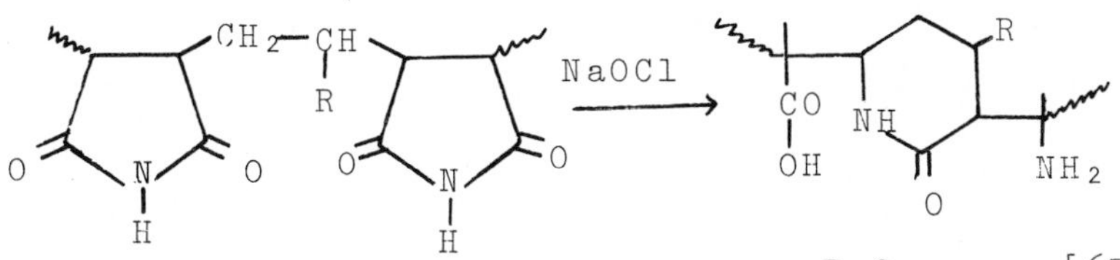

NaOCl

Reference [67]

Studies on intersequence and intrasequence cyclization reactions in copolymers have thus shown that neighboring group interactions play very significant roles in polymer reactions. In the next section the influence of neighboring monomer units on polymer reactivity will be considered.

REACTIVITY OF ESTER GROUPS IN POLYMERS

A large number of factors can influence the reactivity of functional groups in polymers and copolymers. These range from nonspecific effects, that depend to a large degree on overall copolymer composition. to very specific effects that involve interactions of functional groups with neighboring units. It is the latter that are of interest in this paper, but it is difficult to study such effects without interference from less specific ones. The various techniques that have been used to study the influence of copolymer structure or chemical reactivity include: correlation of limiting yields with structural aspects; kinetic studies; double and multiple reaction studies; and spectroscopic analysis of reactive sites. Each of these will be discussed in some detail.

Limiting Conversions

If some of the reactive groups in a copolymer are considerably less reactive than others, it is possible that they may be essentially unreacted under conditions that are adequate for the more reactive groups to react completely. The conversion obtained in such a situation provides a measure of the amount of more reactive units, provided that conversion is not limited by less specific effects, such as polyelectrolyte effects, changes in polymer-solvent interaction during reaction, or to precipitation of partially reacted copolymer.

Bevington and coworkers [61,68-69] have used this approach to study the reactivity of methyl methacrylate units in copolymers with styrene, methyl acrylate and isoprene. The limiting conversions obtained when the copolymers were hydrolyzed in a NaOH-methanol-benzene mixture increased as their methyl methacrylate contents decreased. The limiting conversions obtained agreed reasonably well with the fractions of MMA units in XMX, XMM and MMX triads and in isotactic MMM triads (M=MMA, X=comonomer). They concluded that MMA units in heterotactic and syndiotactic MMM triads are more sterically hindered than those in other triads. Although this is a very reasonable interpretation, it should be noted that styrene/MMA hydrolysis reactions become inhomogeneous when conducted under Bevington's conditions [70] and that the limiting conversions obtained may be influenced by this complication. Von Schreyer and Volker [71], in studies on the alkaline hydrolysis of MMA oligomers, obtained limiting conversion data that indicated the MMA units in the center of such oligomers to be less reactive than those at chain ends. Their results also indicate that MMA units with MMA neighbors are highly hindered.

Glavis [72] found that alkaline hydrolyses of isotactic, syndiotactic and conventional poly MMA proceed at different rates and to different limiting extents, the fastest rates and highest rates being obtained with the isotactic polymer. In contrast, Selegny and Sagain [72] obtained limiting conversions of 85 percent for all forms of poly MMA.

Kinetic Methods

Unless there is a very large difference between the reactivities of functional groups in a copolymer, limiting conversion data can provide only an approximate indication of the proportion of slow reacting groups. Provided that the functional groups react by first order or pseudo first order kinetics, it is possible to obtain the relative amounts and the relative reactivities of the various types of groups present. Thus, if a polymer contains two types of reactive units, which we will designate "fast" and "slow", the relative proportions of which are α and $(1-\alpha)$, respectively, Equation 20 can be written to describe the change in the concentration of reactive units, E, with time, t. Other variables in this equation are k_f and k_s, the specific first order rate constants for the fast and slow reactions, and Eo, the original concentration of reactive units

$$E = Eo\ [\alpha e^{-k_f t} + (1-\alpha)e^{-k_s t}] \qquad (20)$$

After sufficiently long reaction times, the first term in the bracketed expression in Equation 20 approaches zero and Equation 20 can then be rewritten as follows:

$$\ln(E/Eo) = \ln(1-\alpha) - k_s t \qquad (21)$$

Thus, if ln(E/Eo) is plotted versus time, the data at long reaction times define a straight line from which k_s and α can be evaluated. These values can then be used in conjunction with Equation 20 and data obtained at short reaction times to evaluate k_f. Figure 1 shows how this method was used to analyze data obtained in the hydrolysis of vinylpyridine-MMA copolymers

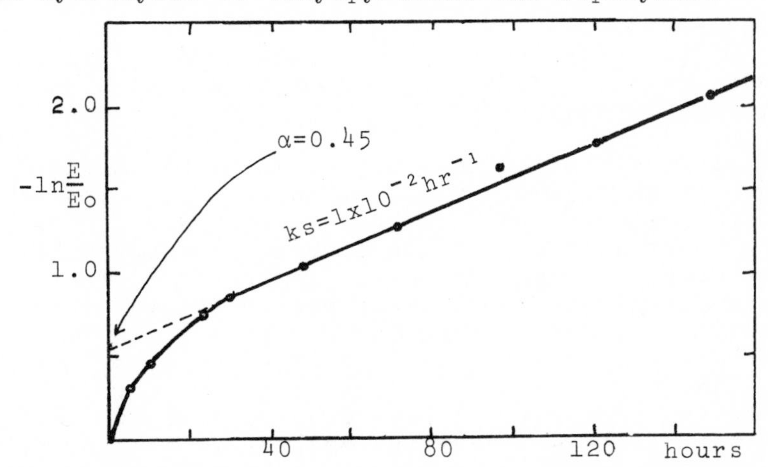

Figure 1. Kinetic Analysis of Data Obtained in the Acid Catalyzed Hydrolysis of a Vinylpyridine-MMA Copolymer

Morawetz [74-75] used the kinetic method to analyze results obtained in the hydrolysis of ester units in methacrylic acid copolymers that contained up to 3 mole percent p-nitrophenyl methacrylate or p-methoxyphenyl methacrylate units. The results indicated the presence of two types of ester groups, differing in reactivity by a factor of 10-15. Approximately 20 percent of the ester groups in each polymer were of the "fast" reacting type. These were considered to be present in favorable configurational environments, since all ester units were present in acid-ester-acid triads in the copolymers. Gaetjens and Morawetz [76] also observed that "fast" and "slow" reacting ester groups are present in p-nitrophenyl acrylate-methacrylic acid copolymers, but that only one type of reactive group is present in copolymers of acrylic acid with p-nitrophenyl acrylate or p-nitrophenyl methacrylate.

Using a similar kinetic approach, Smets and Hesbain [77] found that about 40 percent of the acrylamide units next to acrylic acid units in copolymers hydrolyzed considerably faster than the other acrylamide units. Configurational effects were believed responsible for the differences noted. Such effects are also implicated by results obtained in studes [78-81] on the hydrolysis of methacrylic acid - MMA copolymers and of acrylic acid - ethyl ethacrylate copolymers. Studies on the latter copolymers revealed the existence of a steric effect. Only ester units in acid-ester-acid triads were in the "fast" reacting category. Due apparently to configurational effects, less than 60 percent of these were hydrolyzable, based on limiting conversion data.

Thall investigated the solvolysis of MMA units in copolymers with 2- and 4-vinyl pyridine (VP) using an acetic acid-water-sulfuric acid solvent system. "Fast" and slow reacting MMA units, differing in reactivity by a factor of approximately 20 were indicated by the results and excellent correlations were obtained between the proportions of fast reacting groups and the fractions of MMA units centered in VP-MMA-VP triads. The results, shown in Table VI, are surprising in that the fast reacting groups might have been expected to include the MMA units in VP-MMA-MMA and MMA-MMA-VP triads.

Although kinetic studies can provide valuable information about structure-reactivity relationships in copolymers, they should be used with considerable

TABLE VI

Correlation of the Amounts of Fast Reacting Methyl Methacrylate Units in (Vinylpyridine)-(Methyl Methacrylate) Copolymers with Calculated Methyl Methacrylate Centered Triad Fractions

Sample (Mole % MMA)	Fraction of Fast Reacting MMA Units	Triad Fraction		
		VMV	VMM	MMM
(2-Vinylpyridine)-(Methyl Methacrylate) Copolymers				
83	.07	.04	.30	.66
71	.16	.13	.46	.41
61	.22	.24	.50	.26
54	.30	.36	.48	.16
46	.39	.45	.44	.11
36	.53	.58	.36	.06
(4-Vinylpyridine)-(Methyl Methacrylate) Copolymers				
82	.09	.05	.34	.61
60	.20	.19	.49	.32

caution. Changes in polymer composition during the course of a reaction can lead to polyelectrolyte effects, or to polymer-solvent interaction changes that can affect the reactivity of the remaining groups. Such changes might mislead one to conclude that a polymer contained several types of reactive sites, when in fact, all were the same.

Double and Multiple Reaction Studies

The sensitivity of the limiting conversion and kinetic methods to polyelectrolyte effects, autocatalysis, changes in polymer-solvent interaction during reaction, etc., prompted us to develop a method for studying polymer reactivity that is less sensitive to such effects. The method is based on the fact that acid groups on polymers and copolymers can be cleanly and quantitatively esterified by reaction with diazomethane [86]. Essentially the method involves partially hydrolyzing a polymer containing methyl-C^{14} methacrylate units, remethylation of the product with diazomethane, and then study of the hydrolysis of the resulting polymer.

$$RCOOCH_3^* \longrightarrow RCOOH \xrightarrow{CH_2N_2} RCOOCH_3 \longrightarrow$$

The hydrolyzed-remethylated copolymer contains ester units of differing activity, depending on their reactivity toward hydrolysis and the degree of conversion obtained. The remethylated polymer is then rehydrolyzed and loss of radioactivity is determined as a function of degree of hydrolysis. Analysis of the results enables one to determine the relative amounts and relative reactivities of ester groups in different environments in the polymer.

If the "fast" and "slow" reacting groups and the reacted groups in a polymer are designated by F, S, and X, respectively, and if an asterisk (*) is used to represent the presence of label, then the use of the double reaction technique can be illustrated by the following equations.

```
   ***************************
~~~FFFSSFSFSSSFFSFSSSSFSFFFFSS~~~        I
                ↓
   *  * ** **** * * * **  ****
~~~FXXSXFSXSSSFXSXSXSXFSXXFFSS~~~        II
                ↓
   *  * ** **** * * * **  ****
~~~FFFSSFSFSSSFFSFSSSSFSFFFFSS~~~        III
                ↓
```

Thus, the original polymer (I), which contains all reactive sites uniformly labeled, is partially hydrolyzed to obtain polymer II. Methylation of polymer II with diazomethane yields polymer III in which the "fast" and "slow" reacting groups have 6/13 and 11/14 of the radioactivity present in the groups in polymer I. Polymer III can then be rehydrolyzed and the loss of labeled groups can be measured as a function of conversion. The results obtained can then be analyzed to obtain information about the relative amounts and relative reactivities of fast and slow reacting groups present. Figure 2 shows fractional label loss vs. conversion (p) curves for several limiting cases where the interpretation of the results should be rather straight forward.

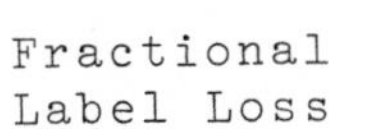

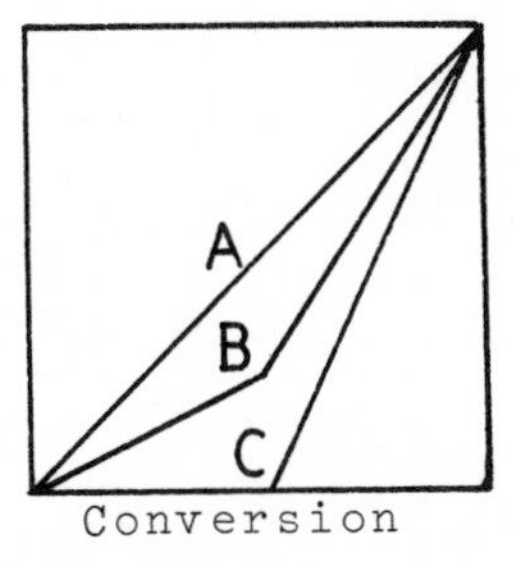

Figure 2. Loss of Labeled Units in Regenerated Polymers as a Function of Conversion for Polymers Having Groups with Different Relative Reactivities

Curve A shows the behavior expected when all groups present in the polymer have the same reactivity. Curves B and C show the behavior expected when F and S differ in reactivity by several orders of magnitude. Whether curve B or C is followed depends on the extent of reaction obtained in the conversion of I to II. Curve B would be obtained when the conversion of I to II was low so that both F and S units were labeled in II. Curve C would be obtained when the conversion of I to II was high so that only a portion of the S units in III were labeled.

Depending on the relative amounts and reactivities of F and S in a polymer, a particular set of label loss vs. conversion curves should be obtained for a given polymer, each curve in the set corresponding to a particular degree of reaction obtained in the conversion of stage I to stage II. In the following section, equations are derived to represent the experimental results. These equations have as parameters, the relative amounts and reactivities of the F and S groups. Those values of these parameters that fit the equations to the experimental results should then represent the relative amounts and the relative reactivities of the fast and slow reacting groups in the polymers.

Theoretical Considerations

If (F) and (S) represent the concentrations of "fast" and "slow" reacting groups present, the simplest expressions that can be written for the rates of disappearance of these species are given by equations 21 and 22, where k_S and k_F are appropriate rate constants and where the factor A includes concentrations of other species involved in the rate determining step. This treatment assumes that both fast and slow reactions are first order with respect to F and S groups and that the order of the reactions with respect to other reactants is the same for both fast and slow reactions.

$$\frac{-d(F)}{dt} = Ak_F(F) \qquad (21)$$

$$\frac{-d(S)}{dt} = Ak_S(S) \qquad (22)$$

Equations 21 and 22 can be combined and integrated to obtain equation 23, where (S_o) and (F_o) are the initial concentrations of slow and fast reacting groups and where $R = k_F/k_S$.

$$(F) = [(S)/(S_o)]^R(F_o) \qquad (23)$$

Now, expressing all concentrations relative to the total initial concentration of reactive groups (both fast and slow) and defining X_o and X as the fractions of slow type groups present initially and when the degree of reaction is p, respectively, the fraction, f, of fast groups unreacted at this degree of reaction is given by equation (24).

$$f = \left(\frac{X}{X_o}\right)^R (1-X_o) \qquad (24)$$

From this, the conversion and the proportions of fast and slow groups (p_F and p_S) reacted can be related through the parameter X by equations (5-7).

$$p = 1 - X - [(X/X_o)^R(1-X_o)] \qquad (25)$$

$$p_F = 1 - (X/X_o)^R \tag{26}$$

$$p_S = 1 - X/X_o \tag{27}$$

Now if a polymer containing labeled groups is reacted to conversion p, and if the reacted groups are replaced by unlabeled groups, the composition of the resulting polymer will be as follows:

Mole fraction unlabeled slow groups = $p_S X_o$
Mole fraction labeled " " = $(1-p_S)X_o$
Mole fraction unlabeled fast groups = $p_F(1-X_o)$
Mole fraction labeled " " = $(1-p_F)(1-X_o)$

If the resulting polymer is then reacted a second time to conversion p', where the proportions of fast and slow groups reacted are p_F' and p_S', the following relationships can be derived for the final polymer:

Mole fraction unreacted labeled slow groups = $(1-p_S)X_o(1-p_S')$

Mole fraction unreacted labeled fast groups = $(1-p_F)(1-X_o)(1-p_F')$

The residual activity of the final polymer will then be given by equation 28, since it should be equal to the mole fraction of unreacted labeled groups present (activity/mole original units).

$$\text{Residual Activity} = \frac{\text{Activity of Final Polymer}}{\text{Activity of Original Polymer}} =$$

$$(1-p_F)(1-X_o)(1-p_F') + (1-p_S)X_o(1-p_S') \tag{28}$$

The fractional label loss, discussed in connection with Figure 2, can be calculated by dividing the residual activity of the final polymer (Eqn. 28) by the residual activity of the intermediate polymer (II), this latter quantity being equal to 1-p.

Since p_F, p_S, p_F' and p_S' are functions of X_o, R, p, and p' and since p and p' can be measured independently, Equation 28 relates the loss in activity of a polymer to conversion through parameters X_o and R This equation can therefore be used to estimate X_o and R by curve fitting techniques, provided that the assumptions on which this derivation is based are appropriate to the situation studied.

Robertson and Harwood [85] applied the double reaction technique in studies on the acidic hydrolysis of PMMA. The technique shows the presence of two types of reactive groups in free radical initiated polymer, even though kinetic results do not. In addition, all the groups in isotactic PMMA were shown by double reaction studies to have the same reactivity, whereas kinetic studies were complicated by autoacceleration.

Multiple Reaction Studies

A useful variation of the double reaction technique is to repeat the hydrolysis-remethylation sequence a large number of times, keeping the conversions constant, and to then monitor the loss in reactivity as a function of the number of cycles. Let x and y be the fractions of fast and slow reacting groups that are hydrolyzed after a given conversion. When the starting, fully labeled polymer is subjected to a hydrolysis-remethylation sequence, the activities of the fast and slow reacting groups will fall, respectively, to (1-x) and (1-y) of that originally present. After each subsequent hydrolysis-remethylation sequence, the activities of the fast and slow reacting groups will also fall by factors of (1-x) and (1-y), respectively. The residual activity (R.A.) of a polymer after n hydrolysis-remethylation sequences will thus be given by Equation 29, where α is the fraction of ester groups that are fast reacting.

$$R.A. = \alpha(1-x)^n + (1-\alpha)(1-y)^n \qquad (29)$$

In evaluating α, x and y, one takes advantage of the fact that as n increases, $(1-x)^n$ approaches zero faster than $(1-y)^n$. After a sufficient number of cycles, the term $\alpha(1-x)^n$ in Equation 29 becomes negligible; α and y can then be determined from several sets of R.A. values. Results obtained in early hydrolysis-remethylation cycles can then be used, in conjunction with α and y values to determine x. Table VII lists R.A. values obtained in a multiple hydrolysis study involving a 28:72 styrene-MMA copolymer, along with R.A. values calculated from α, x and y values that are evaluated from the data. When low conversions are employed in multiple reaction studies, x/y ratios reflect the relative reactivities of the fast and slow reacting groups.

TABLE VII

Comparison of Calculated and Experimental Residual Activities for Multiple Hydrolysis of the 28-72 Styrene-(Methyl Methacrylate) Copolymer Under Acid Conditions

Sample	Residual Activities	
	Calculated*	Observed
After 1st Cycle	0.640	0.640
" 2nd "	0.490	0.492
" 3rd "	0.405	0.408
" 4th "	0.347	0.344
" 5th "	0.299	0.292
" 6th "	0.260	0.264

*Calculated from Equation (29) using values of $\alpha = 0.40$, $x = 0.70$, and $y = 0.13$.

Spectroscopic Identification of Reactive Sites

In those cases where IR or nmr spectroscopy can be used to monitor the appearance or disappearance of functional groups on polymers during reaction, very detailed information about polymer reactivity can be obtained. The environments around groups with different reactivity can be recognized and direct measurements of relative reactivity can be made. Much information about this approach is provided in this volume by Dr. Klesper's paper.

We have tried to use this approach to identify the fast-reacting species in styrene-MMA and related copolymers. The methoxyl proton resonance of such copolymers is observed in three general areas [87-88]. The highest field resonance is due to MMA units in SM*S triads having two <u>meso</u> M-S placements. The central resonance area is due to MMA units centered in MM*S, SM*M and SM*S triads having only one <u>meso</u> M-S placement and the remaining MMA units are responsible for resonance observed in the lowest field area. Our approach was to submit the polymers to the following series of reactions and to monitor the appearance and disappearance of methoxyl proton resonances.

$$\sim\text{COOCH}_3 \xleftarrow{\text{NaOCH}_3} \sim\text{COOCD}_3 \xrightarrow{\text{Hyd.}} \sim\text{COOH} \xrightarrow{\text{CH}_2\text{N}_2} \sim\text{COOCH}_3$$

$$\sim\text{COOCD}_3 \xleftarrow{\text{NaOCD}_3} \sim\text{COOCH}_3$$

ACID CATALYZED SOLVOLYSIS STUDIES

The methods described earlier in this paper were applied to studies on the solvolysis of styrene-MMA and methyl acrylate-MMA copolymers in a mixture of acetic acid, sulfuric acid and water [82,83,89]. The results obtained are shown in Table VIII. For copolymers with high MMA contents, the fraction of slow reacting MMA units, f_{slow}, as determined by double and multiple hydrolysis methods correlates reasonably well with f^*_{MMM}. However, in copolymers with less than 75 percent MMA units, f_{slow} is greater than f_{MMM}. Such copolymers have appreciable XMX triad contents and it seems that their "fast" reacting groups include MMA units in XMX triads plus approximately one-half of those in XMM + MMX triads (X = styrene or methyl acrylate). Examination of molecular models indicates that MMA units in MMX and XMM triads with <u>racemic</u> M-X placements are less hindered than those with <u>meso</u> M-X placements. It is possible therefore that configurational effects are responsible for the results obtained. Slightly higher f_{slow} values were obtained from kinetic studies than from double and multiple reaction studies and the latter are considered the more reliable. It is interesting that the specific rates of the fast and slow reactions occurring during solvolysis of styrene-MMA copolymers are the same as those observed for methyl acrylate-MMA copolymers. This indicates that steric effects are largely responsible for the different reactivities observed.

Our studies included styrene-MMA copolymers and polyMMA derived from methacrylic anhydride. These materials were of interest because they contained large amounts of <u>meso</u> M-M placements as a result of the cyclopropagation step that occurs when methacrylic anhydride units are incorporated into polymers. PolyMMA derived from methacrylic anhydride thus had a high proportion (17%) of <u>isotactic</u> triads. As a result of this, the polymer contained a high proportion of "fast" reacting groups-17 percent as estimated by the kinetic method and 34 percent as estimated by double hydrolysis. Styrene-MMA copolymers derived from methacrylic anhydride seemed to contain three types of reactive groups. Some of these were very reactive--about 20 times as reactive as the fast reacting groups in the conventional copolymers. These are presumed to be present in (MMS + SMM) triads having <u>meso</u> M-M placements and <u>racemic</u> M-S placements.

TABLE VIII

Fractions of Slow Reacting MMA Units Determined for Several Polymers by Kinetic, Double Hydrolysis and Multiple Hydrolysis Methods

Mole % MMA	Fraction of Slow Reacting Units			Triad Distributions		
in Polymer	Kinetic	Double Reaction	Multiple Reaction	S$\overset{*}{M}$S	S$\overset{*}{M}$M	M$\overset{*}{M}$M
Conventional Styrene - Methyl Methacrylate Copolymers						
89	-	0.80	-	0.01	0.21	0.78
83	0.79	0.70	-	0.03	0.31	0.66
72	0.69	0.51	0.60	0.13	0.45	0.42
64	0.69	0.47	0.49	0.26	0.50	0.24
PolyMMA and Styrene - MMA Copolymers Derived from Methacrylic Anhydride						
100	0.83	0.65	-	-	-	1.00
78	0.50	0.53	-	-	0.56	0.44
65	0.46	-	-	-	0.85	0.15
Methyl Acrylate - MMA Copolymers						
86	9.82	-	-	0.02	0.26	0.72
74	0.74	0.60	0.59	0.07	0.39	0.54
65	0.69	0.52	-	0.13	0.46	0.41

Several alternating copolymers containing various combinations of methyl acrylate (MA), MMA, styrene (S) and α-methylstyrene were also studied. Figure 3 compares the relative rates of disappearance of methoxyl groups with time. The alternating S/MA copolymer is the most

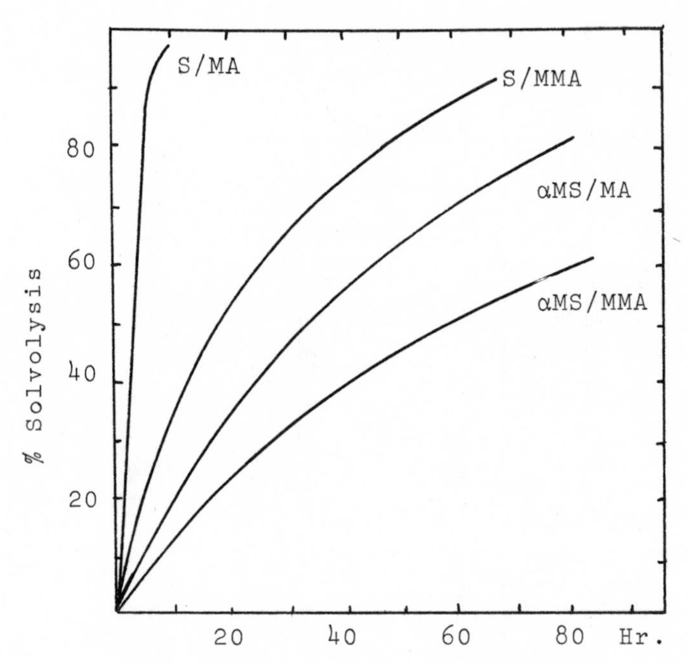

Figure 3. Comparison of Acid Catalyzed Solvolysis Rates of Several Alternating Copolymers Derived from Styrene (S), α-Methylstyrene (αMS), Methyl Methacrylate (MMA) and Methyl Acrylate (MA).

reactive and the alternating α-methylstyrene/MMA copolymer is the least reactive, as might be expected from steric considerations. It is interesting that the S/MMA copolymer is more reactive than the α-methylstyrene copolymer, however. Morawetz [76] has observed a similar effect in studies on the alkaline hydrolysis of p-nitrophenyl esters of acrylic acid or methacrylic acid in copolymers with acrylic acid or methacrylic acid. We thus see that steric effects imposed by nearest neighbors in a vinyl copolymer are at least as effective as those associated with the monomer unit alone.

Multiple hydrolysis and nmr studies on the solvolysis of the alternating S/MMA copolymer indicate all the ester groups to have very similar reactivity. It therefore seems that configurational effects do not influence the reactivity of ester units in SMS triads appreciably, although they are important in the case of MMM triads and are suspected to be important in the case of SMM + MMS triads.

In attempts to conduct multiple reaction studies on the solvolysis of alternating α-methylstyrene/MMA copolymers, considerable difficulty was encountered in remethylation reactions. This was traced to the fact that copolymers cyclized appreciably. Such cyclizations

COOH Ø ⟵ CO OCH$_3$ Ø ⟶ O

have been reported to occur under Friedel-Craft's conditions [64-65], but it was a surprise to learn that they can also occur under our solvolysis conditions. Cyclization was also found to occur during the solvolysis of alternating S/MMA copolymers but at a much slower rate, and it is not believed to have had an appreciable influence on the results previously discussed. The quaternary backbone carbons present in the α-methylstyrene/MMA copolymers apparently restrict the number of conformations available to M-S dyads, and thereby facilitate cyclization. This same phenomenon is believed to be responsible for the fact that vinyl halide/MMA copolymers are much more easily cyclized than vinyl halide/methyl acrylate copolymers [57-58].

TRANSESTERIFICATION STUDIES

It seemed that difficulties due to changes in polymer-solvent interaction, etc., during the course of a reaction, etc., might be avoided by studying "virtual" reactions of the type shown below.

$$RCOOH_3^* + CH_3ONa \longrightarrow RCOOCH_3 + CH_3^*ONa$$

Thus, the rates of loss of C^{14}-labeled methoxyl groups from S/MMA copolymers when allowed to exchange with sodium methoxide were measured. Figure 4 shows the loss in radioactivity of the polymers as a function of time. The rates are seen to increase with the styrene contents of the copolymers, suggesting that MMA units with styrene neighbors are more reactive than those with only MMA neighbors. The data indicate that MMA units in S$\overset{*}{M}$S triads are about 12 times as reactive as those in M$\overset{*}{M}$M triads.

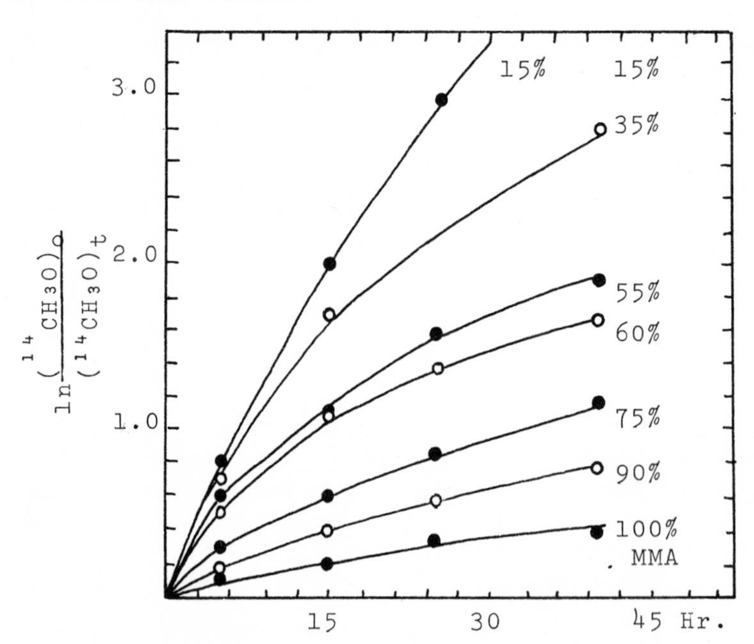

Figure 4. First Order Plots for Exchange of $NaOCH_3$ with Styrene-^{14}C-Methyl Methacrylate Copolymers.

When the rate of uptake of radioactive methoxide ion by S/MMA copolymers was studied, results such as are shown in Figure 5 were obtained. The radioactivity of the copolymers increased and then decreased with time.

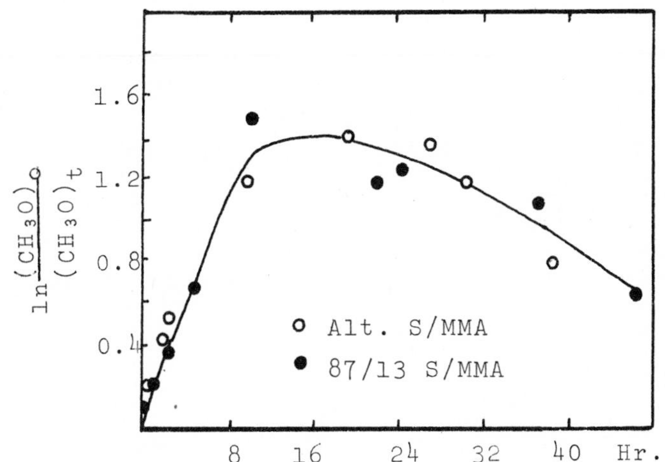

Figure 5. First Order Plot for the Exchange of ^{14}C-Labeled $NaOCH_3$ with S/MMA Copolymers.

The unexpected loss in radioactivity with time was traced to a displacement reaction involving methoxide ion and the methoxyl carbon atoms of the MMA units.

$$RCOOCH_3 + \overset{*}{C}H_3ONa \longrightarrow RCOO\overset{*}{C}H_3 + CH_3ONa$$

$$\downarrow$$

$$RCOONa + CH_3O\overset{*}{C}H_3$$

By comparing the initial rates of radioactivity loss in exchanges of methoxide ion with labeled copolymers with the initial rates of radioactivity gain in exchanges of labeled methoxide ion with unlabeled copolymers, it was estimated that the displacement reaction occurred 1/4 as fast as the exchange reaction.

The observation that a displacement reaction competes with transesterification provided an explanation for a dilemma we had encountered in studies on the transesterification of styrene/phenyl methacrylate copolymers with sodium methoxide [90]. Such reactions seemed to occur quantitatively, yielding S/MMA copolymers, but the products were insoluble in CCl_4, in contrast to the solubility behavior of conventional S/MMA copolymers. We now know that the displacement reaction caused a small amount (∿ 0.5 mole percent) of methacrylic acid units to be present in the derived copolymers and that the presence of such units caused the copolymers to be insoluble in CCl_4. Methylation of the derived copolymers with diazomethane removed such groups and copolymers so treated were soluble in CCl_4.

Acknowledgements

This survey in part covers work done at The University of Akron by T. P. Abbott, G. H. Armstrong, F. A. Blouin, G. Dever, L. J. Guilbault, N. W. Johnston, A. B. Robertson, D. B. Russell, G. O. Schulz, F. Shepherd and E. Thall. The author is indebted to them for their fine experimental work and for many intellectual contributions they have made to this program. Our studies have been supported by grants provided by the National Science Foundation and The Firestone Tire and Rubber Company.

References

1. P.H. Tidwell and G.A. Mortimer, J. Macromol. Sci., Revs. Macromol. Chem. C4, 281 (1970).

2. G. Smets, Makromol. Chem., 34 190 (1959).

3. G. Smets, in "Chemical Reactions of Polymers", edited by E. M. Fettes, Interscience Publishers, Inc., New York, N.Y., 1964, pp 79-91.

4. T. Nakai, H. Kawaoka, and M. Okawa, Bull. Soc. Chem. Japan, 42 508 (1969).

5. P.J. Flory, J. Am. Chem. Soc., 61 1518 (1939).

6. F.T. Wall, ibid., 63, 1862 (1944).

7. F.T. Wall, ibid., 64, 269 (1942).

8. F.T. Wall, ibid., 66, 2050 (1944).

9. E. Merz, T. Alfrey, Jr., and G. Goldfinger, J. Polymer Sci., 1, 75 (1946).

10. T. Alfrey, Jr., C. Lewis and B. Nagel, J. Am. Chem. Soc., 71, 3793 (1946).

11. N.W. Johnston and H.J. Harwood, J. Polymer Sci., Part C, 22, 591 (1969).

12. N.W. Johnston and H.J. Harwood, Macromolecules, 2, 221 (1969).

13. C.S. Marvel, J.H. Sample and M.F. Roy, J. Am. Chem. Soc., 61, 3241 (1939).

14. D.B. Russell, Ph.D. Thesis, University of Akron, 1973.

15. C.S. Marvel, G.D. Jones, T.W. Masten and G.L. Scherty, J. Am. Chem. Soc., 64 235 (1942).

16. T. Alfrey, Jr., H.C. Haas and C.W. Lewis, ibid., 73, 2851 (1951).

17. G.O. Schulz and H.J. Harwood, A.C.S. Polymer Preprints, 7, 356 (1966).

18. G.R. Dever, M.S. Thesis, University of Akron, 1967.

19. C.S. Marvel and C.L. Levesque, J. Am. Chem. Soc., 61, 3234 (1939).

20. D.H. Grant and N. Grassie, Polymer, 1, 125 (1960).

21. J. Semen and J.B. Lando, Macromolecules, 2, 570 (1969).

22. J.C.H. Hwa, W.A. Fleming and L. Miller, J. Polymer Sci., A2, 2385 (1964).

23. J.R. Schaefgen and I.M. Sarasohn, J. Polymer Sci., 58, 1049 (1962).

24. K. Cranwals and G. Smets, Bull. Soc. Chim. Belg., 59, 182 (1950).

25. M. Mullier and G. Smets, J. Polymer Sci., 23, 915 (1957).

26. P.H. Teyssie and G. Smets, ibid., 20, 351 (1956).

27. K. Fujii, J. Ukida and M. Matsumoto, J. Polymer Sci., Part B, 1, 693 (1963).

28. M. Vranchen and G. Smets, J. Polymer Sci., 14, 521 (1954).

29. W.E. Smith, G.E. Ham, H.D. Anspon, S.E. Gebura and D.W. Alwani, J. Polymer Sci., Part A1, 6, 2001 (1968).

30. H.J. Harwood and H.J. Harwood, Jr., J. Appl. Polymer Sci., 15, 1545 (1971).

31. P.J. Flory, J. Am. Chem. Soc., 64, 177 (1942).

32. C.S. Marvel and C.L. Levesque, ibid., 60 280 (1938).

33. C.S. Marvel, G.D. Jones, T.W. Masten and G.L. Scherty, ibid., 64, 235 (1942).

34. J.N. Hay, Makromol. Chem., 67, 31 (1963).

35. N. Grassie and J.N. Hay, ibid., 64, 82 (1963).

36. G.H. Armstrong and H.J. Harwood, A.C.S. Polymer Preprints, 12, 56 (1971).

37. W.J. Burlant and J.L. Parsons, J. Polymer Sci., 22, 249 (1956).

38. R.C. Schulz, Angew. Chem. internat. ed., 3, 416 (1964).

39. R.C. Schulz, H. Chedron and W. Kern, Makromol. Chem., 29 190 (1957).

40. G. Smets and P. Flore, J. Polymer Sci., 35, 519 (1959).

41. L.M. Minsk and W.O. Kenyon, J. Am. Chem. Soc., 72, 2650 (1950).

42. N.L. Zutty and F.J. Welch, J. Polymer Sci., A1, 2289 (1963).

43. M.A. Golub and J. Scanlan in "Chemical Reactions of Polymers", Edited by E.M. Fettes, Interscience Publishers, New York, 1964, pp 107-132.

44. I. Sakurada and K. Kawashima, Chem. High Polymers (Japan), 8, 142 (1951).

45. L.M. Minsk, G.P. Waugh, and W.O. Kenyon, J. Am. Chem. Soc., 72, 2646 (1950).

46. R.J. Kern, Makromol. Chem., 79, 122 (1964).

47. J.A. Powell, J.J. Wang, F.H. Owens and R.K. Graham, J. Polymer Sci., 5A, 2655 (1967).

48. D.B. Denny and J. Giacin, Tetrahedron, 20, 1377 (1964).

49. H. Kwart and M.T. Waroblak, J. Am. Chem. Soc., 89, 7145 (1967).

50. A. Guyot, M. Bert, A. Michel and I.C. McNeill, European Polymer J., 7, 471 (1971).

51. I.C. McNeill, in "Thermal Analysis", edited by R.F. Schwenker, Jr. and P.D. Garn, Academic Press, New York, 1969, Vol 1, pg 417.

52. N.W. Johnston, Macromolecules 3, 566 (1970).

53. F. Shepherd and H.J. Harwood, J. Polymer Sci., Part B, 9 419 (1971).

54. J. Guillot, M. Bert, J. Vialle and A. Guyot, European Polymer J., 8 641 (1972).

55. N.W. Johnston and B.L. Joesten, J. Polymer Sci., Part A, 10 1271 (1972).

56. B.K. Patnaik and N.G. Gaylord, J. Polymer Sci., Part B, 9 347 (1971).

57. N.W. Johnston, Macromolecules, 5, 87 (1972).

58. N.W. Johnston, A.C.S. Polymer Preprints, 13, 1065 (1972).

59. N.W. Johnston and H.J. Harwood, Macromolecules, 3, 20 (1970).

60. F. Shepherd and H.J. Harwood, J. Polymer Sci., Part B, 10 799 (1972).

61. J.C. Bevington and J.R. Ebson, Makromol. Chem., 153, 165 (1972).

62. D.S. Tarbell and P.D. Bartlett, J. Am. Chem. Soc., 59, 407 (1937).

63. L.J. Guilbault and H.J. Harwood, A.C.S. Polymer Preprints, 11 54 (1970).

64. P . Teyssie and G. Smets, J. Polymer Sci., 27 444 (1958).

65. N.G. Gaylord, B.K. Patnaik and Z. Janovic, J. Polymer Sci., Chem. Ed., 11 203 (1973).

66. D.D. Reynolds and W.O. Kenyon, J. Am. Chem. Soc., 72, 1584, 1587, 1591 (1950).

67. E. Goethals and G. Smets, J. Polymer Sci., 40, 227 (1959).

68. F.C. Baines and J.C. Bevington, J. Polymer Sci., Part A-1, 6 2433 (1968).

69. J.C. Bevington, R. Brinson and B.J. Hunt, Makromol. Chem., 134, 327 (1970).

70. E. Thall, Ph.D. Thesis, University of Akron, 1972.

71. G. Von Schreyer and T. Volker, Makromol. Chem., 63, 202 (1963).

72. F.J. Glavis, J. Polymer Sci., 36, 547 (1959).

73. E. Selegny and P. Sagain, J. Macromol. Sci., Chem., A5, 603 (1971).

74. H. Morawetz and E. Gaetjens, J. Polymer Sci., 32, 526 (1958).

75. H. Morawetz and P.E. Zimmering, J. Phys. Chem., 58, 753 (1954).

76. E. Gaetjens and H. Morawetz, J. Am. Chem. Soc., 83, 1738 (1961).77

77. G. Smets and A.M. Hesbain, J. Polymer Sci., 40, 217 (1959).

78. W. DeLoecker and G. Smets, J. Polymer Sci., 40, 203 (1959).

79. G. Smets and W. DeLoecker, J. Polymer Sci., 40, 375 (1959).

80. G. Smets and W. DeLoecker, J. Polymer Sci., 40, 461 (1959).

81. G. Smets and W. Van Humbeek, J. Polymer Sci., Part A, 1, 1227 (1963).

82. E. Thall and H.J. Harwood, Preprint III-21 of Papers Presented at the International Symposium on Macromolecules, Helsinki, Finland, July 2-7, 1972.

83. E. Thall, Ph.D. Thesis, The University of Akron, 1972.

84. H.J. Harwood and A.B. Robertson, Preprint 10/9 of papers presented at the IUPAC International Symposium on Macromolecular Chemistry, Budapest, 1969.

85. A.B. Robertson and H.J. Harwood, A.C.S. Polymer Preprints, 12, 620 (1971).

86. A. Katchalski and E. Eisenberg, J. Polymer Sci., 6, 145 (1951).

87. K. Ito, S. Iwase, K. Umehara and Y. Yamashita, J. Macromol. Sci., Part A, 1, 891 (1967).

88. H.J. Harwood in "NMR-Basic Principles and Progress", ed. by P. Diehl, E. Fluck and R. Kosfeld, Springer Verlag, New York, 1971, pg 71.

89. F.A. Blouin, R.C. Chang and H.J. Harwood, A.C.S. Polymer Preprints, 14 25 (1973).

90. R.G. Bauer, H.J. Harwood, and W.M. Ritchey, A.C.S. Polymer Preprints, 7(2), 973 (1966).

DISCUSSION SESSION

Discussion Leader - M.Lewin

Moore: Did you check for changes in the stereochemical integrity of the polymer during hydrolysis?

Harwood: We compared the NMR spectra of copolymers which had been partially hydrolyzed and remethylated with the parent copolymer spectra. They were the same. We would have seen different methoxyl proton resonance patterns if the configurational structure had changed. Methyl acrylate units might epimerize during the reaction and this would have to be considered if very precise data were attainable. At present we seem to be separating only gross steric effects.

Challa: The average reactivity of the isotactic units in long sequences of isotactic units will be greater than that of isotactic units in short sequences if there is a tendency for sequential reaction. Did you observe this effect in your kinetic analyses? Would your method be sensitive to the distribution of isotactic sequences in a highly isotactic polymer?

Harwood: I don't think so. This would lead to a distribution of reactivities and I don't believe that the method will be sensitive enough to separate them. It would probably indicate all groups to have equal reactivity, and this is what we observe.

Rivin: Have you investigated other solvent systems? The steric effects you observe might be somewhat solvent dependent.

Harwood: No. The solvent systems employed were selected to enable us to study as wide a range of copolymers as possible and to insure that the reaction mixtures were homogeneous. It is difficult to find suitable solvent systems. We are currently using crown ethers to solubilize inorganic reagents and this is providing some flexibility in solvent selection. I do not anticipate observing significant changes in the steric effects. The MMA units in MM*M triads are dineopentyl systems at best and are highly hindered, whereas those in SMM and SMS triads are considerably less hindered.

Salamone: Could the high reactivity of ester units in VMV triads in the MMA-vinyl pyridine copolymers be due to reaction of an MMA unit with a vinyl pyridine neighbor to form an acyl pyridinium intermediate and that the other vinyl pyridine unit associates with a molecule of water which then attacks the acyl intermediate, causing hydrolysis?

Harwood: Perhaps, but I don't believe that this can explain the high reactivity of MMA units in VMV triads in the 4 vinyl pyridine copolymers. The pyridyl nitrogen atoms are not suitably placed for this.

St. Pierre: Is it possible that pyridine acts as a general base operating through several water molecules? This would enable the 4-pyridyl units to act specifically on neighboring MMA units.

Harwood: Yes.

St.Pierre: Would it be useful to study the reaction of PMMA with pyridine?

Harwood: I don't think it would help in this case. The fast reaction seems to be restricted to VMV triads. We seem to be dealing with combined steric-specific solvation effects that would be difficult to duplicate in a pyridine-PMMA system.

Loucheux: How do the actual rates of the reactions compare?

Harwood: The slowesr rate is about the same in all copolymers. In acid catalyzed hydrolysis of S/MMA copolymers the "fast" groups are about 8 times as reactive as the "slow" groups. The ratio of reactivities is about 12 in the case of transesterification. The ratio was about 20 for the acidic hydrolysis of vinyl pyridine-MMA copolymers.

Dale: Is there any precedent for the ether-forming reaction you have observed and have you observed this reaction with less sterically hindered molecules?

Harwood: I am not sure about what precedent there is for such reactions, but I would think that they could be observed if methyl esters were equilibrated with sodium methoxide methanol mixtures for long periods. Acetate ion is not such a bad leaving group.

Daly: When the carbonyl group is hindered, S_N2 attack by methoxide ion at the methyl carbon atom can compete effectively with attack at the carbonyl groups.

Cho: If you use the multiple reaction technique and the reactions are not clean, the products of side reactions will accumulate on the polymer. This would seem to be a limitation of the method.

Harwood: That is correct and it is possible that cyclization may be causing some trouble in studies of the acidic hydrolysis of styrene-MMA and related copolymers.

Lewin: In that connection, what is the accuracy of the method and how reproducible is it?

Harwood: The fraction of slowly reacting groups and the relative reactivity can be estimated to ± 5-10% and 10-20%, respectively. The method requires that one be able to reproduce conversion-time behavior as precisely as possible.

Kelly: I wonder if ^{13}C studies would provide clearer information in multireaction studies than proton spectra? You might methylate partially hydrolyzed polymers with ^{13}C-enriched diazomethane and monitor the appearance of various ^{13}C resonances.

Harwood: Dr. Klesper has already described some experiments in which ^{13}C spectra were used. I believe it was the carbonyl carbon resonance which was most sensitive to environment, however.

Klesper: The carbonyl carbon resonance is most sensitive to structure. It is interesting that the α-methyl carbon resonance is sensitive only to the relative configurations of neighboring acid or ester units and not to composition. This is a rather unexpected result.

Smets: As a comment on the use of ^{13}C measurements for the elucidation of internal copolymer structure, I would like to present some results on data from the copolymerization of acrylic acid/styrene. These copolymers were prepared at 60° using AIBN at a total monomer concentration of ~2 moles/liter; the copolymer precipitates when formed in benzene. It remains soluble in dimethyl sulfoxide. The copolymers were analyzed for their acid content, and the acrylic acid triads sequences (^{13}C of the carbonyl carbon atom).

				Triads	
Solvent	AA/S	Wt%AA	AAA	AAS	SAS
φH	1.2/1	50	-	27	73
DMSO	"	34	-	30	70
φH	1.5/0.7	54	~10	37	53
DMSO	"	38	12	33	55

Though the overall composition is noticeably different, nevertheless the triads for acrylic acid incorporation are identical in both solvents. It suggests that the copolymer obtained in DMSO contains longer stryrene sequences than that precipitated from benzene.

Salamone: Is there any correlation of tacticity with the rates you obtained?

Harwood: There does not appear to be an effect in the case of SM*S triads, because all the groups have the same reactivity in alternating copolymers. of course, the MMA units in isotactic MM*M triads are more reactive than those in heterotactic or syndiotactic triads, as seen from studies on the homopolymers. We don't know much about SM*M triads, but the result obtained with high styrene copolymers indicate that the differrnt configurations have different reacti vities.

PHOTOCHEMICAL REACTIONS OF POLYMERS IN SOLUTION

C.H. Bamford

Department of Inorganic, Physical and Industrial Chemistry, University of Liverpool, England.

Although the thermal and photochemical reactions of an isolated group in a polymer molecule resemble those of a similar group in a low-molecular-weight compound, some influences of the polymer environment are often apparent. Reaction rates and quantum yields may be affected by restrictions in movements imposed by the molecular chains or by steric or shielding effects arising from chain coiling; for example, the quantum yield of a photochemical process will be low if achievement of the transition state requires unlikely changes of chain conformation. If the groups in the polymer molecule are not effectively isolated their interaction can lead to a wide range of interesting phenomena. Reactivities may be increased or decreased or changes in mechanism may be brought about by neighbouring group participation; the high local concentrations of functional groups in the neighbourhood of a polymer molecule containing one group per repeat unit is one factor responsible for the behaviour of polymers as catalysts. Many of these effects are of course apparent in suitably chosen low-molecular-weight models.

Photochemical reactions, with which we are primarily concerned in this review, involve both photophysical and chemical processes. A chromophore electronically excited by light absorption is confronted by several possibilities other than chemical reaction : deactivation may ensue as a result of light emission or a radiationless transition involving energy transfer. Electronic energy transfer, in which relatively large amounts of energy are transferred, is a very important photophysical process both in systems composed of small molecules and in those containing polymers. It may be inter- or intra-molecular; in polymer systems the latter implies migration of energy along the polymer chains. These processes are not only of theoretical interest but they have important practical implications, notably in providing methods of stabilising polymers against photodegradation and of achieving controlled degradation. Incorporation of a quenching group in a polymer chain provides an "energy sink" which in suitable circumstances dissipates the absorbed energy before photolysis can occur. Such stabilisation is potentially more effective than that obtainable by the incorporation of inhibitors which suppress chemical reactions following photolysis.Energy transfer is also important in polymers exposed to high energy radiation, and when the materials are used as scintillators or phosphors.

Photophysical processes in small molecules have received much study and four types of electronic energy transfer are recognised (1). (i) The so-called "trivial" type, in which radiation emitted by the excited donor is absorbed by the acceptor. This requires overlap between the relevant emission and absorption spectra and may operate over large distances. It is not now thought to be of general importance in condensed systems. (ii) A non-radiative process involving exchange interaction between donor and acceptor. This is a collisional process and operates over short distances (1-1.5nm); its rate is limited by the collision frequency of the interacting species. Quantum restrictions (Wigner's spin rule) are important : if S_1 and S_2 are the initial spin quantum numbers, the resultant spin quantum number of the two systems taken together must have one of the values $S_1 + S_2$, $S_1 + S_2 - 1$ ...... $S_1 - S_2$. (iii) A non-radiative process operating over relatively large distances (5-15nm). A theoretical basis for this process based on dipole-dipole interaction has been given by Förster (2). Transfer of this type is forbidden if there is a change of spin in either partner. In fact, however, triplet – singlet transfer has been observed; although the process is theoretically very slow it can compete with deactivation of the donor if this is also forbidden. This type of energy transfer is not dependent upon collisions and takes place in rigid media as well as in solution. (iv) Exciton migration. An "exciton" is an excited state delocalized over a group of molecules; it may be either singlet or triplet in character. The molecules must have high local order; exciton phenomena are therefore most frequently encountered in crystals although there is evidence that they participate in liquid aromatic systems. The exciton diffusion length may be of the order of 50nm.

We conclude this section with a brief summary of the type of evidence which indicates that electronic energy can migrate along polymer chains.

Cozzens and Fox (3) found that poly(1-vinyl naphthalene) in a rigid glass at $77^{\circ}K$ irradiated by light of $\lambda = 290nm$ shows a delayed emission spectrum corresponding to mean radiative lifetimes of 80ms in the fluorescence region and about 1.9s in the phosphorescence region. Delayed fluorescence did not appear in the spectrum of the low-molecular-weight model 1-ethyl naphthalene. Similar delayed emission spectra were also obtained from a mixture of the polymer with benzophenone on irradiation with λ = 366nm, at which wavelength the polymer is transparent. The triplet quencher piperylene quenched the delayed fluorescence and phosphorescence of poly(1-vinyl naphthalene) but not the fluorescence or phosphorescence of 1-ethyl naphthalene at equivalent concentrations. It was concluded that the delayed fluorescence involves triplet excitation (see also p. 9). To account for these findings it was postulated that two photons may be absorbed by the same polymer molecule, forming excited singlet states which undergo intersystem crossing to the triplet manifold; migration of triplet energy along the chain then occurs until the two triplets become sufficiently close to undergo triplet-triplet annihilation (equation (1))

$$T + T \longrightarrow S_1 + S_0, \quad S_1 \longrightarrow S_0 + h\nu \qquad (1)$$

Fluorescence from the resulting excited singlet will appear to be longer-lived than that from an excited singlet produced by direct absorption since the excitation has resided temporarily in the triplet. Further, the intensity of the delayed fluorescence was found to be proportional to the square of the incident intensity, in conformity with equation (1). Experiments with copolymers containing methyl methacrylate indicated that a minimum sequence-length of naphthalene chromophores is necessary for detectable delayed fluorescence, since methacrylate units would be expected to act as barriers to energy transfer. It was pointed out that triplet migration would be expected to be very sensitive to the presence in the polymer chains of end-groups possessing triplet levels lower than those of the 1-vinyl naphthalene units since trapping of the energy could occur in such groups; processes leading to delayed fluorescence would thus be reduced in importance. Subsequently Fox and Cozzens (4) reported detailed studies of 1-vinyl naphthalene-styrene copolymers and mixtures of the two homopolymers under similar conditions. Comparison of the two systems revealed that the copolymers excited by light absorbed by polystyrene show quenching of phosphorescence arising from styrene segments accompanied by sensitization of phosphorescence of the 1-vinyl naphthalene segments. Delayed fluorescence is significant from polymers containing more than about 5 mol % naphthyl groups, but with decreasing concentration of these groups below this value the delayed fluorescence rapidly decreases in importance. These experiments show that intramolecular transfer of triplet energy from styrene to 1-vinyl naphthalene segments accurs in the copolymers, the latter segments acting as "energy sinks". The decrease in delayed fluorescence observed at low proportions of vinyl naphthalene units indicates that under these conditions the naphthyl sequences are too short to allow triplet-triplet annihilation. It was concluded that energy transfer must be substantially intramolecular, since intermolecular quenching of styrene segment phosphorescence should occur in mixtures of the homopolymers; the observations show that such quenching is relatively small. In a later paper Fox, Price and Cozzens (5) reported that while piperylene quenches the phosphorescence of pure polystyrene (in a glassy matrix at 77°K) it cannot compete with the intramolecular energy transfer to naphthyl segments in the copolymer under the same conditions. Energy migration through chains having as many as 140 styrene units was observed. The interesting conclusion was drawn that in the copolymers containing styrene the origin of much of the phosphorescence is in the region of the chain where the two types of unit are joined; further, the natures of the groups at the ends of a 1-vinyl naphthalene sequence exercise an important influence on the photophysical behaviour. The intramolecular transfer of singlet excitation from styrene to 1-vinyl naphthalene units was also reported. The authors draw the analogy between energy migration in polymer systems and exciton diffusion in organic crystals. In organic crystals containing small quantities of guest molecules or defects, energy absorbed by the host is delocalized and is ultimately observed as emission from the guest molecules. A polymer molecule through which similar delocalization of energy occurs may be likened to a one-dimensional crystal, especially if some part of the chain acts as an energy trap. Many kinds of such traps of differing depths may be envisaged, for example double bonds, branches and initiator fragments, and interaction between trapped and migrating energy must then be considered. Phenomena of this type give rise to non-exponential growth and decay curves for fluorescence and phosphorescence. Evidently a study of light emission provides a highly sensitive technique for detecting certain kinds of impurities in polymer molecules.

Emission spectra frequently reveal the existence of excimers, i.e. excited dimers formed by the interaction of a singlet excited species with an unexcited molecule:

$$S_1 + S_0 \rightleftharpoons \underset{\text{excimer}}{(S_0 S_1)^*} \longrightarrow 2S_0 + h\nu \tag{2}$$

Pyrene solutions in ethanol show this phenomenon; at a pyrene concentration of 3×10^{-3} mol l^{-1} excimer emission dominates the luminescence. The dimerization in (2) occurs only after excitation (i.e. there is no ground state of an excimer) and so is not normally observable in the absorption spectrum. Excimer formation depends critically on the geometrical arrangement of the interacting molecules (6), consequently its existence in polymer systems can provide information about the conformation of the polymer chains. Vala, Haebig and Rice (7) showed that certain emission bands observed in polystyrene and in poly(1-vinyl naphthalene) arise from excimers formed by interaction of nearby chromophores on the same polymer chain; one conclusion reached by these workers was that a rigid helical isotactic polystyrene chain in solution would give a much smaller excimer emission than is observed in practice, so that, in fact, the coherence length of helical portions of the chains in solution is short. Fox and his colleagues (8,9) have also studied excimer emission in several polymers. It appears that in solutions of vinyl aromatic polymers excimer formation is probably an intra-chain phenomenon except in the most concentrated solution. Excimer sites can act as efficient singlet traps for migrating electronic energy.

DIRECT PHOTOCHEMICAL REACTION OF POLYMERS

(1) Simple Bond-breaking Processes, without Degradation.

The simplest systems are probably those in which the chromophore is a C-halogen group in the polymer. On irradiation fission of the C-halogen bond occurs with formation of a carbon radical attached to the polymer chain and a halogen atom (equation (3)).

$$\sim\sim CH_2Br \xrightarrow{h\nu} \sim\sim \dot{C}H_2 + \dot{B}r \qquad (3)$$

By carrying out this reaction in a polymerizable vinyl monomer Melville and his colleagues (10) were able to synthesise block copolymers. The liberated bromine atoms also initiate, so that the final product consists of a mixture of polymers. If the halogen atoms are not terminal, a similar series of reactions leads to graft copolymers. An example is provided by some unpublished work of Mullik and Norrish on polyvinyl bromide. Primary photolysis in tetrahydrofuran solution (λ = 253.7nm) proceeds according to equation (4) with unit quantum yield.

$$\sim CH_2-\underset{\displaystyle Br}{\underset{|}{C}H}\sim + h\nu \longrightarrow \sim CH_2-\underset{\displaystyle \dot{B}r}{\underset{+}{\dot{C}H}}\sim \qquad (4)$$

The bromine atoms attack the solvent, and also enter into hydrogen abstraction reactions with the methylene groups of the polymer, which as a consequence develops unsaturation. Ultimately conjugation becomes sufficiently extensive to produce brown or black colorations in the polymer. Polyacrylonitrile and polyethylene have been grafted on to polyvinyl bromide with the aid of reaction (4). Clearly, if during grafting the chains terminate by combination of radicals, cross-linking and ultimately network formation are possible in suitable circumstances.

(2) Degradation

Many polymers are degraded by exposure to visible or ultraviolet light and a great deal of effort has been expended in attempts to design light-stable polymer systems. Much of this work is of a utilitarian kind, and does not elucidate the detailed mechanism of the photoreactions. It seems to be generally realised that, although the utilitarian approach is undoubtedly useful, major advances are more likely to be made by more fundamental work, for example, on the relevant photophysical processes described earlier. We shall now consider some recent investigations of chemical processes involved in degradation.

Grassie and his colleagues (11) have carried out a systematic study of homopolymers and copolymers of methyl methacrylate (MMA) and methyl acrylate (MA) under a wide variety of conditions. On ultraviolet irradiation at ordinary temperatures, polymethyl methacrylate undergoes chain scission and a rapid reduction in molecular weight, but polymethyl acrylate becomes insoluble on account of cross-link formation. Chemical evidence suggests that the primary acts are similar in both cases, and consist of scission of an ester group (reactions (5) and (6)).

$$\sim CH_2-\underset{\underset{COOCH_3}{|}}{\overset{\overset{H}{|}}{C}}\sim \xrightarrow{h\nu} \underset{(I)}{\sim CH_2-\overset{\overset{H}{|}}{\dot{C}}\sim} + \dot{C}OOCH_3 \qquad (5)$$

$$\sim CH_2-\underset{\underset{COOCH_3}{|}}{\overset{\overset{CH_3}{|}}{C}}\sim \xrightarrow{h\nu} \underset{(II)}{\sim CH_2-\overset{\overset{CH_3}{|}}{\dot{C}}\sim} + \dot{C}OOCH_3 \qquad (6)$$

The difference in behaviour of the polymers is therefore ascribed to differing reactions of radicals (I) and (II), the latter tending to undergo chain scission (equation (7)) while the former mainly dimerize.

$$\sim\!\sim C(CH_3)(COOCH_3)-CH_2-\dot{C}(CH_3)-CH_2\sim\!\sim \longrightarrow \sim\!\sim \dot{C}(CH_3)(COOCH_3) + CH_2=C(CH_3)-CH_2\sim\!\sim \quad (7)$$

An interesting difference between the behaviour of copolymers rich in MMA and those rich in MA was apparent in studies of the materials as films and in solution in methyl acetate. The rates of chain scission of the former copolymers are much greater in solution, while the methyl acrylate-rich copolymers undergo scission in films and in solution at comparable rates. The two types of copolymers have glass transition temperatures above and below ambient temperatures, respectively, and it was suggested that, in films, cage recombination of the primary products arising from reactions (5) and (6) are much more important for the more rigid methacrylate-rich copolymers. In solution such recombination would be expected to be greatly reduced. Supporting evidence for these views was adduced from experiments on a methacrylate-rich copolymer at different temperatures, which showed that the rate of chain scission in films increases rapidly with increasing temperature in the vicinity of the glass transition.

Purely thermal degradation of polymethacrylates, which begins near 200°C, generally involves depolymerization to the monomer and also ester group decomposition, which leads to the formation of carboxylic acid residues in the polymer and the corresponding olefine. The latter reaction proceeds through a cyclic transition state depending on interaction between the carbonyl group and a β hydrogen atom; it therefore becomes more important as the number of β hydrogens in the molecule increases and is the almost exclusive mode of decomposition with the t-butyl ester. At somewhat lower temperatures ($\sim$160°C) photolysis produces almost quantitative yields of monomer even from t-butyl methacrylate, presumably by the unzipping of radicals such as those arising in reaction (7). Apparently the unzipping process cannot pass freely through acrylate units in copolymers containing methyl acrylate.

These results demonstrate that physical conditions can exert a predominating influence on the secondary processes occurring in polymer photodegradation and thus control the rate of the overall decomposition.

The photochemistry of carbonyl compounds has received extensive attention. It is well-known that simple ketones containing aliphatic groups may undergo four types of primary process (1):

$$R\overset{O}{\overset{\|}{C}}CH_2CH_2CH_3 \xrightarrow{h\nu} R\dot{C}O + \dot{C}H_2CH_2CH_3 \quad (8)$$

$$R\overset{O}{\overset{\|}{C}}CH_2CH_2CH_3 \xrightarrow{h\nu} R\overset{O}{\overset{\|}{C}}CH_3 + CH_2{=}CH_2 \quad (9)$$

$$R\overset{O}{\overset{\|}{C}}CH_2CH_2CH_3 \xrightarrow{h\nu} \text{(cyclobutanol with OH and R)} \quad (10)$$

$$RR'CO \xrightarrow{h\nu} RR'CO^{*} \xrightarrow{SH} RR'\dot{C}{-}OH + \dot{S} \quad (11)$$

Reactions (8) and (9) are generally referred to as Norrish Types I and II, respectively. In hydrocarbon solution at room temperatures reaction (8) is strongly quenched (12), most probably through operation of a cage effect. The type II process, which leads to a methyl ketone and an olefine, is not quenched in solution (12) and is most likely to involve a cyclic 6-membered transition state (III) (equation (12)) which breaks down into the enol form of a methyl ketone (13,14). A biradical intermediate of the form (IV),

$$R{-}\overset{O}{\overset{\|}{C}}{-}CH_2{-}CH_2{-}CH_2{-}H \xrightarrow{h\nu} \text{(III)} \longrightarrow R{-}C(OH){=}CH_2 + CH_2{=}CH_2 \quad (12)$$

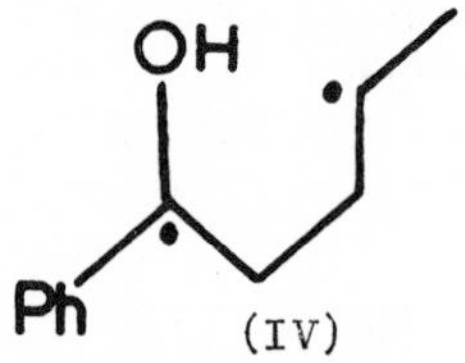

OH

Ph (IV)

which may partly revert to the ground state of the original ketone, has also been suggested (15,16). The type II process is naturally restricted to ketones which possess a hydrogen atom on the γ-carbon atom; it occurs with aryl alkyl as well as dialkyl ketones. Type I may proceed through either singlet or triplet (n,π^*) excited states of the ketone; Type II involves similar states with dialkyl ketones, but mainly the (n,π^*) triplet with aryl alkyl ketones. Reaction (10), which like Type II requires the presence of a γ- H, often proceeds through the (n,π^*) triplet state. Reaction (11) occurs with arylalkyl or diaryl ketones; benzophenone is a very familiar example. The excited benzophenone singlet (n,π^*) arising from light absorption in the region 340nm undergoes intersystem crossing to the triplet with almost unit efficiency; this may then abstract hydrogen from a donor (SH) or transfer its energy to a reactant or quencher molecule (X). These processes are summarized below.

$$Ph_2CO \xrightarrow{h\nu} {}^1(Ph_2CO)^* \longrightarrow {}^3(Ph_2CO) \quad \begin{cases} \xrightarrow{SH} Ph_2\dot{C}OH + \dot{S} \\ \xrightarrow{X} Ph_2CO + {}^3X \end{cases} \tag{11a}$$

By virtue of these reactions benzophenone functions as an efficient photosensitizer.

The occurrence of Norrish types I and II processes in polymer systems was demonstrated by Guillet and Norrish (17) and by Wissbrun (18) in their studies of the photolysis of polymethyl vinyl ketone in solution. The primary reactions are shown in equation (13)

$$\sim CH_2-\underset{\underset{CH_3}{|}}{\underset{CO}{|}}{CH}-CH_2-\underset{\underset{CH_3}{|}}{\underset{CO}{|}}{CH}\sim \;\xrightarrow{\;I\;}\; \sim CH_2-\underset{\underset{CH_3}{|}}{\dot{C}O}\!\!\!\!\!\!\!\!\overset{\bullet}{C}H-CH_2-\underset{\underset{CH_3}{|}}{\underset{CO}{|}}{CH}\sim$$

$$\xrightarrow{\;II\;}\; \sim CH_2-\underset{\underset{CH_3}{|}}{\underset{CO}{|}}{C}=CH_2 \;+\; \underset{\underset{CH_3}{|}}{\underset{CO}{|}}{CH_2}\sim \qquad (13)$$

Clearly only type II leads to main chain scission; type I in the simplest case would result in a small reduction in number-average molecular weight due to loss of acetyl groups. Similar reactions occur in solid films of the polymer.

The quantum yield of photolysis of polymethyl vinyl ketone in solution is low- 0.025 - about an order of magnitude lower than that for low-molecular-weight ketones. Amerik and Guillet (19) found that in copolymers of methyl vinyl ketone and methyl methacrylate containing minor amounts of the former the quantum yield is approximately 0.2. This is a surprising result since a ketone unit flanked by methacrylate units has no γ-hydrogen atom available to form the cyclic intermediate for the type II reaction. Amerik and Guillet suggested the participation of a 7-membered ring, involving a δ-hydrogen in this case. Further, it was thought that energy migration between neighbouring carbonyl groups in the homopolymer would increase the probability of conversion of electronic to vibrational energy, and hence lead to a low quantum yield (see also p. 1). On this view, quantum yields for main-chain scission near 0.2 would be expected for vinyl copolymers containing pendant methyl ketone groups if the carbonyl groups are isolated from similar groups along the chain and provided there is adequate chain mobility.

The occurrence of energy transfer in polymethyl vinyl ketone and other aliphatic polyketones has been further examined by Guillet and his colleagues (20,21). The quenching efficiency of biacetyl for the fluorescence of the polymeric ketones was found to be significantly greater than that for low-molecular-weight ketones; at the same time the quenching of naphthalene fluorescence is more efficient with small ketones than with the polymers. These results suggest that the (singlet) excitation energy in the polymer chain is delocalised along the chain thereby increasing the probability of an encounter with a quenching molecule. Associated spectral changes in the polymers indicate some interaction of ketone groups (20). Reference (20) also quotes other examples of the migration of singlet energy along polymer chains. Determinations of the quenching of phosphorescence of polymethyl vinyl ketone and polymethyl isopropenyl ketone by 1-cis, 3-cis cyclo-octadiene reveal that the sizes of the quenching spheres are similar for the polymers and small ketones, indicating that no intramolecular migration of triplet energy occurs.

A related polymer-polyphenyl vinyl ketone has been studied by David, Demarteau and Geuskens (22). Irradiation in benzene solution by light of $\lambda = 365$nm brings about a decrease in the viscosity of the solution indicating polymer degradation. No cross-linking occurs during irradiation, so that photoreduction of the ketone groups by reaction (11), which would lead to radical formation and subsequent cross-linking, is unimportant. The occurrence of a type I reaction is insignificant. The main reaction occurring is therefore the type II scission:

$$\sim CH(COC_6H_5)-CH_2-CH(COC_6H_5)-CH_2 \sim \xrightarrow{h\nu} \sim CH_2(COC_6H_5) + CH_2=C(COC_6H_5)-CH_2 \sim \qquad (14)$$

The number of chain scissions per polymer molecule was shown to be linear in the reaction time, and the quantum yield for scission is approximately 0.3. It is of interest that these findings show the similarity between the photolysis of the polymer and that of butyrophenone, which may be regarded as a low molecular weight model for the polymer. According to Baum, Wan and Pitts (23) the quantum yield for the type II decomposition of butyrophenone in benzene solution at 25°C is 0.43, while the sum of the quantum yields for processes other than type II is less than 10^{-2}.

In a subsequent publication David, Demarteau and Geuskens (24) have reported that the presence of naphthalene reduces the rate of chain scission, indicating that the excited state of the ketone responsible for scission is a triplet. Deactivation of the excited polymer chromophores is accompanied by the appearance of the sensitized phosphorescence of naphthalene and the intensity of this was measured in thin films of the polymer at 77°K. It thus became possible to estimate the transfer efficiency f, defined as the ratio of the observed quantum yield for naphthalene phosphorescence to the maximum quantum yield obtainable by complete transfer. The transfer efficiency increases with the naphthalene concentration until a saturation value is reached; at this point every C_6H_5CO-group has a naphthalene molecule within its sphere of action. The saturation value is 0.3 mol l^{-1}, approximately, and corresponds to a quenching sphere of radius 2.6nm and containing 108 neighbours. This value is greater than that found for the benzophenone -

naphthalene system (1.3nm) and may be taken to indicate the existence of triplet energy migration in the polymer (which would lead to an increase in the interaction distance). A similar conclusion has been reached by Dan,Somersall and Guillet (21). The dependence of f on the concentration of naphthalene is in good quantitative agreement with the theory of Inokuti and Hirayama (25) for energy transfer by exchange interaction.

Irradiation of films of polyphenyl vinyl ketone produces a decrease in molecular weight without any gel formation.Type I processes are insignificant, and the photochemical behaviour closely resembles that observed in solution.

Golemba and Guillet (26), in studies of the photolysis (λ= 313nm) of polyphenyl vinyl ketone and copolymers of phenyl vinyl ketone with styrene in benzene solution at 25°C, confirmed the predominance of the type II process, although benzaldehyde was identified at high light doses. These workers used 1-cis, 3-cis-cyclo-octadiene as triplet quencher and discussed their observations in terms of the Stern-Volmer relation:

$$\frac{\Phi_o}{\Phi} = 1 + k_q \tau [Q] \qquad (15)$$

In this equation φ_o, φ are the quantum yields for the process under consideration (the type II reaction in this case) in the absence and presence of the quencher Q, respectively, k_q is the bimolecular rate coefficient for the quenching process and τ the lifetime of the excited state in the absence of quenching agent. The relation (15) is a general one, applying to light emission and chemical reaction, and is readily derived from a simple scheme in which the observed process (first order) is in competition with quenching (second order). According to (15), a plot of φ_o/φ against [Q] should be linear and of slope $k_q\tau$, hence the latter quantity may be evaluated. Quenching under normal conditions is often diffusion-controlled and in such circumstances values of k_q in the range $10^9 - 10^{10}$ mol^{-1}l s^{-1} might be expected for typical organic solvents at 25°C. Data for polyphenyl vinyl ketone, plotted according to equation (15) (the φ's referring to chain scissioning) gave the expected straight line. Golemba and Guillet took k_q = 2.5x10^{9} mol^{-1}l s^{-1} and hence obtained a value for the lifetime of the excited triplet state of 1.7x10^{-8}s.

On the other hand, data for the polyphenyl vinyl ketone - styrene copolymers do not give linear Stern-Volmer plots. The slope is greatest in the lower part of the concentration range ([Q] < 0.05 mol l^{-1}, approximately) but with increasing [Q] it steadily decreases and approaches that found for the homopolymer of phenyl vinyl ketone. In the copolymer there exists a distribution of sequences of various lengths of phenyl vinyl ketone units, and Golemba and Guillet point out that a curved Stern-Volmer plot would be expected if there is a difference in triplet lifetime between isolated and sequential units. On this view, the form of the plot indicates that the triplet lifetime is greater for isolated units than for sequences, or, alternatively, the efficiency of quenching might be greater for sequences. A difference of this kind might arise from delocalization of excitation energy, or excimer formation, in sequences.

The lifetimes deduced for the excited chromophores in the copolymers fall within the range of values obtained by Wagner and Kemppainen (27) for a series of low molecular weight phenyl alkyl ketones. In the latter series it appears that factors leading to a short lifetime are the presence of a tertiary γ-hydrogen, or a secondary γ-hydrogen which is α to a phenyl group. Thus the ease of γ-hydrogen abstraction (equation (12)) may be rate-determining. However, the polymeric ketones are less reactive, perhaps as a result of stiffness of the polymer chains - the ease of formation of the cyclic intermediates in equation (12) rather than the ease of abstraction of the γ - H controlling the lifetime.

Lukáč et al (28) have reported an interesting study in which a quenching molecule (naphthalene) is built into a polymer molecule. Copolymers of phenyl vinyl ketone with 1- or 2-vinyl naphthalene containing up to 10% (w/w) of vinyl naphthalene were prepared and photolysed in benzene solution at 30°C. Quantum yields for type II chain scission were determined and the ratio φ_0/φ plotted against the concentration of naphthalene structural units (cf. equation (15)). In this way it was found that $k_q\tau = 1.4\times10^3$ mol^{-1} l, a value about 21 times as great as that for the quenching of the type II decomposition of polyphenyl vinyl ketone by naphthalene in solution. As would be anticipated, intramolecular energy transfer in this type of system is very efficient. The work clearly demonstrates that the concept of "built in" triplet stabilisers is of potential practical importance.

Copolymers of ethylene and carbon monoxide consist of polyethylene chains with randomly situated carbonyl groups and would therefore be expected to resemble dialkyl ketones in their photochemistry. This has been demonstrated by extensive studies of Guillet and his colleagues (29,30,31). On irradiation in hydrocarbon solution by light with λ = 313nm reactions of types I and II occur; in these systems both lead to chain scission and consequent reduction in molecular weight. The participation of singlet and triplet excited states was examined by use of 1-cis, 3-cis-cyclooctadiene as triplet quencher. For a polymer containing 1% CO groups in decalin at 80°C about 45% of the total reaction can be quenched and therefore arises from the triplet state. The limiting slope (low [Q]) of the Stern-Volmer plot (equation (15)) for triplet quenching shows that $k_q\tau = 20\ mol^{-1}l$. An estimate of k_q was made indirectly from measurements of the quenching of naphthalene fluorescence by the copolymers; for details the reader should consult the original paper (30). The derived value of τ for the copolymer in decalin solution at 80^oC is $1.4x10^{-8}s$; a slightly smaller value, $1x10^{-8}s$, was obtained for a solid polymer film. The authors point out that these short lifetimes probably account for the inability of atmospheric oxygen to quench the reaction. For similar reasons, stabilization of polymers of this type against photodegradation by the use of energy-transfer processes would require uneconomically large concentrations of stabiliser.

Guillet and coworkers (30,32,33) also investigated the photochemistry in solution of some relatively low-molecular-weight ketones, including a series of linear 2-alkanones from C_6 to C_{19} and 6-undecanone. One object of this work was to determine the effect of chain-length on the course of photolysis and to compare the reactions of these model compounds and the ethylene-carbon monoxide copolymers. Type I was found to be a minor component of the photolysis of the 2-alkanones at temperatures up to 120^oC. With increasing chain-length the type I quantum yields decrease but the type II yields remain relatively unchanged (they decrease somewhat in the range 0.25-0.15). Further, the quantum yields of type I increase with rising temperature and decrease with increasing solvent viscosity, while those of type II are little affected. Some of these findings are reminiscent of early results with ketones of lower molecular weight (12); cage recombination of type I intermediates is favoured by increases in viscosity and low temperatures. Quenching experiments with 1-_cis_, 3-_cis_-cyclo-octadiene showed that the contribution of triplet states of the type II reaction is about 64%, independent of chain-length.

These and other observations show that the overall photochemical quantum yield (reactions (9) and (10)) for methyl alkyl ketones and symmetrical dialkyl ketones possessing secondary γ-hydrogens decreases with increasing chain-length. Ease of rotation around bonds in the neighbourhood of the carbonyl chromophore must influence the attainment of the cyclic transition state (III) and may account in part for the changes in the type II quantum yield. Reversion to the parent ketone of the postulated biradical intermediate (IV) may also increase in rate with increasing chain length for a similar reason, and thus give rise to decreasing overall quantum yields. The latter effect should be more important with symmetrical dialkyl ketones, in conformity with observation .

3. Cross-linking and Grafting

Photo-cross-linking of polymers finds many industrial applications in image formation, including the production of printing plates and relief images and printed and integrated circuits particularly for use in micro-electronics. A discussion of some of these topics has been published by Williams (34). Photosensitive polymers which cross-link on irradiation are the basis of "negative" photoresists; image formation is based on the insolubilization of portions of the polymer which have been exposed to light and the subsequent removal of unexposed material by appropriate solvent treatment. The industrial importance of photosensitive polymers has stimulated a great deal of research which cannot be completely reviewed here, so we shall select a few typical examples.

One of the simplest examples from the mechanistic view-point is provided by the photolysis of polyvinylbenzophenone (V), studied by David, Demarteau and Geuskens (35).

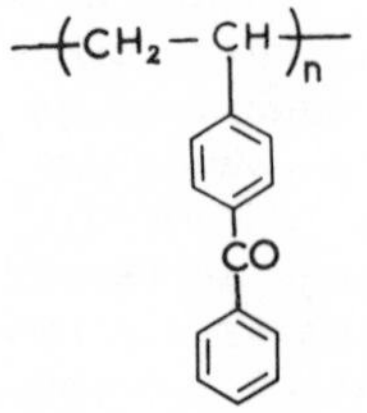

$$-(CH_2-CH)_n- \text{ with side group } -C_6H_4-CO-C_6H_5 \qquad (V)$$

This polymer, unlike polyphenyl vinyl ketone discussed earlier, has no labile hydrogen attached to γ-C, so that type II decomposition of (V) is excluded. The low-molecular-weight analogue p-propyl benzophenone was also studied.

Photolysis (λ = 365nm) of the polymer and p-propyl benzophenone in benzene solutions containing isopropanol (10^{-1}mol l^{-1}) produces acetone by reactions summarised below.

$$R\,Ph\,CO^{*} + CH_3CH(OH)CH_3 \longrightarrow R\,Ph\dot{C}OH + CH_3\dot{C}(OH)CH_3 \qquad (a)$$

$$R\,Ph\,CO + CH_3\dot{C}(OH)CH_3 \longrightarrow R\,Ph\dot{C}OH + (CH_3)_2CO \qquad (b) \qquad (16)$$

$$2R\,Ph\,\dot{C}\,OH \longrightarrow R\,PhC(OH)\,C(OH)R\,Ph \qquad (c)$$

Here $R = -C_6H_4-C_3H_7$ or $-(CH_2-CH)_n-$ with $-C_6H_4-$ attached

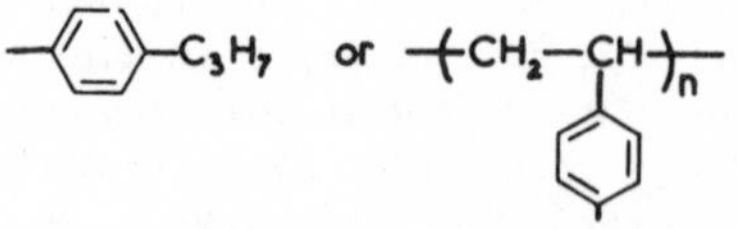

Reaction (16c) may be inter- or intra-molecular; the former would result in cross-linking while the latter would lead to a decrease in intrinsic viscosity of the polymer by reducing the hydrodynamic volume of the polymer coil. Insignificant quantities of volatile products were obtained from p-propylbenzophenone, showing that no rupture of the side-chains occurs; thus it may be inferred that rupture of the main chains of polyvinyl benzophenone is unimportant.

With dilute solutions of the polymer no appreciable amount of gel is formed, and a decrease in viscosity of the solution occurs on irradiation. Under these conditions, intramolecular cross-linking (16c) is the predominant reaction. On the other hand, in concentrated solutions and solid films, a gel fraction approaching 100% is obtainable, showing that cross-linking in these circumstances is mainly intermolecular.

Both cyclisation in dilute solutions and cross-linking in concentrated solutions or films are retarded by the presence of naphthalene, as would be anticipated if these reactions proceed through triplet (n,π^*) intermediates.

It is interesting that, under comparable conditions, the quantum yield of acetone formation is greater for p-propyl benzophenone that for polyvinyl benzophenone. This difference probably arises from the reduced accessibility of the excited carbonyl group inside the macromolecular coil.

Recently Sumitomo, Nobutoki and Susaki (36) have reported the photo-grafting of methyl methacrylate on to styrene-vinyl benzophenone copolymers. In some of the experiments methanol or toluene was used as hydrogen donor. The results appear to be consistent with a mechanism basically similar to (16); grafting originates from ketyl radicals $(RPh\dot{C}OH)$ arising in a reaction analogous to (16a), with methanol or toluene as hydrogen donor.

Polymers containing cinnamic acid residues find wide application in negative photoresist processes (34). Cinnamic acid, even in the solid state, undergoes photodimerization with formation of cyclobutane derivatives (37,38). In the crystalline state the reaction is controlled by topochemical factors, the reaction occurring with the minimum of atomic or molecular movement. Thus <u>trans</u>-cinnamic acid exists at room temperatures in a stable (α) form and a meta-stable (β) form; on irradiation these give α-truxillic and β-truxinic acids, respectively, (provided the temperature does not rise sufficiently for the $\beta \rightarrow \alpha$ transformation to be significant) (see equation (17)). Many related compounds also photodimerize in the solid state. Reactions of this type bring about cross-linking if the active groups are attached to a polymer chain and have been thoroughly studied from this point of view. For example, polyvinyl alcohol may be esterified with cinnamic acid, or cinnamic residues may be introduced into cellulose or epoxy resins. Equation (18) represents the basic cross-linking process schematically for polyvinyl cinnamate. Photodimerization is brought about either by direct light absorption or photosensitization. Curme, Natale and Kelley (39) concluded from spectroscopic studies that in both cases excitation of the double bond to the (n,π^*) triplet state is involved. Since the triplet, having some degree of radical character, may also abstract hydrogen from suitable donors, free-radical polymerization would also be expected and indeed evidence

HOOC H H Ph $\xrightarrow{h\nu}$ α-truxillic acid

α-form (crystal) (17)

$\xrightarrow{h\nu}$ β-truxinic acid

β-form (crystal)

for the existence of radicals in the irradiated polymers is provided by esr observations of Nakamura and Kikuchi (40). However, Sonntag and Srinivasan (41) decmonstrated the presence of α-truxillic residues in the products of hydrolysis of the cross-linked polymer, so that there is no doubt about the participation of photodimerization.

Photosensitization is of practical importance since it extends the spectral sensitivity range of the polymer. For efficient sensitization the triplet energy of the sensitizer should be higher than that of the cinnamate; further, the triplet state of the sensitizer should be formed in high yield by intersystem crossing and have a relatively long life ($>10^{-2}$s) (42). The presence of a heavy atom such as bromine raises the probability of intersystem crossing and so improves the efficiency of sensitization. Typical sensitizers are benzophenone, Michler's ketone, p-dimethylamino-nitrobenzene, 2,6-dibromo-4-nitro-1-dimethylaminobenzene and N-acetyl 4-nitro-1-naphthylamine. As would be anticipated from the mechanism described, a high sensitizing efficiency is correlated with high phosphorescent intensity of the sensitizer (43).

Polymers carrying pendant chalcone and coumarin groups which also cross-link through photodimerization are illustrated in (VI) and (VII), respectively. Other examples have been listed by Schryver and Smets (43a). Photodimerization of coumarin gives different products, (VIII) or (IX), according to the conditions. (VIII), (IX) probably arise from reactions of the ground state with the lowest excited singlet and the lowest triplet, respectively (44).

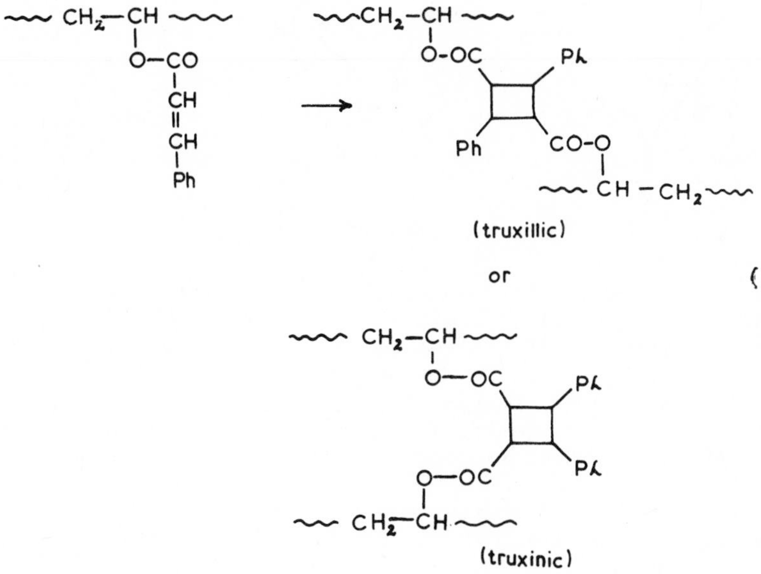

(truxillic)

or (18)

(truxinic)

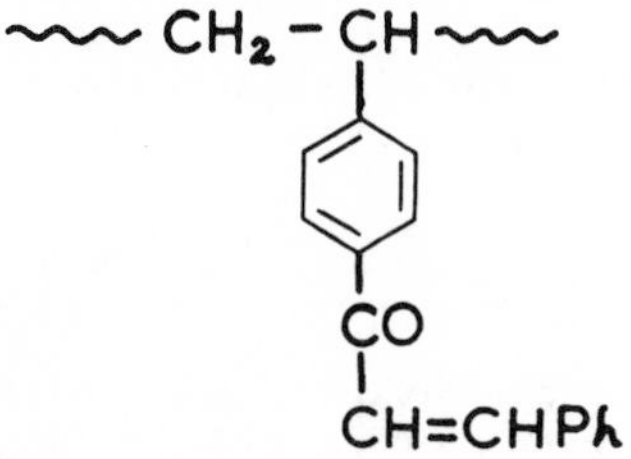

(VI)

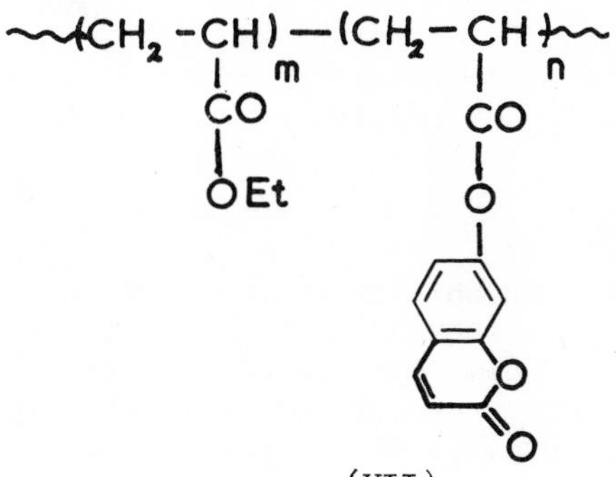

(VII)

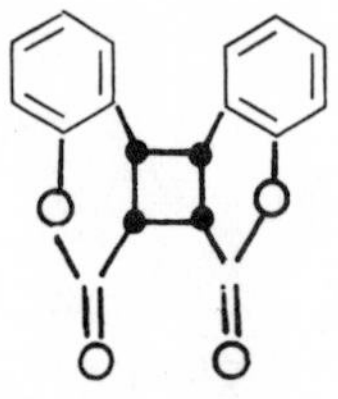

cis-dimer (VIII)   trans-dimer (IX) $CH=CH_2$ (X)

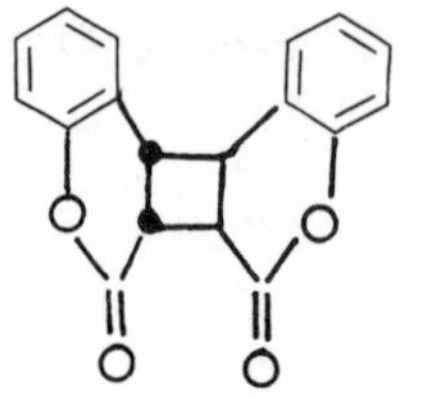

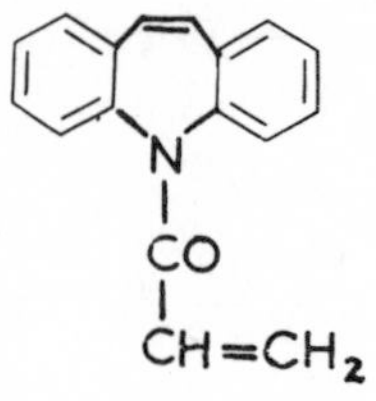

Recently Hyde, Kricka and Ledwith (45) in our laboratories have shown that N-acryloyl dibenz[b,f]azepine (X) when homopolymerized, or polymerized with styrene, gives a polymer which photocross-links by cyclobutane ring formation. Cross-linking may be sensitized by benzophenone and is retarded by oxygen, and so presumably proceeds through triplet states. The photodimerization resembles the stepwise photo-polymerization of bis-maleimides (46).

Photocross-linking systems may also be based on polymers containing azide groups (47); of the many compounds available examples are shown in (XI) and (XII).

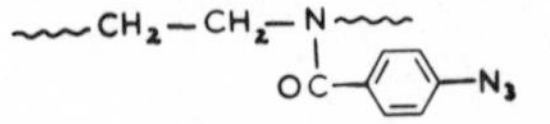

(XI)

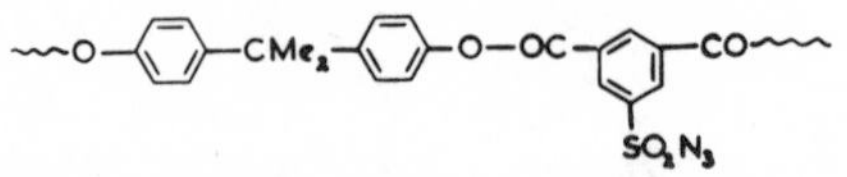

(XII)

Photolysis of arylazides often proceeds through elimination of nitrogen and formation of nitrene intermediates (48):

$$Ar{-}N_3 + h\nu \longrightarrow Ar{-}N + N_2 \qquad (19)$$

Nitrenes (which show analogies to carbenes in their behaviour) are highly reactive and enter into different types of reactions. Reiser et al (49c)have demonstrated by flash photolysis studies that triplet nitrenes derived from aromatic azides may recombine to form azo derivatives, or partake in the processes shown below.

Hydrogen abstraction from solvent,

$$\begin{aligned} R{-}N + HR' &\rightarrow R\dot{N}H + \dot{R}' \\ R\dot{N}H + HR' &\rightarrow RNH_2 + R'^{\bullet} \end{aligned} \qquad (20)$$

followed by recombination of radicals.

Hydrogen abstraction from azide, or a second nitrene,

$$HRN + HRN_3 \longrightarrow HRNH - RN_3 \longrightarrow \text{polymer}$$
$$HRN + HRN \longrightarrow HRNH - RN \longrightarrow \text{polymer} \qquad (21)$$

Reaction with azide with N_2 elimination,

$$RN + RN_3 \longrightarrow RN = NR + N_2 \qquad (22)$$

The relative importance of these processes depends on the conditions; thus high nitrene concentration favours recombination, high azide concentration enhances polymer formation and so forth. Insertion of nitrenes into C - H and some other bonds can also occur (50). In general, the precise mechanisms of nitrene reaction is less well understood than those of the carbenes. However, it is clear that the processes mentioned are capable of leading to cross-linking in polymer systems. The decomposition of azides has been shown to be photosensitized both by transfer of singlet and triplet energy (51), e.g. from excited singlet naphthalene and triplet benzophenone, respectively.

A review of polymer reactions involving nitrenes has been given by Imoto and Nakaya (52) and Delzenne (53) has published a more general account of the synthesis and properties of photoactive polymers.

4. Photoisomerization Processes

It has been known for some time that _cis_ / _trans_ rearrangements of polymers containing double bonds can be effected photochemically. Thus Voigt (54) showed that rearrangement occurs when unsaturated polyesters containing fumaric and maleic units are irradiated by ultraviolet light with eventual establishment of a photostationary state favouring maleic acid.

The photoisomerization of _cis_- to _trans_- 1,4-polybutadiene was observed by Golub (55); this process may be sensitized by a variety of organic bromine compounds, sulphides and mercaptans and then probably proceeds by a free-radical mechanism as was demonstrated with model compounds (56). Participation of charge-transfer complexes between double bonds and the sensitizers has also been suggested (57). Direct irradiation of films of 1,4-poly-butadiene and 1,4-polyisoprene induces _cis/trans_ isomerization; with the latter polymer irreversible reactions leading to the formation of cyclopropyl groups also occur (58,59). The mechanism of _cis/trans_ isomerization in a number of low-molecular-weight compounds has been shown to involve triplet excitation by Hammond et al (60).

The well-known photo-induced trans → cis isomerization of azo compounds has also been shown to occur in polymers both for pendant azo-groups (61) and azo-groups located in the main chain. Schulz and his colleagues (62) have investigated several systems of the latter type, for example the polymer with units of type (XIII) below.

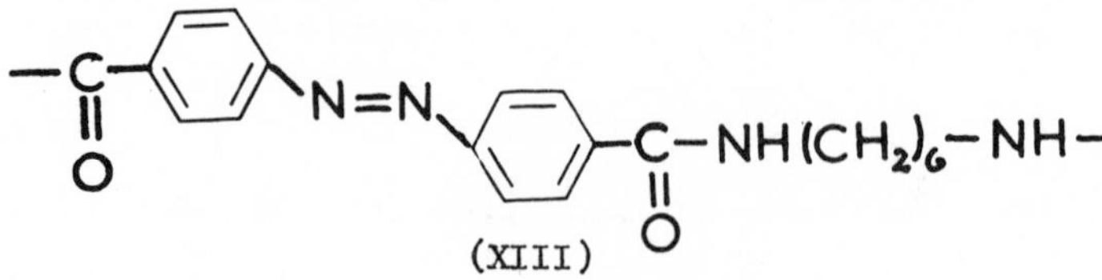

$$-\underset{\underset{O}{\|}}{C}-C_6H_4-N=N-C_6H_4-\underset{\underset{O}{\|}}{C}-NH(CH_2)_6-NH-$$

(XIII)

Irradiation of solutions of the polymer in hexafluoroisopropanol by light with λ = 365nm brings about a great decrease in absorption corresponding to the trans form; a subsequent dark reaction exhibiting first-order kinetics causes reversion to the all-trans polymer. Examination of low-molecular weight models under the same conditions revealed no significant differences in rate coefficients or activation energies between the polymer and model systems. It was therefore concluded that each azo group in the polymer chain behaves independently; probably the distance between such groups in the polymer is too great to allow significant interaction.

The cis/trans isomerization of azo compounds is an example of photochromism - a reversible change in absorption spectrum produced by irradiation. The interesting question is to what extent photochromic phenomena are influenced by incorporation of the photochromic group in a polymer chain. Clearly, in general, it would be expected that differences between low-molecular-weight systems and polymers would be greatest when the photochromic change requires segmental movements in the polymer chains. Smets and his colleagues (63) have synthesized polymers containing spiropyran groups, a typical unit being shown in (XIV). Irradiation causes rupture of the C-O pyran bond, and subsequent rotation of a part of the molecule leads to the planar form (XV). Polymers based on (XIV) do not behave very differently from model compounds containing the same group, although the polymer decolorizes more slowly on irradiation. On the other hand, marked differences have been observed with polytyrosine containing spiran substituents in the side chains, the photochromic behaviour of the polymer being much less affected by the polarity of the solvent.

An interesting photochemical rearrangement in a polymer - N-chloronylon 66 (XVI)-has been reported by Schulz and his colleagues (62). On irradiation by the unfiltered light of a high pressure mercury arc in solution or in a solid film the content of N-chloroamide decreases and unsubstituted amide groups appear. At the same time the total chlorine content remains almost constant. The chlorine atoms are therefore shifted into the main chain,

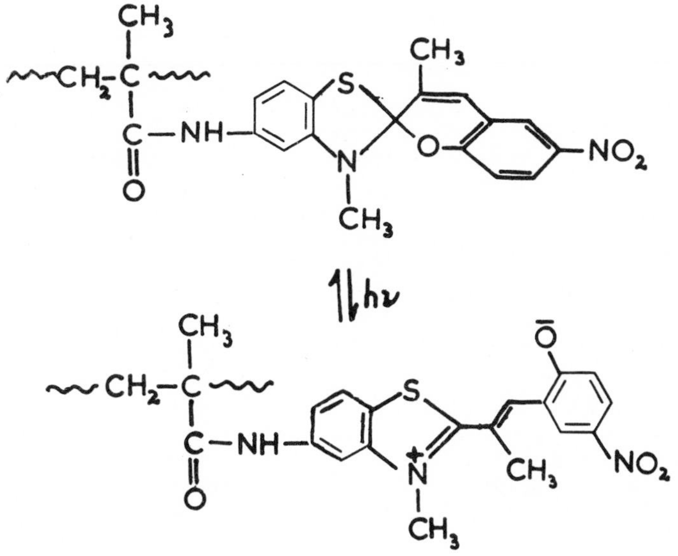

(XIV)

(XV)

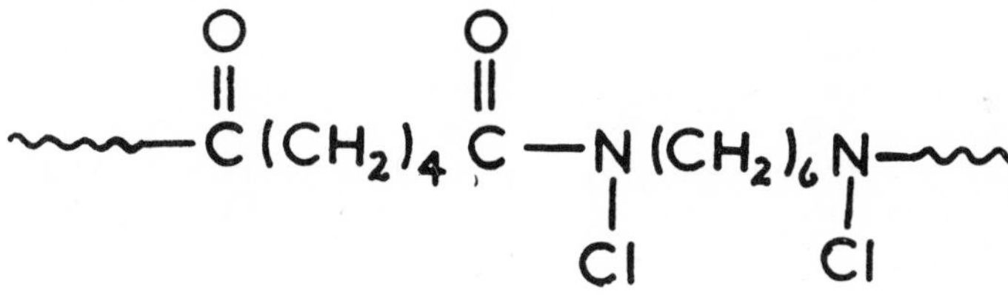

(XVI)

probably mainly into the diamine segments, although a mixture of products is formed. When the photorearrangement is carried out in styrene or acrylonitrile polymerization occurs, suggesting the intervention of free-radical intermediates. Two points of practical interest arise. The rearrangement is accompanied by a strong decrease in polymer solubility, since the regenerated amide groups form strong intermolecular hydrogen bonds. Further, the decrease in N-chloroamide content implies a reduction in the ability of the polymer to oxidize potassium iodide. Schulz (62) has demonstrated that the latter change can be the basis of an image-forming process, since, after "development" of the exposed film with aqueous potassium iodide, iodine appears only in the exposed areas and there produces a deep brown or black coloration.

5. Photosensitization by Polymers

In the preceding discussion we have encountered numerous examples of the tranfer of excitation energy from a polymer molecule to a molecule of a quenching agent. Such findings naturally suggest the possibility of photosensitization by polymers, in which excitation energy of the polymer molecule is transferred to a suitable reactant. There do not appear to be many investigations of this phenomenon. Searle et al (64) reported the isomerization of stilbene photosensitized by polyvinyl benzophenone; they compared the quantum yields for the *trans* $\rightarrow$ *cis* and *cis* $\rightarrow$ *trans* reactions in benzene with those of the same processes photosensitized by the model compound 4-methyl benzophenone and found no significant differences. One suggested explanation of this result was that the large size of the polymeric sensitizer was exactly compensated by a reduction in the diffusion-controlled collision rate. More recently, Hammond et al (65) in the same laboratories have studied the stilbene isomerization photosensitized by polystyrenes in which approximately two-thirds of the phenyl rings were substituted with α - or β-npahthoyl residues. Again the polymer-sensitized reactions were found to have the same quantum yields as reactions sensitized by low-molecular-weight models, except for the *trans* $\rightarrow$ *cis* reaction sensitized by the α -naphthylated polymer. In this case the polymer was less efficient as a photosensitizer than the model, and moreover its efficiency increased somewhat with decreasing molecular weight. The results were interpreted in terms of steric considerations, with the suggestion that the excited *trans*-stilbene molecules are confined in a polymer cage which restricts the isomerization. The difference between the reactions of the *cis*-and *trans*-derivatives was connected with the existence of a positive volume of activation for isomerization of the latter.

PHOTOSENSITIZED REACTIONS OF POLYMERS

1. Grafting and Cross-linking

Generation of free radicals in a system containing macromolecules may lead in principle to the formation of radical sites on the polymer backbones on side chains by transfer reactions and so may bring about grafting or cross-linking. Cross-linking by chain propagation through reactive double bonds attached to polymer molecules is a familiar and important process, for example with unsaturated polyesters. Since free-radical formation may be photosensitized by a variety of processes, photosensitized grafting and cross-linking are frequently encountered. However, reactions of this type are not properly the subject of the present review in the absence of specific features associated with photosensitization; such features essentially involve reaction of the macromolecules either with the photosensitizer or species formed directly from it by light absorption.

Examples of the latter processes are provided by the grafting of acrylamide chains to natural rubber photosensitized by benzophenone (66) and the grafting of methyl methacrylate to natural rubber by the use of rubber latex containing 1-chloranthraquinone as photosensitizer (67). In these systems grafting is presumably initiated from radical sites generated by hydrogen abstraction from the rubber molecules by the excited sensitizer. Geacintov et al (68) demonstrated that dyes which are active in phototendering can induce grafting to cellulose. By use of anthraquinone-2,7-disulphonic acid as photosensitizer they obtained grafts of methyl methacrylate, styrene and vinyl acetate containing weights of these monomers over three times the weight of the cellulose.

In our laboratories we have studied the initiation of polymerization by organometallic derivatives of transition metals in association with organic halides (e.g. CCl_4) and have used these reactions in synthesizing graft and block copolymers and networks. For details of the potentialities and chemistry of these initiating systems the reader may consult recent reviews by Bamford (43a,69), which also give earlier references. Here we shall present only an outline of the main features.

Radical formation involves electron-transfer from the metal $\underline{M}$ to the halide R-X; as a result the metal assumes a higher oxidation state and the halide is cleaved into an ion and a radical fragment:

$$\underline{M}^{0} + X - R \longrightarrow \underline{M}^{I}X^{-} + R\cdot \qquad (23)$$

The overall reaction (23) does not normally occur directly between the organometallic component and halide under the conditions normally holding in polymerization studies. Kinetic investigations show that the first step is ligand exchange between the organometallic derivative and monomer (M), or other electron-donating species present, resulting in the formation of an active complex (I') which participates in a redox reaction with the halide. In thermal reactions the ligand exchange may be S_N1 or S_N2 in character, depending on the nature of the system; typical examples are illustrated in equations (24) and (25).

$$S_N1: \ Ni\{P(OPh)_3\}_4 \rightleftharpoons Ni\{P(OPh)_3\}_3 + P(OPh)_3$$

$$Ni\{P(OPh)_3\}_3 + M \longrightarrow \underset{(I')}{Ni\{P(OPh)_3\}_3 M} \qquad (24)$$

$$S_N2: Mo(CO)_6 + M \rightleftharpoons \underset{(I')}{M-Mo(CO)_5} + CO \quad (25)$$

Many systems are photochemically active. Thus with metal carbonyls of Group VIA formation of (I') follows the course exemplified in equation (26)

$$\begin{aligned} Mo(CO)_6 + h\nu &\rightleftharpoons Mo(CO)_5 + CO \\ Mo(CO)_5 + M &\longrightarrow (I') \end{aligned} \quad (26)$$

The carbonyls of Group VIIA, particularly those of manganese ($Mn_2(CO)_{10}$) and rhenium ($Re_2(CO)_{10}$) are of particular interest as photoinitiators. The long-wave limits of absorption are approximately 460 and 380nm, respectively, and photosensitization with unit quantum efficiency occurs up to these wavelengths in the presence of relatively low halide concentrations (e.g. $[CCl_4] > 10^{-2}$ mol l^{-1} for $Mn_2(CO)_{10}$). A considerable amount of evidence supports the view that the primary dissociation is unsymmetrical, as represented by equation (27), and that radical generation follows reaction of one type of fragment with the halide,

$$Mn_2(CO)_{10} + h\nu \longrightarrow Mn(CO)_4 + Mn(CO)_6 \quad (27)$$

$$Mn(CO)_4 + CCl_4 \longrightarrow \dot{C}Cl_3 + Mn(CO)_4Cl \xrightarrow{CO} Mn(CO)_5Cl \quad (28)$$

Formation of CCl_3 radicals from CCl_4 has been established by polymerization experiments with carbon tetrachloride enriched with $^{14}CCl_4$; the expected concentration of labelled endgroups is found in the polymer. Other halides behave analogously. If the halide group is part of a preformed polymer molecule these reactions lead to the formation of radicals attached to the chains, and, in the presence of a polymerizable monomer, to grafting or cross-linking (equation (29)). Equation (29a) represents the initiation of a propagating chain; termination by combination (equation (29b)) or disproportionation (equation (29c)) yields crosslinks or grafts. Polymer networks are readily synthesized by these reactions. If the halide group occupies a terminal position of the preformed polymer chain similar reactions may be employed to synthesize block copolymers.

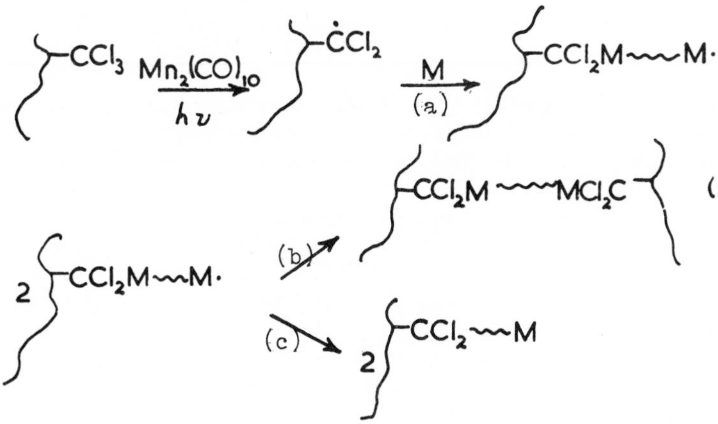

$$\sim CCl_3 \xrightarrow[h\nu]{Mn_2(CO)_{10}} \sim \dot{C}Cl_2 \xrightarrow[(a)]{M} \sim CCl_2M \sim\sim M\cdot$$

$$2 \sim CCl_2M \sim\sim M\cdot \xrightarrow{(b)} \sim CCl_2M \sim\sim MCl_2C \sim \qquad (29)$$

$$\xrightarrow{(c)} 2 \sim CCl_2 \sim\sim M$$

A wide range of composite polymers may be prepared in this way. We have been particularly interested in the properties of network and block copolymers containing chains with widely differing properties, for example combinationsof hard and rubbery polymers, or of hydrophobic and hydrophilic chains. Examples of preformed polymers containing suitable halide residues are shown in (XVII) - (XIX). Polyamides containing N-chloro units such as (XVI) are also

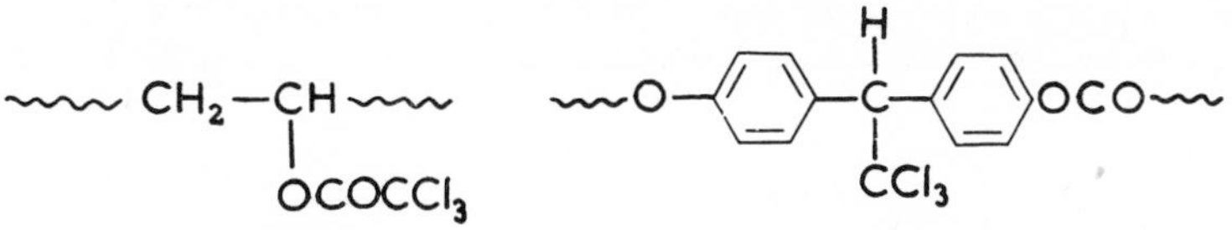

$$\sim CH_2-CH(OCOCCl_3) \sim$$

(XVII)

$$\sim O-C_6H_4-CH(CCl_3)-C_6H_4-OCO \sim$$

(XVIII)

$$\sim(CH_2-CH(C_6H_5))_m-(CH_2-CH(C_6H_4OCOCCl_3))_n\sim$$

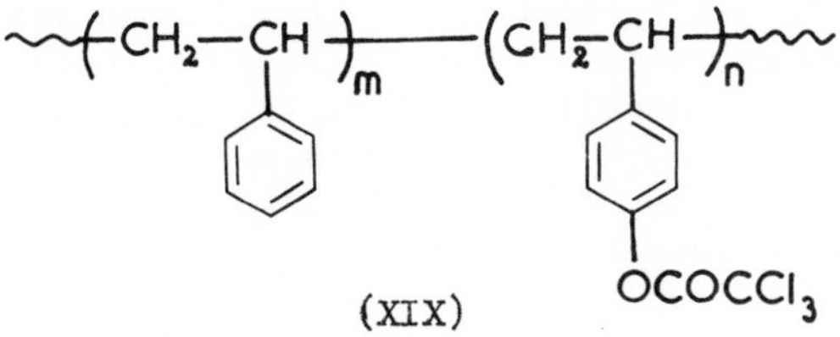

(XIX)

very reactive prepolymers. Methyl methacrylate, styrene and chloroprene have been employed as cross-linking monomers in networks.

From the practical aspect these systems generally have the advantage of yielding products effectively free from homopolymers of the second component (M). In networks the mean cross-link length of the second component is controllable by adjustment of the rate of initiation or the concentration of M and the cross-link density is determined inter alia by the duration of reaction. If chain-transfer is effectively absent the relative numbers of cross-links and branches depend on the nature of the termination reaction as indicated in equations (29b,c). Addition of a sufficiently high concentration of transfer agent suppresses network formation and the reaction products then contain the simple graft copolymer together with homopolymer of M. Lower concentrations of transfer agent give rise to network copolymers containing both cross-links and branches of M. Further, the presence of a low-molecular-weight halide during polymerization produces unattached radicals which enter into the termination reaction and yield branches. The cross-link : branch ratio is thus controllable.

It is clear that statistical control of the network structure is obtainable. Obviously the chain-length distribution is broad compared to that in simple anionic polymerizations but the synthesis has the advantage of being very versatile.

Finally we note that in suitably designed experiments measurement of the gel time provides a method for determining the ratio of combination to disproportionation in free-radical polymerizations.

2. Photosensitized additions

Schulz and his colleagues (62) have provided a number of examples of addition reactions involving polymers with reactive double bonds, for example unsaturated polyesters. These workers showed that tetrahydrofuran, tetrahydropyran, 1,3-dioxolane, 1,4-dioxane and trioxane can be added to a polyester (prepared from maleic anhydride and ethylene glycol) with the aid of acetophenone as photosensitizer. The reaction probably follows the free-radical mechanism elaborated by Rosenthal and Elad (70) for the benzophenone-photosensitized addition of cyclic ethers to double bonds, the first step being hydrogen abstraction from the ether by the excited sensitizer (cf. equation (11a)). The products from the polymer reactions were soluble, except in the case of tetrahydropyran addition where same cross-linking occurred.

Schulz has also reported the addition of a carbene to polyester double bonds; the former was prepared by photolysis of diazofluorene and the product contained spirocyclopropane groups.

CONCLUSION

We have endeavoured to illustrate by typical examples rather than to review exhaustively the diversity of direct and sensitized photochemical reactions involving macromolecules. Developments in the theoretical aspects of electronic energy transfer and our knowledge of the photochemical behaviour of small molecules enable many of the phenomena encountered to be interpreted and rationalised. The practical importance of polymer photochemistry is clear, not only in the further development of industrial processes, but also in efforts to eliminate the familiar disadvantages of polymer systems, notably their sensitivity to radiation and their deleterious effects on our environment.

REFERENCES

1. An account of these and related matters has been given by A. Cox and T.J. Kemp in Introductory Photochemistry, McGraw-Hill, London, 1970.

2. Th Förster, Disc. Faraday Soc. 27, 7 (1959).

3. R. F. Cozzens and R.B. Fox, J. Chem. Phys. 50, 1532 (1969).

4. R.B. Fox and R.F. Cozzens, Macromolecules 2, 181 (1969).

5. R.B. Fox, T.R. Price and R.F. Cozzens, J. Amer. Chem. Soc. 54, 79 (1971).

6. F. Hirayama, J. Chem. Phys. 42, 3163 (1965).

7. M.T. Vala, Jr., J. Haebig and S.A. Rice, J. Chem. Phys. 43, 886 (1965).

8. R.B. Fox, T.R. Price, R.F. Cozzens and J.R. McDonald, J. Chem. Phys. 57, 534 (1972).

9. J. R. McDonald, W.E. Echols, T.R. Price and R.B. Fox, J. Chem. Phys. 57, 1746 (1972).

10. A.S. Dunn, B.D. Stead and H.W. Melville, Trans. Faraday Soc. 50, 279 (1954).

11. N. Grassie, Pure and Appl. Chem. 34, 247 (1973).

12. C.H. Bamford and R.C.W. Norrish, J. Chem. Soc. p.1531 (1938) idem ibid p.1544 (1938).

13. W. Davis, Jr. and W.A. Noyes, Jr., J. Amer. Chem. Soc. 69, 2153 (1947); F.O. Rice and E. Teller, J.Chem.Phys. 6, 489 (1938).

14. G.R. McMillan, J.G. Calvert and J.N. Pitts, Jr., J. Amer. Chem. Soc. 86, 3602 (1964); R. Srinivasan, J. Amer.Chem. Soc. 81, 5061 (1959).

15. P.J. Wagner and G.S.Hammond, Advan. Photochem, 5, 98 (1968).

16. N.C. Yang and S.P. Elliott, J. Amer. Chem. Soc. 91, 7550 (1969).

17. J.E. Guillet and R.G.W. Norrish, Proc. Roy. Soc. A 223, 153 (1955).

18. K.F. Wissbrun, J. Amer. Chem. Soc. 81, 58 (1959).

19. Y. Amerik and J.E. Guillet, Macromolecules 4, 375 (1970).

20. A.C. Somersall and J.E. Guillet, Macromolecules 5, 410 (1972).

21. E. Dan, A.C. Somersall and J.E. Guillet, Macromolecules 6, 228 (1973).

22. C. David, W. Demarteau and G. Geuskens, Polymer 8, 497 (1967).

23. E.J. Baum, J.K.S. Wan and J.N. Pitts, Jr., J. Amer. Chem. Soc. 88, 2652 (1966).

24. C. David, W. Demarteau and G. Geuskens, European Polymer J. 6, 1405 (1970).

25. M. Inokuti and F. Hirayama, J. Chem. Phys. 43, 1978 (1965).

26. F.J. Golemba and J.E. Guillet, Macromolecules 5, 212 (1972).

27. P.J. Wagner and A.E. Kemppainen , J. Amer. Chem. Soc. 99, 5896 (1968).

28. I. Lukáč, P. Hrdlovič, Z. Maňásek and D. Belluš, J. Polym.Sci. A1 9, 69 (1971).

29. (a) G.H. Hartley and J.E. Guillet, Macromolecules 1, 165 (1968); (b) ibid 1, 413 (1968).

30. M. Heskins and J.E. Guillet, Macromolecules 3, 224 (1970).

31. J.E. Guillet, Naturwiss 59, 503 (1972).

32. F.J. Golemba and J.E. Guillet, Macromolecules 5, 63 (1972).

33. P.I. Plooard and J.E. Guillet, Macromolecules 5, 405 (1972).

34. J.L.R. Williams, Topics in Current Chemistry 13.2, 227 (1969).

35. C. David, W. Demarteau and G. Geuskens, Polymer 10, 21 (1969).

36. H. Sumitomo, K. Nobutoki and K. Susaki, J. Polymer Sci. A-1 9, 809 (1971).

37. M.D. Cohen and G.M.J. Schmidt, J. Chem. Soc. 1996 (1964).

38. J. Bregman, K. Osaki, G.M.J. Schmidt and F.I. Sonntag, J. Chem. Soc. 2021 (1964).

39. H. G. Curme, C.C. Natale and D.J. Kelley, J. Phys. Chem. 71, 767 (1967).

40. K. Nakamura and S. Kikuchi, Bull Chem. Soc. Japan 41, 1977 (1968).

41. F. I. Sonntag and R. Srinivasan, Technical Papers, p.163 Regional Technical Conference, Society of Plastics Engineers 6 Nov. 1957.

42. N.M. Moreau, Polymer Reprints, Amer. Chem. Soc. 10, 362 (1969).

43. M. Tsuda, Bull Chem. Soc. Japan 42, 905 (1969).

43a. In reactivity, Mechanism and Structure in Polymer Chemistry, ed. A. D. Jenkins and A. Ledwith (Wiley, New York, in press)

44. G.S. Hammond, C.A. Stout and A.A. Lamola, J. Amer. Chem. Soc. 86, 3103 (1964).

45. P. Hyde, L.J. Kricka and A. Ledwith, Polymer Letters, in course of publication.

46. F.C. Schryver, N. Boeus and G. Smets, J. Polymer Sci. A-1 10, 1687 (1972).

47. S.H. Merrill and C.C. Unruh, J. Appl. Polym. Sci. 7 , 273 (1963).

48. P.A. Smith, in Nitrenes ed. W. Lwowski, Interscience, 1970.

49. (a) A. Reiser, G. Bowes and R.J. Horne, Trans. Faraday Soc. 62, 3162 (1966); (b) A. Reiser, H.M. Wagner, R. Marley and G. Bowes, Trans. Faraday Soc. 63, 2403 (1967); (c) A. Reiser, F.W. Willets, G.C. Terry, V. Williams and R. Marley, Trans. Faraday Soc. 64, 3265 (1968).

50. M. Takebayashi and T. Shingaki, Kagaku (Kyoto) 26, 602 (1971).

51. J.S. Swenton, T.J. Ikeler and P.H. Williams, J. Amer. Chem. Soc. 92, 3103 (1970).

52. M. Imoto and T. Nakaya, J. Macromol. Sci. Rev. Macromol.Chem. 7, 1 (1972).

53. G.A. Delzenne, European Polymer J. Supplement 55 (1969).

54. J. Voigt, Z. Phys. Chem. 209, 255 (1958).

55. M.A. Golub, J. Polymer Sci. 25, 373 (1957).

56. C. Sivertz, J. Phys. Chem. 63, 34 (1959).

57. W.A. Bishop, J. Polymer Sci. 55, 827 (1961).

58. M.A. Golub and C.L. Stephens, J. Polymer Sci. C 16,1, 765 (1967).

59. M.A. Golub and C.L. Stephens, J. Polymer Sci. A-1 6, 763 (1968).

60. G.S. Hammond, J. Saltiel, A.A. Lamola, N.J. Turro, J.S. Bradshaw, D.O. Cowan, R.C. Cownsell, V. Vogt and C.Dalton, J. Amer. Chem. Soc. 86, 3197 (1964).

61. R. Lovrien and J.C.B. Waddington, J. Amer. Chem. Soc. 86, 2315 (1964) : R. Lovrien, Proc. Nat. Acad. Sci. Wash. 57(2), 236 (1967).

62. R.C. Schulz, Pure and Appl. Chem. 34, 305 (1973).

63. P. H . Vandewyer and G. Smets, J. Polymer Sci. A-1 8, 2361 (1970); G. Smets, Pure and Appl. Chem. 30, 1 (1972).

64. R. Searle, J.L.R. Williams, J.C. Doty, D.E. DeMeyer, S.H. Merrill and T.M. Laakso, Makromol. Chem. 107, 246 (1967).

65. H.A. Hammond, J.C. Doty, T.M. Laakso and J.L.R. Williams, Macromolecules 3, 711 (1970).

66. G. Oster and O. Shibata, J. Polymer Sci. 26, 233 (1957).

67. W. Cooper and M. Fielden, J. Polymer Sci. 28, 442 (1958).

68. N. Geacintov, V. Stannett, E.W. Abramson and J.J. Hermans, J. Appl. Polymer Sci. 3, 54 (1960).

69. C. H. Bamford, European Polymer J- Supplement 1 (1969)

70. I. Rosenthal and D. Elad, Tetrahedron 23, 3193 (1967).

DISCUSSION SESSION

Discussion Leader - R. B. Fox

Turner: Would it be possible to prepare block copolymers by generating free radicals from polymers terminated with $-C\,Cl_3$ groups by treatment with manganese carbonyl in the presence of a second monomer?

Bamford: Yes, we have prepared block copolymers by this method. Either $-C\,Cl_3$ or $-CBr_3$ terminal groups may be employed, the latter having the advantage of much higher reactivity. Polymers with a single reactive terminal group per molecule yield the ABA type of block copolymer with monomers in which termination occurs by combination of radicals and the AB type with monomers in which exclusive disproportionation occurs. If the preformed polymer carries reactive groups at both ends of the molecule, more complex block structures may be formed, e.g. those of type (AB)n. We have synthesized ABA copolymers in which A=polystyrene and B= polychloroprene and are examining these materials as thermoplastic elastomers.

Turner: Are homopolymers formed?

Bamford: Under suitable conditions no homopolymers are formed, since all the initial radicals generated from the halide groups are attached to chains of the preformed polymer. Homopolymer will arise if chain-transfer is significant, or if low molecular weight halides which act as initiators are present. An exception to the former statement arise when nickel derivatives (e.g. $Ni[P(OPh)_3]_4$) are used with halides such as ethyl trichloroacetate; in this case some unattached radicals, perhaps chlorine atoms, are produced. The effective absence of homopolymer formation in suitable systems is a great advantage by this method of synthesis.

Challa: Regarding excimer emission in polystyrene: you said in isotactic polystyrene having high helix content, excimer emission was less than in amorphous polystyrene because of decreased interaction between rings in the helix. Then a less stereoregular isotactic polystyrene, having less helical structure, should also give excimer emission?

Bamford: Yes.

Challa: Can you have intermolecular excimers when polystyrene is highly isotactic and its helix content prevents intramolecular excimer formation ? At 50° C phenyl groups begin to rotate more extensively. Above 50° will excimer emission begin to decrease in polystyrene?

Bamford: Except in very concentrated solutions, intramolecular excimers are strongly favored. This has been emphasizid by Dr. Fox in his work on polyvinylaromatic systems. I would expect the excimer emission to decrease when excessive rotation sets in.

Fox: You need a generally face-to-face arrangement of the aromatic groups for excimer formation. As a bimolecular interaction, it is temperature dependent. With motion, in fluid solution there will be at any given time some face-to-face conformations. At room temperature with vinylaromatic polymers like polystyrene and poly-(1 vinyl naphthalene) where energy can be rapidly delocalized along the chain to a potential excimer site, the bulk of prompt emission will be excimer fluorescence, although some molecular fluorescence may also be seen. As the temperature of the solution is lowered excimer fluorescence will decrease until none can be observed. Now, if a solid solution (e.g. poly(vinyl naphthalene) in poly (methylmethacrylate))is formed at a temperature at which excimer fluorescene can be observed, then a decrease in temperature may not change this emission since the excimer-forming conformation are rigidly fixed. Under these conditions ordinary fluorescene may increase because its quantum yield generally increases as temperature is lowered. In bulk polystyrene, not in solution, one may observe intermolecular as well as intramolecular excimer formation.

Challa: Do you expect higher excimer emission in atactic than in isotactic polystyreme?

Fox: Not in solution, since you have sufficient motion to produce excimer-forming sites in both polymers. The rate of energy delocalization is apparently sufficiently high that if a site is present, even for a very short time, it will generate excimer emission.

Moore: Dr. Fox, how do you differentiate experimentally among all these photophysical process?

Fox: We utilize the emission spectrum itself, i.e. the specific bands observed and the lifetime measured from the decays of specific emissions. Initially, excitation usually generates an excited singlet state, which is short-lived (~100nsec or less). Its emission is fluorescence. Excimer emission just discussed, can also form, and too is short-lived; it appears as a broad band at wavelengths longer than for fluorescence, since the eximer is a lower energy species than is the corresponding singlet. Fluorescence is observed at wavelengths slightly longer than the longest wavelength absorption band: the o-o bands of the absorption and fluorescence should overlap, although they may not appeas as strong bands in the spectra. In principle, fluorescence would be the mirror image of absorption, but this is not always observed for various reasons. The singlet, after intersystem crossing yields an excited triplet species whose emission is called phosphorescence. The triplet as a longer-lived species with lifetimes varying from 10^{-5} to 10 or even 100sec; depending on the molecule. Its energy is less than that of the excimer and its emission spectrum will appear at longer wavelengths than that of the excimer. Emissions originating from a singlet or a triplet species are resolved by placing a rotating sector in the light path and with the short-lived or "prompt" singlet emissions thus eliminated, the long-lived or "delayed" triplet emissions alone can be observed. If an observed emission is from a single species, an exponential decay curve can be measured and the lifetime of the species can be determined. If the emission is generated from more than one species with differing rates of decay, or bimolecular quenching or other bimolecular interactions are involved, a non-exponential decay curve will often result. Emissions thus far discussed are monophotonic. Two triplet species, resulting from the absorption of two photons, can interact to yield an excited singlet and a ground state singlet. These can combine in place to yield an excimer (a complex between an excited and a ground state singlet of the same molecular species), or the excited singlet can generate fluorescence. In both cases, the emission will now be long-not short-lived, because its lifetime is dependent on that of the triplets from which it was formed. These emissions are called delayed excimer fluorescence and delayed fluorescence, respectively. Their intensities are proportional to the square of the intensity of the exciting light, whereas the monophotonic emission intensities are proportional to the first power of the exciting light intensity. When the excited singlet and the ground state singlet are different chemical species, yet another emission is produced. It is called exciplex fluorescence, and can be prompt or delayed, depending on the manner in which it originated. All of these emissions have been observed with polymers. We can distinguish among energy-trapping impurities not attached to polymer chains, attached to a few polymer chains or impurities, such as end-groups, that are part of every polymer chain, if we follow the emission spectra as a function of polymer concentration. This can be done wherever energy migration along polymer chains occurs, as it does in the poly (vinyl aromatic) polymers. Where energy is trapped in a chain, the locus of trapping or delocalization might be called a "defect". Physical conformations, including chain folding, may well determine the kind of emission that will be observed.

Moore: Can glassy polymers be used as matrices in which to observe emission, instead of the low-temperature glass-forming solvents?

Fox: This is a common technique; polymethylmethacrylate is the usual polymer.

Lewin: Can emissions due to aggregations be detected? I am interested in possible applications of this technique to the study of heterogeneous nucleation, chain folding, etc..

Fox: In organic crystals and in liquids or concentrated solutions in which crystals are forming, spectra have been compared, and particularly decay curves of the emissions have been analyzed in this way. In principle it should be possible to do what you suggest.

NEW METHODS OF PREPARATION OF TAILOR MADE POLYMER NETWORKS

Paul Rempp

Centre de Recherches sur les Macromolecules - CNRS
6,rue Boussingault - 67083 STRASBOURG CEDEX - France

INTRODUCTION

Polymer networks have been devoted much interest for a number of years. Swelling behavior, stress strain properties, stress birefringence, elastomosis, as well as diffusion properties of small molecules in gels, and gel permeation chromatography phenomena have been largely investigated in many laboratories. However most of the experimental work was carried out on gels prepared either by polycondensation processes, or by radical copolymerization of two monomers one of which being bifunctional. The samples thus obtained are far from being structurally well defined for a number of reasons:

- Copolymerization is a random process. The length of the linear segments between two successive branch points fluctuates around its average value. These fluctuations are often quite large, especially in those cases in which the difunctional monomer is consumed more rapidly than the other one.

- In some cases syneresis may occur: if the amount of solvent present exceeds the equilibrium swelling degree of the gel when conversion is quantitative, some of the solvent is expelled in the form of small droplets, which make the gel turbid and inhomogeneous. Syneresis is sometimes desired, especially for networks meant for GPC, since the macroporosity of the gels has been assumed to be a condition for satisfactory chromatographic separation.

- Even when there is no syneresis, it should be recalled that the gel-point occurs usually at low conversion, i.e., when only a small proportion of the monomers present have reacted. At this stage the obtained gel is swollen by both solvent and residual monomers. As conversion proceeds the gels become more and more tight. One can say therefore that the gel-point has no physical significance. On the other hand the "memory-term" which appears in all theoretical treatments of the swelling behavior of the gels cannot be related with any measurable quantity.

THEORETICAL ASPECTS OF THE EQUILIBRIUM SWELLING DEGREE OF A GEL

All theoretical approaches of the equilibrium swelling degree of a gel in a pure solvent require a few hypotheses concerning the structure and the homogeneity of the network:

- No syneresis should occur: the segment concentration should be constant within the whole gel.

- The network should be gaussian, which means that all its linear chain elements should obey gaussian statistics.

- The network should be ideal, which means that all its linear chain elements should be attached by their two ends to two different branch points. There should be neither pendent chains, nor loops, nor bifunctional branch points. If this is true the total number of elastic chains per ml of dry network is given by (1):

$$\nu_e = \frac{1}{M\, v_3^0}$$

where M is the average molecular weight of the linear elements and v_3^0 is the specific volume of the dry gel.

To evaluate the equilibrium swelling degree of a gel one can calculate the free energy of the system, which comprises one "dilution term" involving the thermodynamic interaction constant between solvent and polymeric segment χ, and one term characterizing the elastic deformation of the network (2,3). The chemical potential of the solvent in the gel can be written:

$$\mu_1 - \mu_1^0 = RT\left|\ln(1-v_3)+v_3+\chi v_3^2\right| + RT\nu_e\,\overline{V}(Ah_3^{2/3}\,v_3^{1/3} + Bv_3)$$

when equilibrium is reached the chemical potential of the solvent $\mu_1 = \partial\Delta F/\partial N_1$ is the same in the gel and in the surrounding solvent. Then it becomes:

$$\nu_e = \frac{\ln(1 - v_3) + v_3 + \chi v_3^2}{V_1(Bv_3 - Ah_3^{2/3}\, v_3^{1/3})}$$

In this expression:

v_3 : is the volume concentration of polymeric segments, i.e., Q_3^{-1}

χ : the interaction parameter

V_1 : the molar volume of the solvent

h_3 : is the so-called memory-term, i.e., the volume fraction of the segments in the gel when the chains are in the relaxation state

A and B are constants. According to DUSEK and PRINS (3), the best fitting values are

$$A = 1 \qquad \text{and} \qquad B = \frac{2}{f}$$

where f is the average functionality of the branch points.

To test the validity of these theoretical treatments it is therefore of interest to synthesize model-networks in which the molecular structure is unambiguously known. This was the starting point of several recent research projects carried out in several laboratories, and especially at the CRM in Strasbourg. The following goals were attempted:

- To make gels in which the length of the linear chain elements fluctuates as little as possible around its average value.

- To make gels of great homogeneity in which no syneresis happens to occur, and in which the segment density may be considered to be constant.

- To make gels under such conditions that the memory-term may have a physical significance, and may be related thus to the concentration of polymeric segments at which crosslinking occurred.

- To make gels in which the functionality of the branch points is constant all over the gel.

If the obtained gels have all these structural properties together, it may be possible to check for the validity of the expected swelling behavior, and to relate their properties with their structure.

PREPARATION OF TAILOR MADE NETWORKS BY ANIONIC BLOCK COPOLYMERIZATION

As everybody knows, anionic polymerization carried out in aprotic solvents, using organometallic initiators may yield polymers of known chain length and of narrow molecular weight distribution curves.

It was tried a few years ago to make star shaped polymers by anionic block copolymerization (4). Starting from a monofunctional anionic polystyrene (BuLi + styrene), and using this 'living' polymer to initiate the polymerization of a small amount of divinylbenzene, it was possible to get such star polymers. Each molecule is made of a small nodulus of poly-DVB, connected with p branches which are, in first approximation identical in length. It was shown by ZILLIOX and WORSFOLD (5) that the average value of p is a function of the overall concentration of polymer at the time of addition of DVB. It was shown also that, if adequate mixing was attained, the fluctuations over p were rather small within a sample (6).

If instead of using a monofunctional initiator one takes a bifunctional initiator for the first step of this process one gets a bifunctional living polymer, fitted at the two ends of its chain with organometallic sites, able to initiate further polymerization. If again a small amount of DVB is then added to the solution the polymerization of DVB will induce gellation, because each precursor chain will be connected with two DVB noduli, and it thus becomes an elastic chain (7). The network thus consists of N linear chain elements connected with 2 N/f, f-functional branch points (the DVB noduli).

The experimental conditions have to be chosen such as to allow adequate mixing to occur prior to the gellation. Also, care has to be taken to choose the proper conditions to allow rapid initiation of the DVB, as compared with propagation, in order to avoid pendent chains to be formed.

As a matter of fact, the gels prepared thus, by anionic block copolymerization of styrene and divinylbenzene, are not, rigorously speaking, model-networks:

- A few of the chain-ends may have been accidentally protonated; these chains are connected only by one end to a branch point; they are called pendent chains and do not contribute therefore to the elasticity of the network.

- It may also occur that the two ends of a given chain react with the same DVB nodulus, and form a loop. A loop is not an elastic chain, and it does not contribute to the elasticity of the network.

- It may also happen that two-functional branch points are formed: a DVB nodulus may be connected to two chain ends only, or at least to only two independent chain-ends (loops apart). Such a branch point is in fact merely a coupling site, and it cannot be counted as an elastic nodulus. Moreover, the elastic chain has twice the average length, which introduces a new source of polydispersity.

- Finally, it should be pointed out that this method does not allow adequate knowledge of the functionality of the branch points, f, nor of their number. It is not well controlled by the experimental conditions of preparation, and its value is not experimentally accessible.

By functionality f we mean, of course, the average number of elastic chains attached to one single poly-DVB nodulus, loops being excluded, as well as pendent chains.

The first experiments were devoted to the system styrene-DVB and many gels were made, in various solvent media, with precursors of various molecular weights, and at various overall conditions (the amount of DVB added was also varied, but it turned out that the amount of DVB involved has not a great influence on the swelling behavior of the gel).

In a second step we tried to generalize the procedure to other anionically polymerizable monomer systems. In doing that it is necessary to choose the bifunctional monomer in such a way that its polymerization may be initiated rapidly and totally by the living anions of the other monomer. This feature has been investigated thoroughly, by many authors who tried anionic copolymerizations. The systems we have studied are the following (8):

styrene	: DVB	
isoprene	: "	
vinylpyridine	: "	or (DMG)
methylmethacrylate	: DMG	
butylmethacrylate	: "	
isopropylidene glyceryl methacrylate	: "	

A few words of comment can be given here on some special problems which require some attention:

- Concerning gels of polyisoprene it is of course useful to have 1,4 cis steregular chain elements between branch-points, in order to make comparisons with vulcanized rubber. But this is not easy to achieve, since bifunctional initiators are not soluble in non-polar solvents; it is well known that to get 1,4 cis polyisoprene it is necessary to use Li initiators in non-polar solvents, only. As soon as polar solvents are added to help solubilize the initiator the stereoregularity is lost. Recent attempts to initiate polymerization with lithium metal (finely divided) have been successful.

- Water soluble polymers cannot be made anionically. But it was possible to synthesize isopropylideneglycerylmethacrylate, a rather bulky monomer, but which undergoes rather nicely anionic polymerization, and which may be block-copolymerized with DMG. Thereafter the acetal grouping may be destroyed to regenerate the two OH groupings of glycerol, thus making the gel compatible with water.

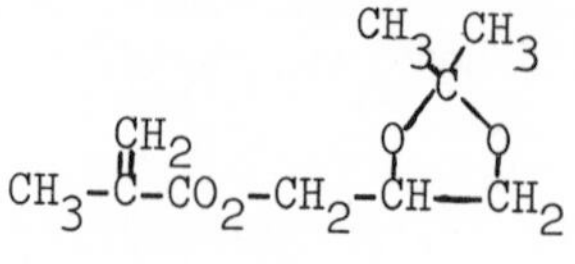

$$CH_3-\overset{\overset{CH_2}{\|}}{C}-CO_2-CH_2-\underset{O}{CH}-CH_2(O)-C(CH_3)_2$$

iPGMA

$$CH_3-\overset{\overset{CH_2}{\|}}{C}-CO_2-CH_2-CH_2-O_2C-\overset{\overset{CH_2}{\|}}{C}-CH_3$$

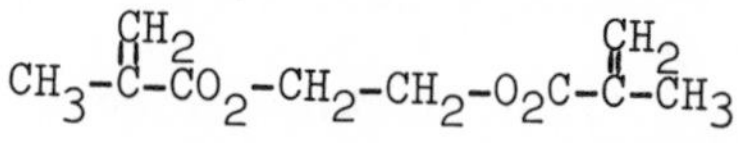

DMG

- Some attempts have also been made to chemically transform model-networks, in order to fit them with selected functions, namely ionic ones. No special problems arise with poly-2-vinylpyridine networks which can be treated with HCl or RCl to quaternize it.

Further experiments were carried out with polystyrene in order to sulfonate it. Two methods were attempted: sulfonation with sulfuric acid in the presence of silver sulfate, or by addition product of dioxane with SO_3. It appears that this latter method which was expected to be smooth and selective (in the para position) does not yield the desired products. The degree of swelling of the sulfonated species is much less than expected, owing to SO_2-bridge formation, which tightens up the network. But the sulfuric acid method yielded sulfonated polystyrene networks which did not seem to have undergone either chain scission, or additional bridge formation. It is interesting to notice however, that the degree of swelling of the sulfonated gels is much higher in water than that of the initial gel in benzene. This is due to the repulsive effect of the sulfonic anions along the chain.

EQUILIBRIUM SWELLING OF THE TAILOR-MADE NETWORKS

To investigate the swelling behavior of the obtained model gels we carried out equilibrium swelling measurements on a homologous series of model gels, prepared under identical conditions, the only parameter changing from gel to bel being the average length of the linear chain elements. Let us first consider the limitations imposed by the hypothesis we have made concerning the structure of these gels:

- Ideality of the gels - Since up to 2% of linear polystyrene can be extracted from the gels, it appears that about 12% of the anionic sites have been deactivated during the process. This means that 21% of the chains are linked by one end only, i.e., pendent chains.

- Memory-term - It was first thought that h_3 could be identified with v_3^c, the concentration at which crosslinking occurs. But DUSEK and PRINS (3), and JAMES and GUTH (10) have shown that, if account is taken of the influence of a crosslinking process on the radius of gyration of an individual linear chain, the concentration of h at which the chains are in the relaxed state should be connected with the concentration v_3^c at which crosslinking has occurred by the following relation:

$$h_3 = \left[\frac{\ell_r^2}{\ell_\ell^2}\right]^{3/2} v_3^c$$

Since $\frac{\ell_r^2}{\ell_\ell^2}$ is of the order of 0.5 it follows that:

$$h_3 = 0.354\ v_3^c$$

- Segment-segment interaction parameter χ - One could use the χ parameter obtained on studying the corresponding linear polymer in the same solvent, arguing that, no matter whether the segments are part of a network or part of a linear chain, the interaction in a given solvent should be the same. But recently it was shown that in branched polymers (star-shaped or comb-shaped (12)), owing to the high segment density in the polymer coil, the χ parameter cannot be considered anymore as a constant. In other terms, the two-parameter theory based on the single contact approximation has to be modified, owing to the non-zero probability of multiple contacts (13), when the segment density becomes high. Similar results have been obtained on studying concentrated solutions of linear polymers (14).

If these considerations are applied to the case of gels (9), it should follow that χ should become a function of the swelling degree. For high Q the χ value should be rather close to the χ value observed for linear polystyrene in benzene (0.456), and for low values of Q, χ should increase further

$$\chi = \chi_0 + a\, v_3 + b\, v_3^2$$

- <u>Functionality of the branch points</u> - As already indicated the functionality of the branch points is neither known from the preparation conditions of the network, nor experimentally accessible. The swelling formula may however help in getting the average f value, provided all the other quantities are known, namely, V_1, $Q = v_3^{-1}$, h_3, , M. The swelling equation can be written as follows:

$$f = \frac{2\, V_1\, v_3}{V_1\, h_3^{2/3}\, v_3^{1/3} + M\, v_3^0\, |\ell n(1 - v_3) + v_3 + \chi\, v_3^2|}$$

Prior to calculating f it is necessary to check whether is or is not a function of v_3. And this may be done by plotting as a function of $Q = v_3^{-1}$ according to:

$$\chi = \frac{V_1\, Q}{M\, v_3^0} \left|\frac{2}{f} - (h_3\, Q)^{2/3}\right| - \left|Q + Q^2\, n\, (1 - \frac{1}{Q})\right|$$

Assuming that for our series of gels h_3 and f are constant we can thus plot χ versus v_3. We get thus straight lines for each couple of values of f and h_3. The best fitting curves are those in which $h_3 = 0.354\, v_3^c$, (this confirms the assumption of DUSEK and PRINS) and in which f = 4 and f = 10. When f = 4 the extrapolated value χ_0 is very close to 0.45, i.e., the value measured for dilute solutions of polystyrene in the same solvent. From the slope of this line the following relationship can be established:

$$\chi = \chi_0 + 0.9\, v_3$$

Going back now to the functionality of the branch points we have investigated the variation of Q with M, the average length of the linear chain elements, and compared the curve with theoretical expectations for various plausible values of f. The swelling formula can be written:

$$M = \frac{V_1\left[\frac{2}{fQ} - \frac{h_3^{2/3}}{Q^{1/3}}\right]}{\bar{v}_3^0 \left|\ln(1 - \frac{1}{Q}) + Q^{-1} + \chi_0 Q^{-2} + a Q^{-3}\right|}$$

We have chosen f = 3,4,5 and we have compared these 3 theoretical curves with the experimental one. It can thus be seen that the best fitting curve is that with f = 4, though for tight networks f might be somewhat above and for loose networks, f might be be-tweem 3 and 4.

It should be remembered that here loops and pendent chains are not taken into account, since f is the number of elastic chains connected with one branch point. As a matter of fact, if we assume that there is on the average one pendent chain and one loop per branch point the average total functionality of a DVB nodulus in the gel would amount to 7, which seems reasonable.

STATIC AND DYNAMIC PARTITION COEFFICIENTS

Besides swelling of the gels in a pure solvent, the partition coefficient of a linear polymer in such gels was also investigated. This can be done in two ways:

- Static partition coefficients (15) are defined by the ratio of the concentration of linear polymer inside the gel to its concentration in the surrounding solution, at equilibrium. Account has to be taken of the fact that the degree of swelling is not the same in a pure solvent and in a solution of polymer in that solvent, owing to osmotic effects of deswelling. It was found that below a certain molecular weight K is almost equal to 1, and that above another value of M, $K_S = 0$. Partition occurs usually over about a decade or a little more. This can be interpreted by the fact that small molecules fit into all pores of the gel, whereas very big molecules are excluded from all pores. In the intermediate region K_S is a measure of the proportion of pores in which the linear polymer molecules may fit.

- Dynamic partition coefficients (16) are connected with gel permeation chromatography. It was interesting to establish whether macropores are necessary for a gel meant to be used for GPC. The tailor made polymer networks we have are homogeneous, with no macropores in them, and the pore size can be related with the molecular weight of the precursor. Since chromatographic separation was easily detected, using these gels as fillings in the chromatographic columns, it is obvious that the process is linked with penetration of the pores by the linear polymer in solution. However it must be indicated that there is a very big difference between K_S and K_D, the latter being defined as:

$$K_D = \frac{V_e - V_i}{V_t - V_i}$$

with V_t : total available volume

V_i : interstitial volume (between gel particles)

V_e : measured elution volume

It follows that the GPC process with our model networks proceeds very far from equilibrium, and that it is therefore difficult to compare measurements which are not carried out at constant flow rate, T, pressure, and with identical gel particle size.

It should also be pointed out that the mechanical behavior of these gels is not satisfactory and that for this reason the experiments are not easy to perform.

PREPARATION OF TAIL-MADE NETWORKS BY CHEMICAL REACTIONS ON TERMINAL GROUPS

Until now we have considered preparation of tailor-made networks in which the functionality of the branch points remains unknown; new efforts were devoted therefore to produce homogeneous networks in which both length of the chains and functionality of the branch points are known from the preparation procedure. Anionic block copolymerization of styrene and DVB cannot be used here, and we switched to deactivation processes of ω-functional polymers using plurifunctional electrophillic deactivators.

Here again we checked first the formation of star-polymer molecules, by killing a monofunctional living polymer with such a deactivator. Many authors have published similar experiments, using:

$SiCl_4$ — 1,2,4-tris(chloromethyl)benzene (CH_2Cl, ClH_2C, CH_2Cl) — 1,2,4,5-tetrakis(chloromethyl)benzene (ClH_2C, ClH_2C, CH_2Cl, CH_2Cl) — $(PNCl_2)_3$

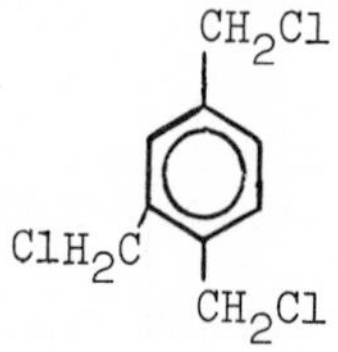

but in all cases the obtained products were mextures of star-molecules exhibiting p, p-1, p-2, ... branches, and the yields calculated for the stars with p branches remained quite low.

Two additional difficulties have been encountered:

- elimination reactions take place every time the CCl linkage is in α-position with respect to a hydrogen bearing carbon atom. This is the reason why benzylic chlorides have been widely used. They cannot give rise to elimination.

- exchange reactions are known to take place as side reactions when a halogenide (especially benzylic) is reacted with an organometallic site (especially when Li is the metal).

Our own attempts were carried out with the following 3 compounds (17,18):

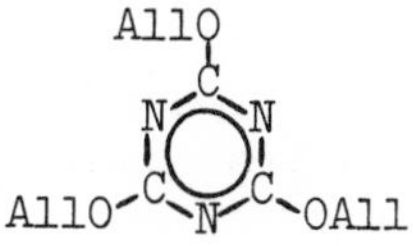

AllO, AllO, OAll (triazine ring: C, N)

I

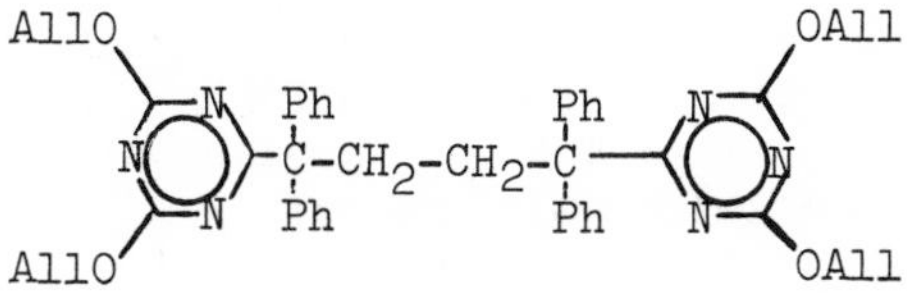

AllO, AllO — triazine — C(Ph)(Ph)-CH_2-CH_2-C(Ph)(Ph) — triazine — OAll, OAll

II

$C(CH_2X)_4$

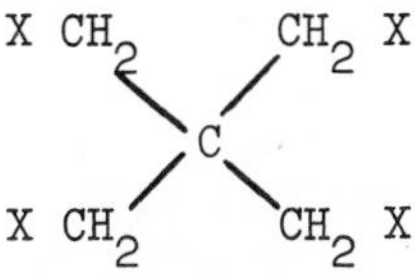

III

Trialloxytriazine (I) is a commercial product. Compound II is obtained by reaction of I in excess with the disodium derivative of diphenylethylene-dimer, in THF. Compound III was made in two steps: pentaerythritol and HBr in acetic acid yielded the tribromo compound; the latter was treated with PBr_3 to form III. The iodo compound III was made from the tetrabromo derivative, by treatment with LiI in butanone at 100°C.

All these compounds react stoichiometrically with the styryl anion of living polystyrenes. Stars with 3 and 4 branches were thus formed with good yields, and characterized as such by various techniques. It does not seem that the reactivity of an allyloxy function is decreased when the two other functions have already reacted.

If instead of reacting monofunctional living polystyrene with any of these compounds one uses a bifunctional living polymer one gets a gel. There are however two difficulties to overcome:

- Exact stoichiometry must be observed, which is not always easy, owing to the uncertainty on the molecular weight of any anionic polymer.

- Adequate mixing must be provided, and temperature has to be chosen such as to avoid the reaction occurring before a homogeneous mixture has been obtained.

Besides these difficulties of an experimental nature, it is necessary to consider two other points:

- Even under the best possible experimental conditions it is still possible that some pendent chains remain in the gel, as well as a few unused deactivating sites. It is not possible therefore to claim that all branch points are p functional, though most of them definitely are so.

- The possibility of loop formation cannot be excluded here, as it could not be in the anionic copolymerization process. It cannot be prevented that some polystyrene chains reacted by their two living chain ends with two deactivating sites of the same molecule.

Let us finally consider another method of tailor-made network formation, which does not proceed anionically, and which has led to very interesting results (19). It is based on the reaction of an α,ω-difunctional polymer with the p functions of a suitable reagent.

It is well known that - SiH linkages are reactive, especially towards allylic double bonds.

$$-SiH + CH_2 = CH - CH_2 - R \rightarrow - Si - CH_2-CH_2-CH_2-R$$

We were able to obtain a set of polydimethylsiloxane samples fitted with α,ω terminal SiH bonds. We have reacted each of these polydimethylsiloxane samples (mw ranging from 1700 to 20,000) with the following compounds:

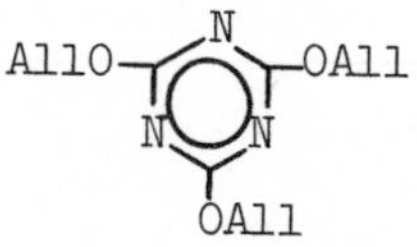

I: AllO, OAll, OAll substituted triazine ring (N, N, N)

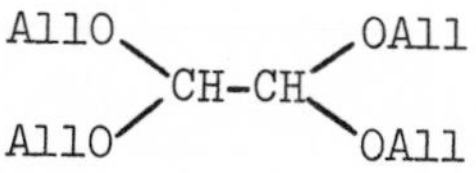

II: (AllO)(AllO)CH-CH(OAll)(OAll)

III: $C(OAll)_4$

I and II are commercial products; III is obtained in two steps: the triallyl derivative is obtained by reaction of pentaerytritol with allylbromide in concentrated aq. NaOH. This compound is reacted with Na, and then with allylbromide to yield III.

The crosslinking reaction is carried out either in bulk or in toluene solution, using strictly stoichiometric amounts of polydimethylsiloxane and of the allylic derivative (taking account of its functionality) in the presence of a small amount of H_2PtCl_6 which acts as a catalyst for the reaction.

Crosslinking takes place readily, at temperatures ranging from 20° to 75°C. It is noteworthy that when the concentration chosen is too low no gellation of the whole medium takes place.

When compound I is used the network first obtained spontaneously demolished, caused by slow acidic hydrolysis of the triazine rings (due to the presence of H_2PtCl_6). But when compound II is used the obtained gel is perfectly stable and can be studied as such. The network is rubberlike and its swelling degree in a pure solvent depends largely upon the concentration of the reactants during the crosslinking process.

SWELLING BEHAVIOR OF THE POLYDIMETHYLSILOXANE-MODEL GELS

It was shown that for a homologous series of networks prepared under identical conditions, with precursor polydimethylsiloxanes of different molecular weights, M, the equilibrium swelling degree is proportional to M . This can be explained from the swelling formula:

$$\frac{1}{M} = \frac{\bar{v}_3^0(\ln(1-v_3) + v_3 + \chi v_3^2)}{V_1 (\frac{2}{f} v_3 - h_3^{2/3} v_3^{1/3})}$$

If it is assumed that χ is equal to 1/2 and does not show any variation with the segment density, if the first three terms of the development of the log are taken, and if it is assumed that $h_3 = 0.354\ v_3^c$ one gets the expression (for f = 4):

$$Q = \left[\frac{2}{3\,V_1}\right]^{3/8} \frac{[v_3^o]^{3/8}}{{v_3^c}^{2/8}\left[1 - \left(\frac{v_3}{v_3^c}\right)^{2/3}\right]^{3/8}} M^{3/8}$$

The term in [ ] of the denominator still contains $v_3 = Q^{-1}$, but its variation is rather slow for the Q interval covered by these samples (Q = 3 to 8), and it can be neglected in first approximation. The value of the slope was found experimentally to be 0.21, whereas the theoretical value is about 0.215, in excellent agreement.

Thus it appears that the swelling behavior of these gels is rather different from that of the anionically made polystyrene gels. This is due to the fact that the latter are much less tight and that they consequently swell much more. For the polystyrene gels a linear function of Q versus $M^{3/5}$ approximates the results satisfactorily. Since $v_3 \ll 1$, one can write $\ln(1 - v_3) = - v_3 + v_3^2/2$, and also neglect, to a first approximation, the variation of χ with v_3.

CONCLUSION

This presentation of some recent results obtained at our Institute on preparation of tailor-made networks and on investigation of their swelling behavior was mainly meant to show that several methods are now available to prepared model gels, and that it is absolutely necessary to use such networks of well-known structure to test the validity of the theoretical expectations concerning their swelling behavior, the partition coefficients, etc. The first comparison made between theory and experiment yielded a rather satisfactory agreement, showing that the main hypothesis made in the theoretical treatment of DUSEK and PRINS seem to rest on solid ground.

But it should be emphasized once more that the limitations of preparative methods have not been entirely overcome, and that account has to be taken of inevitable defects in tailor-made networks, such as pendent chains, loops, and fluctuations over p and over M.

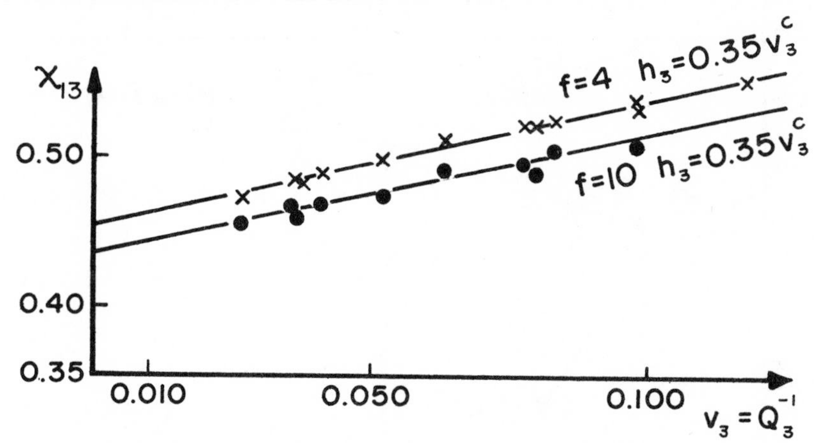

Fig. 1. Variation of χ versus v_3 for a homologous series of polystyrene gels, assuming f=4 and f=10. χ is calculated from the swelling data.

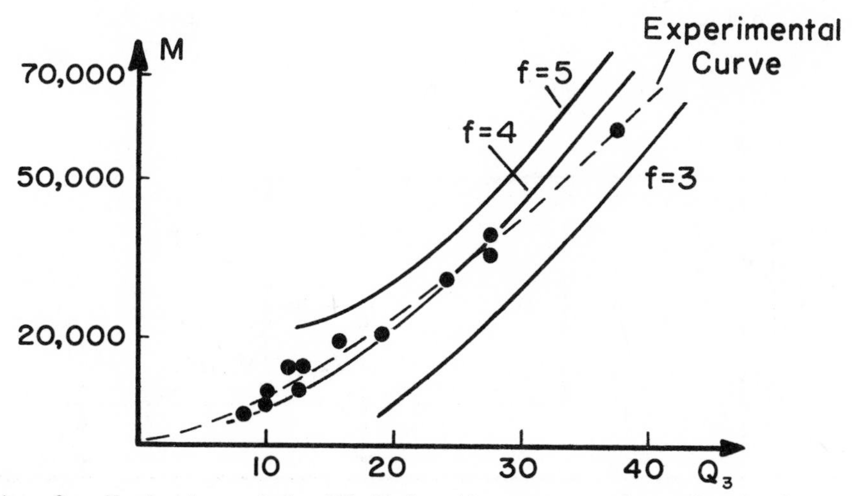

Fig. 2. Variation of Q with M for the same series of polystyrene gels. Dotted line : experimental curve. Full lines : calculated curves assuming f=3, 4, 5.

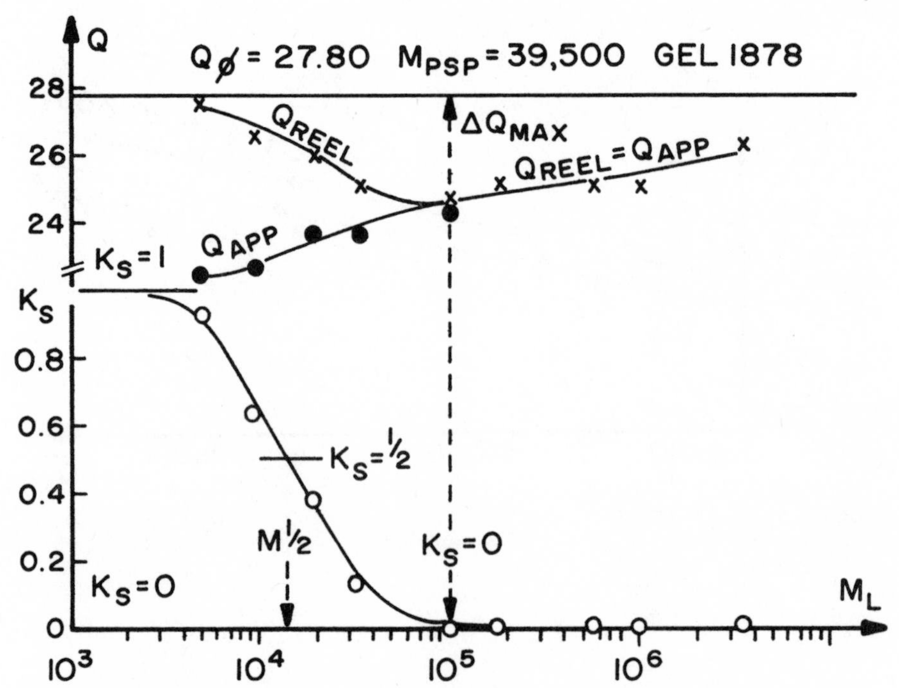

Fig. 3. Example of a static partition curve : K versus molecular weight of the linear polymer used. Upper curve shows the deswelling effect of the gell, due to osmotic effect.

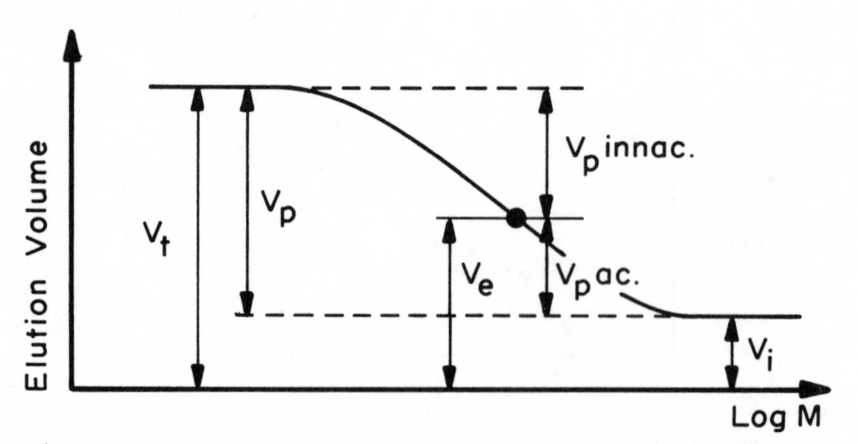

Fig. 4. Elution volume versus molecular weight ; K_D is the ratio of the accessible porous volume, $V_{p\ ac.}$, to the total volume of the pores.

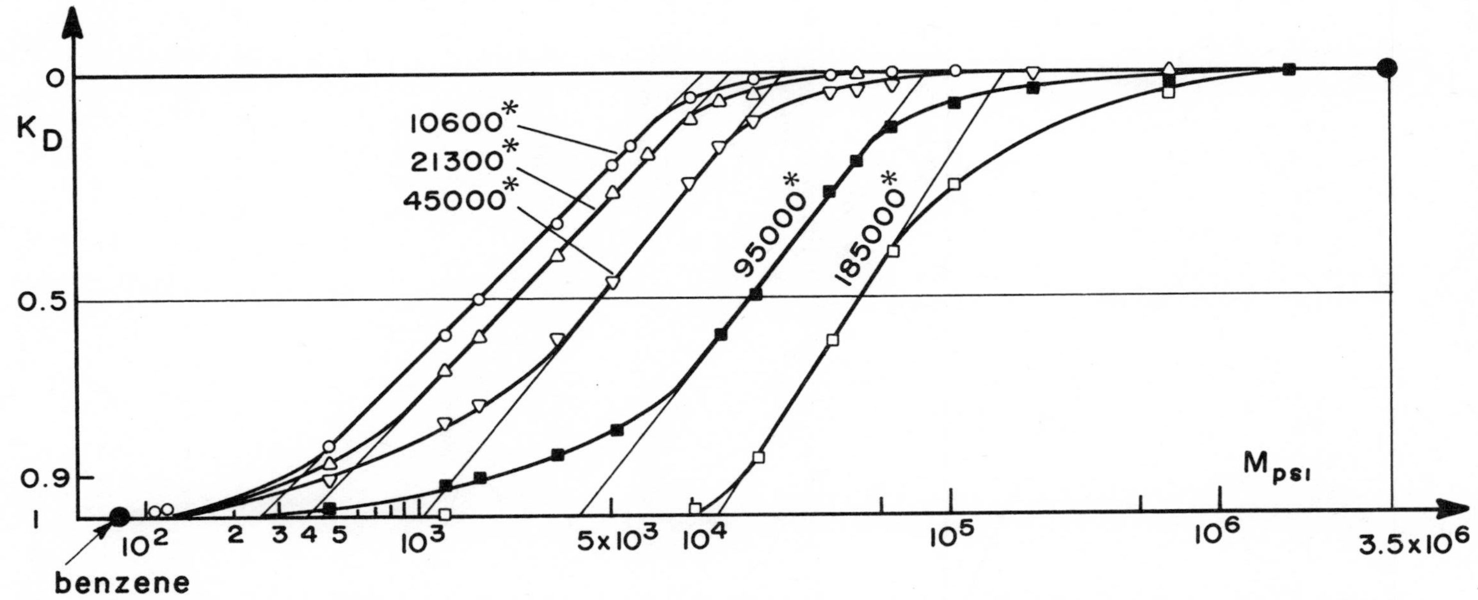

Fig. 5. G P C partition coefficients for a homologous series of polystyrene gels, plotted versus the molecular weight of the eluted linear polystyrenes. Numbers indicate the molecular weights of the precursor of each gel.

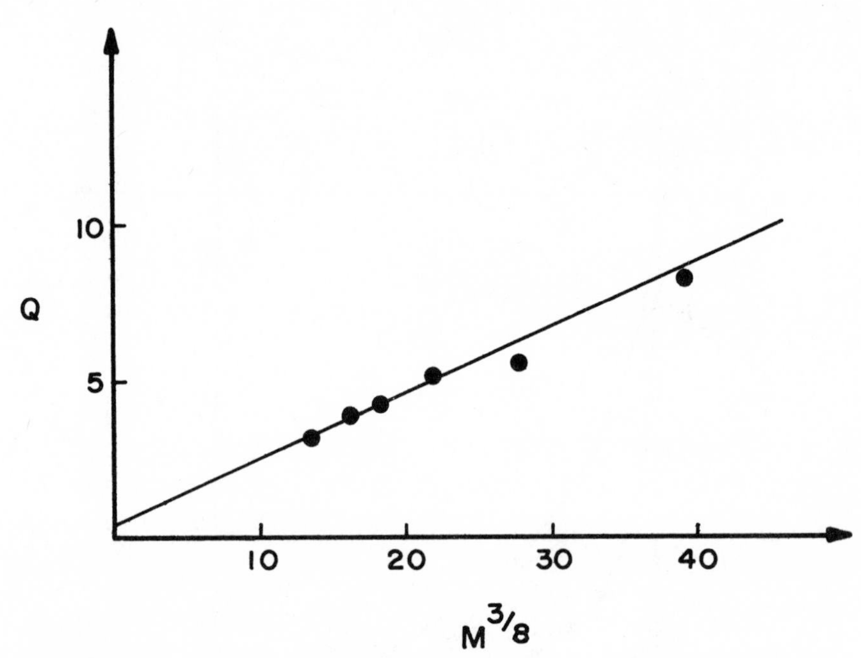

Fig. 6. Variation of the equilibrium swelling degree of the P D M S gels with the molecular weight of the precursors used.

REFERENCES

1) P. WEISS, P. REMPP and J. HERZ, Makromol. Chem. 141, 145 (1971).
2) P. J. FLORY, J. Chem. Phys. 18, 108 (1950).
3) K. DUSEK, W. PRINS, Adv. Polym. Sci. 6, 1 (1969).
4) J. G. ZILLIOX, P. REMPP and J. PARROD, J. Polymer Sci. C-22, 145 (1968).
5) D. J. WORSFOLD, J. G. ZILLIOX and P. REMPP, Can. J. Chem. 47, 3379 (1969).
6) A. KOHLER, J. POLACEK, I. KOSSLER, J. G. ZILLIOX and P. REMPP, European Polymer J. 8, 627 (1972).
7) P. WEISS, G. HILD, J. HERZ and P. REMPP, Makromol. Chem. 135, 249 (1970).
8) G. HILD, P. Rempp, C. R. Acad. Sci. Paris 269, 1622 (1969).
9) A. HAERINGER, G. HILD, P. REMPP and H. BENOIT, Makromol. Chem. (1973).
10) H. M. JAMES, E. G. GUTH, J. Chem. Phys. 15, 669 (1947).
11) J. G. ZILLIOX, Makromol. Chem. 156, 121 (1972).
12) F. CANDAU, P. REMPP and H. BENOIT, Macromolecules 5, 627 (1973).
13) H. BENOIT, D. DECKER, A. DONDOS and P. REMPP, J. Polymer Sci. C-30, 27 (1970).
14. see e.g., C. PICOT, H. BENOIT, Pure and Appl. Chem. 12, 545 (1966).
15) G. HILD, D. FROELICH, P. REMPP and H. BENOIT, Makromol. Chem. 151, 59 (1972).
16) P. WEISS, J. HERZ, Z. GALLOT and H. BENOIT, ibid. 145, 105 (1971).
17) J. HERZ, M. HERT and C. STRAZIELLE, ibid. 160, 213 (1972).
18) M. HERT, C. STRAZIELLE and J. HERZ, C. R. Acad. Sci. Paris C-276, 395 (1973).
19) M. A. BELKEBIR-MRANI, J. HERZ and P. REMPP, European Polymer J. (1973).

DISCUSSION SESSION

Discussion Leader - W. H. Daly

Moore: You had some trouble using bi-functional lithium compounds because of insolubility. Would you expect that crown ethers might be useful in dissolving them?

Rempp: We are presently working with crown ethers and cryptates, but we haven't done the work you are thinking of, yet. I'm not sure that it will be very useful because these compounds solvate the cation and will strongly increase the reactivity of the corresponding anion with respect to the tight ion pairs usually found in non-polar solvents. My fear is that the system will act as if it were in tetrahydrofuran or any polar solvent, so we would not gain anything.

Goethals: Heterocyclic compounds can be polymerized by cationic means giving "living" polymers, e.g., tetrahydrofuran. Have you considered this method to synthesize networks?

Rempp: We are presently studying this approach, but the results are not yet completed.

Carraher: I can envision a number of interesting "star"-type networks formed by the chelation of metal ions by polymers "capped" with Lewis base ligands, e.g., nylons capped with amine groups.

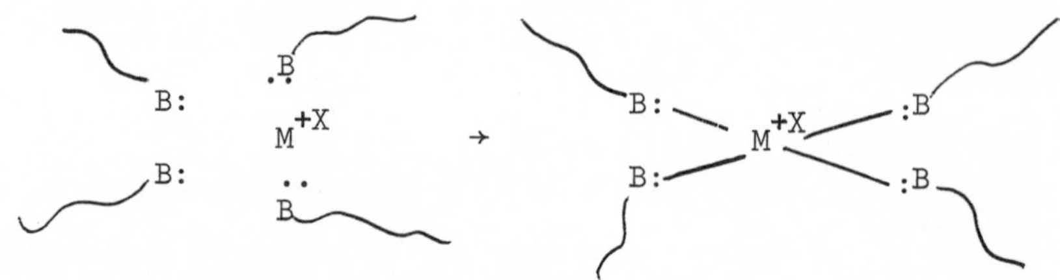

As an aside, we know that the addition of metal ions to polymer chains containing Lewis base groups, under proper conditions, causes the formation of presently uncharacterized precipitates. As our technology and knowledge increases, I hope we will return to these reactions to better describe the products and better tailor reaction systems. These reactions are easily carried out and may offer straight-forward routes to interesting and useful products.

Liu: The χ value is a function of v_3 but is actually inherently different. If you have a sequence of five units or twenty units their interactions with solvent or other segments should be different. When you involve entanglements and crosslinking, the difference will be enhanced. A linear molecule with a small amount of crosslinking can be treated as a dilute solution, as we have shown experimentally. When you have either crosslinked polymer or very concentrated polymer solutions, the solvent and the polymer will experience different environments. We can also show when the segment knows it is a particular type of polymer segment and when it is not. For example, if you imagine the crosslinked chain lengths between segments are for units of amino-acids, helices cannot form. However, if you make them longer helices will be able to form and with different types of structures the segment-segment or segment-solvent interactions will be very different. I think this suggests that a great number of interesting experiments which may be derived by controlling the segmental differences to study either the reaction kinetics or other types of interactions.

Rempp: I agree. However, we are working with distances between branch points of about 30 monomer units, never four.

Liu: I think it would be very interesting to study segment sizes below 30 because that will be a rather sensitive region to show how the segments can gradually recognize their origin.

Bamford: I think Prof. Rempp was being a little hard on free radical reactions and I fear I have to defend them. He was talking about the least favorable conditions which can be employed. It is possible to use better techniques, such as those we have developed. In our systems the gel point has real physical significance and the difficulties of monomer conversion are overcome. We use a preformed polymer as part of the network. I would agree with Prof. Rempp that for thermodynamic studies it is highly desirable to use the most precise network which can be made, but I think for many other things less precise things will probably do. If one, in addition, has to produce a network with polymers of very different types it would be very difficult to accomplish this by anionic means, but the free radical method is very versatile.

Rempp: I quite agree that anionic polymerization can only be performed with a limited number of monomers. We are also doing quite a lot of radical polymerization and I was just trying to point out schematically what are, in the less favorable cases, the limitations of radical crosslinking processes.

CHEMICAL AND BIOCHEMICAL REACTIONS IN THE GEL PHASE

R. Epton

Department of Physical Sciences,
The Polytechnic, Wolverhampton, (England)

Currently, there is an upsurge of interest in the immobilisation and regimentation of biologically active molecules, notably enzymes, anti-bodies and various bio-specific inhibitors, by conjugation with synthetic and semi-synthetic carriers. The resulting materials are of practical importance because they are recovered readily from heterogeneous reaction mixtures by filtration and, in certain cases, may be used continuously in perfused or fluidised beds. More important from a theoretical viewpoint, choice of a suitable carrier for the immobilisation process enables the researcher to define precisely the microenvironment of the constrained biological molecule. Simulation of the native, cellular environment of the molecule and, further, the creation of new, artificial environments, calculated to modify and perhaps improve, biological activity become distinct possibilities.

Typical semi-synthetic carriers available for the immobilisation process are those based on agarose, cellulose, cross-linked dextrans and certain proteins. Synthetic carriers include poly-amino acids, polyacetals, polyacrylamides, polymethacrylic acids, polymaleic acids,polystyrenes, silastic resins and porous glass. Numerous functional derivatives of these polymers are available. Also several chemical and physical binding techniques may be used. Consequently, the biological molecule may be immobilised in a variety of novel situations.

The obvious criterion of success for a given immobilisation procedure is the retention of biological activity. In this context, it is important to remember that *in vivo* biological reactions take place in an aqueous environment even though the reactants are constrained within a cellular membrane. Consequently, an essential condition for the retention of biological activity in synthetic conjugates is that the immobilised molecules should remain effectively in solution. Successful immobilisation involves constraining the biological molecule within a microenvironment approximating to that within an aqueous gel.

Many carriers are effective because, to a greater or lesser extent, they undergo gelation in water. This property also facilitates preliminary chemical modification of the carriers and the subsequent preparation of synthetic conjugates. In relatively few instances need we consider chemical and biochemical reactions at a simple interface between solid and solution. Even in these circumstances reactions take place involving molecules or molecular sites which are effectively in solution even though the molecule as a whole is anchored to the solid phase.

CLASSIFICATION AND MICROARCHITECTURE OF CARRIERS

It is a basic premise of the lecture that reactions involving immobilised biological molecules take place within the gel phase. By this I mean the gel phase in its very broadest sense. With this in mind it is convenient to adopt the classification commonly used by the gel chromatographer. Potential support matrices may be subdivided into three categories: xerogels, aerogels and xerogel-aerogel hybrids.

Xerogels

Xerogels are gels in the classical sense. They consist partly of pure solvent and partly of individual polymer chains in solution. A schematic representation is given in Fig. 1a. The solvated chains comprise thread-like volumes of solution whose movement, relative to one another, is restricted by cross-linking or physical interaction. Removal of the solvent from the xerogels results in collapse of the three-dimensional matrix of hitherto solvated polymer chains. Although the gels are apparently soft solids, diffusion of various solutes may occur within the gel just as in the case of a normal liquid.

Since the gelation process is one of restricted dissolution, a given polymer will only form xerogels in the limited range of solvents in which the individual polymer chains are inherently soluble. Consequently, polymers capable of forming aqueous xerogels are, of necessity, hydrophilic in character. This is fortuitous because most biologically active macromolecules are compatible only with hydrophilic matrices.

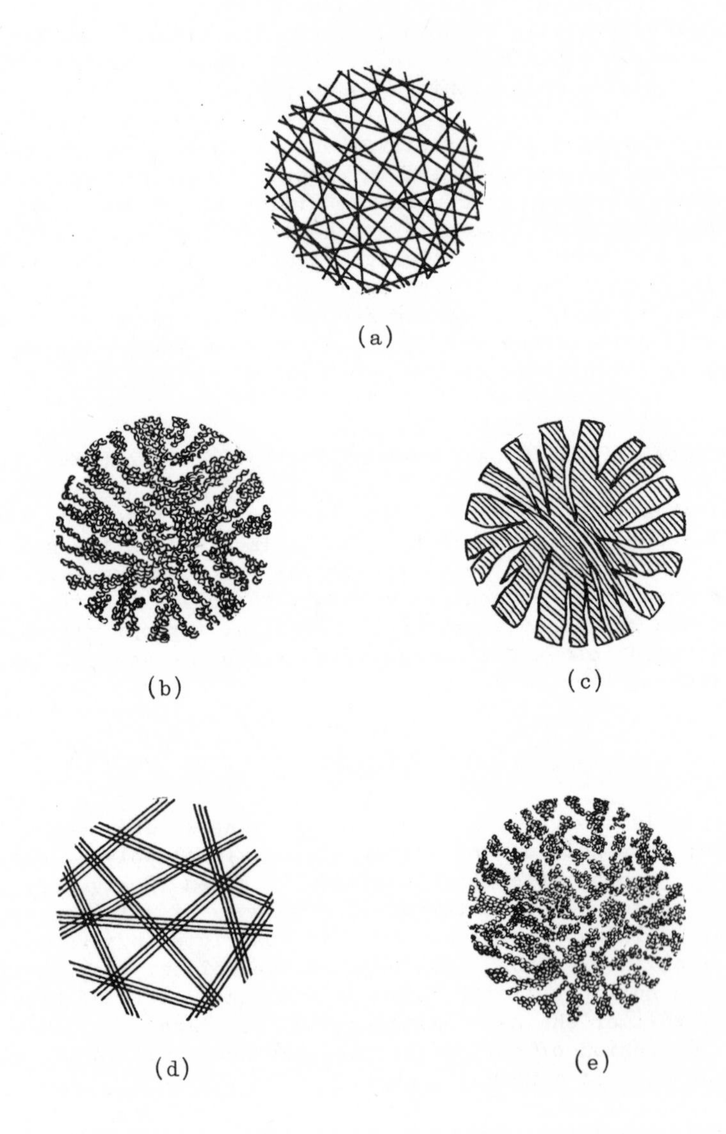

Fig. 1. Micro-architecture of (a) typical xerogel, (b) porous silica (aerogel), (c) porous 'silica rich' glass (aerogel), (d) agarose gel (xerogel-aerogel hybrid) and (e) typical macroreticular polymer (xerogel-aerogel hybrid).

Examples of xerogels which, after suitable derivatization, are commonly used to covalently bind biological molecules are cross-linked dextrans (1) and cross-linked polyacrylamides (2,3). Cross-linked polyacrylamide is probably the most suitable matrix for the immobilisation of biological macromolecules by entrapment (4,5). A number of ionic conjugates of the xerogel type are derivatives of ethylene-maleic acid copolymers (6). In these materials the ionic groups responsible for xerogel character are derived from anhydride residues during the immobilisation procedure

Aerogels

Aerogels are formed by introducing solvent into a rigid preformed macroporous matrix. This does not undergo constrained dissolution and consequently it cannot collapse or change in porosity on removal of the liquid component. The porous silica of Le Page _et al_ (7) and the porous, 'silica rich', alkali borosilicate glass of Haller (8) are examples of suitably hydrophilic matrices of this type (Fig. 1, b and c). Porous silica is produced by calcination of spheres of silica gel containing a small amount of sodium oxide. This results in particles built up to smooth, almost crescent shaped microstructures fused together in such a way as to form micro-cavities and channels. In the production of porous glass, suitable alkali borosilicate materials are prepared and tempered such that phase separation takes place. This leads to 'silica rich' and 'silica poor' regions. The microheterogeneous matrix is then leached with dilute hydrochloric acid to remove the 'silica poor' phase leaving a continuous system of uniform channels.

From a structural viewpoint the analogy between aerogels and xerogels is superficial. However, the microenvironments provided by two types of gel bear some resemblance insofar as an immoblised biological molecule is concerned. A molecule constrained at the interface between the solvent and solid aerogel matrix may interact with the latter in much the same way as with the solvated polymer chains of a typical xerogel. Porous glass, suitably derivatized, has been shown to be a particularly effective support for the immobilisation of enzymes (9,10).

Xerogel-Aerogel Hybrids

As the name implies, xerogel-aerogel hybrids have structural features common to both xerogels and aerogels. Many have a macroporous structure. This may arise either by cha[illegible] clustering during the actual gelation process or as a result of [illegible]ctural features already present in a preformed, semi-rigid ma[illegible]ix.

Usually, the latter are prepared by macroreticular polymerisation techniques.

Agarose gels are the only common examples of macroporous, aqueous xerogel-aerogel hybrids produced by chain clustering. (see Fig. 1d). Agarose is a seaweed polysaccharide composed of alternating 1,3-linked β-D-galactose and 1,4-linked 3,6-anhydro-α-L-galactose residues. When an aqueous solution of agarose is cooled below 50° gelation occurs. This involves hydrogen bonding between the individual polysaccharide chains and results in the formation of microscopic strands. Grouping of these strands into micro-crystallites produces a cross-linking effect. Chain clustering results in agarose being much more mechanically stable and having a much greater pore size than aqueous xerogels of comparable matrix dilution. As a result, agarose is a useful support for the covalent binding of species whose biological activity is directed towards macromolecules.

Unfortunately, agarose gels have some disadvantages. Because they "melt" on heating, sterilisation by heat is impracticable. This is serious because, being a natural polymer, agarose is very susceptible to microbial attack. Further, once prepared, the gel must be stored hydrated. In spite of these disadvantages agarose is the matrix of choice both for the immobilisation of enzymes active against macromolecular substrates (11) and also for the preparation of column packings destined for use in affinity chromatography (12).

Macroreticular polymers are prepared by copolymerisation of monomers in a solvent in which the polymer chains are only sparingly soluble. The polymer precipitates in the form of sub-microscopic spherical particles which fuse to form micro-agglomerates. These in turn are grouped together to give a system of pores and crevices within the material which generally has a semi-rigid structure. On introducing a solvent to constitute the gel, little dissolution of the matrix occurs and consequently swelling is minimal. Common polymers, such as polystyrenes and polymethacrylate esters, which may readily be prepared in macroreticular form are usually too hydrophobic to be of use in the preparation of biologically active conjugates. No matter what coupling procedure is employed, hydrophobic polymers are reluctant to react with hydrophilic biological molecules in aqueous solution (13). Furthermore , conjugation with a hydrophobic surface often results in impaired biological activity (14).

To date, there have been relatively few attempts to utilise hydrophilic, macroreticular materials for the preparation of biological conjugates. Reactive derivatives of ionic, hydrophilic polystyrenes and polystyrene-methacrylic acid copolymers have been prepared in macroreticular form and used to covalently bind enzymes (15). Recently, the preparation of hydrophilic, macroreticular, hydroxyethyl methacrylate polymers has been described (16). Following activation, these materials were used to prepare synthetic conjugates of enzymes and enzyme specific inhibitors (17). These materials proved excellent supports for affinity chromatography.

The term xerogel-aerogel hybrid, although originally coined to classify a group of macroporous chromatographic packings, also describes a number of heterogeneous materials which contain regions able to imbibe solvent. These materials may be considered to undergo restricted gelation. Cellulose and collagen are important examples of hydrophilic materials of this type.

Of all the materials available for the preparation of synthetic, biologically active conjugates, cellulose and its derivatives are most popular (18,19). Cellulose consists of linear chains of 1,4-linked β-D-glucose residues. In native cellulose the glucopyranose rings are orientated so that all three free hydroxyls may be involved in intercatenate hydrogen bonding (20). This results in regions of high crystallinity. Most of the individual cellulose chains are grouped into fibrils which, in turn, are aligned to form whole fibres. These consist of layers of fibrils spiralling in opposite directions along the fibre axis. Because alignment of the individual cellulose chains is frequently imperfect, irregularities in fibril structure occur. This is reflected in the structure of the whole fibre which consequently consists of poly-crystalline aggregates separated by amorphous regions.

Swelling of the cellulose matrix in water involves hydration and consequently a degree of constrained dissolution of the cellulose chains in the amorphous regions. This is important because it is in these regions of low three-dimensional molecular order that chemical reaction most readily occurs. Conjugation with biological molecules, particularly macromolecules, is most likely within these accessible, gelatinous sites rather than with the highly ordered, crystalline regions. It follows that any factor leading to a decrease in crystallinity will lead to enhanced binding capacity. For example, alkali treatment (mercerisation) of native cellulose leads to rotation of D-glucose residues about the glycosidic link. This results in the C-3 hydroxyl becoming unfavourably placed for intercatenate hydrogen bonding thereby reducing cohesion in the crystal lattice and, coincidentally, increasing accessibility. The amorphous character of native cellulose may also be increased by treatment with certain organic solvents, notably ethylamine.

The popularity of cellulose and its derivatives as carriers for the immobilisation of biological molecules probably arises by reason of tradition. Much of the successful early work was based on cellulosic materials. In practice, cellulose is one of the more difficult carriers to use. Covalent coupling of reactive cellulose derivatives with biological molecules and subsequent washing procedures are complicated by adsorption effects (21). Furthermore, the physical form of the resulting conjugates is frequently not ideal for use in packed beds or stirred reactors. Paradoxically, some of the more satisfactory biologically active cellulose conjugates are those prepared from micro-crystalline cellulose (22,23). This is a particulate material produced by controlled acid digestion of some of the more amorphous sites of the native material. Diminution in the number of amorphous sites is probably offset by increased porosity.

Collagen, the major protein of tendon and dermis, is much underestimated as a support matrix for biological molecules. It may be processed into a variety of physical forms and shapes including coarse fibres, fine filaments and transparent sheets. After glycine, the most abundant amino acid residues in collagen are proline and hydroxyproline, the latter contributing significantly to the overall hydrophilicity of the protein. The basic structural unit (24) is tropocollagen, a rigid rod composed of a triple helix formed from three, slightly dissimilar polypeptide chains. Linear aggregation of these rods leads to microfibrils which are formed in such a way as to leave regular gaps between the tail of one unit and the head of the next. This results in the formation of "holes" which are regularly spaced along the axial length of the microfibril. These "holes" are intensely hydrophilic in character and are responsible for the aqueous swelling of the polymer. Both covalent coupling and physical binding of biological molecules take place at these sites and in this respect they are analogous to the amorphous regions of cellulose.

It is recognised that the classification of potential carriers according to their ability to form xerogels, aerogels, or xerogel-aerogel hybrids has limitations. A number of useful materials do not fit into any of these catagories. Nevertheless it is often possible to identify structural features which undergo constrained dissolution or which contribute to overall hydrophilicity. Such materials include various polyaminoacids (25,26), various inorganic adsorbents (27,28) and nitrocellulose (collodion) membranes (29).

IMMOBILISATION METHODS

The general methods which have been applied to the immobilisation of biological molecules fall into four catagories, each depending on an essentially different principle. These are:-

(a) physical interaction (adsorption or ionic binding) with a suitable carrier
(b) entrapment within a xerogel matrix
(c) covalent bonding with a chemically reactive carrier
(d) covalent cross-linking either alone or in the presence of an inert support.

The approach selected depends primarily on the eventual use to which the conjugate will be put. Being prone to dissociation, conjugates dependent on adsorption or ion exchange are unsuitable for application in affinity chromatography or as immunoadsorbents. Dissociation may also be serious in the case of enzymes active against macromolecular substrates. Inaccessibility becomes a problem when the biological activity of the conjugate is directed towards macromolecules. In these circumstances molecular entrapment is not applicable and it is best to resort to covalent binding to a macroporous carrier. In the case of enzymes active against small substrates, accessibility and dissociation are not such problems and all four technqiues are often applicable.

Physical Interaction

Synthetic conjugates prepared by physical adsorption frequently involve binding interaction between the biological molecule and a bulk surface in contact with the aqueous phase. Examples include such carriers as activated charcoal (28) or alumina (30) in conjugation with β-D-fructofuranosidase and cephalin coated silica or carbon in conjugation with phosphoglucomutase (31). The nature of the binding interaction in these conjugates and its effect, adverse or otherwise, on the bound enzyme has received scant attention.

Conjugates prepared by interaction of the biological molecule with cellulose differ in that sorption occurs at sites where partial dissolution of the matrix has occurred. Relatively weak binding of a variety of biological macromolecules takes place at these accessible sites. Much stronger binding is observed with glycoside hydrolases, notably γ-amylase (32) and α-amylase (21). This is probably due to structural resemblances between the polysaccharide carrier and the substrate resulting in bio-specific interaction (affinity binding) with sites on the enzyme normally involved in the enzyme substrate complex. In this context it is interesting to note that enzymes immobilised by binding to purpose synthesised affinity supports frequently retain their characteristic catalytic activity (33).

A number of successful enzyme conjugates have been prepared by ion exchange using CM and DEAE cellulose (34). Ion exchangers in general may be used to immobilise enzymes in active form. However, a major disadvantage of synthetic bio-polymer conjugates prepared by either adsorption or ion exchange is their tendency to dissociate on varying such factors as ionic strength, pH or temperature. Synthetic conjugates involving enzymes may also dissociate on contact with substrate solution (35). With small substrates this dissociation may be so slight as to be negligible but if the substrate is a soluble macromolecule substantial partitioning of the enzyme with the bulk solution surrounding the conjugate usually occurs. For example, cellulose conjugates of both α-amylase and γ-amylase depending on adsorption may be dissociated by repeated washing with starch solution (21). Resolubilisation on washing with a solution of the macromolecular substrate is a useful way of differentiating between physically and covalently bound enzymes.

Although physical interaction is often the mechanism by which enzymes are located in their natural, cellular environment this method has been, until recently, the least fashionable for the preparation of synthetic conjugates. This has arisen because of their tendency to dissociate. Consequently, practical problems arise from the progressive diminution of the amount of enzyme actually immobilised. Also, in theoretical studies, it is difficult to estimate the proportion of activity due to resolubilised enzyme.

An early insight into the potential of physically bound enzymes was provided by a study of a ribonuclease conjugate derived from the ion exchanger, Dowex 50, a microporous sulphonated polystyrene. This conjugate was atypical in that it could be used repeatedly against the macromolecular substrate, sodium ribonucleate without dissociation (36). One of the first viable commercial applications of immobilised enzymes was in the selective hydrolysis of L-acetylamino acids in D,L-acetylamino acid mixtures using ionically immobilised acylase. The DEAE derivatives of both cellulose and cross-linked dextran (Sephadex) are used as ionic carriers (37,38) to constrain the enzyme. The commercial potential of γ-amylase ionically bound to DEAE cellulose has also been demonstrated (39). Providing low ionic strength buffers were employed to maintain pH the conjugate could be used continuously for several weeks without deterioration. It is apparent that physical conditions, notably pH and ionic strength may be selected such that most enzymes will remain tightly ionically bound even when they are used to process large volumes of substrate solution.

Recently, an important series of membranous collagen-enzyme conjugates have been described (24). Collagen conjugates of β-D-fructofuranosidase, lysozyme, urease, glucose oxidase and penicillin amidase were prepared either by direct impregnation of a preswollen collagen membrane or by electro-deposition from a collagen dispersion containing dissolved enzyme. The resulting materials were applied in a capillaric coil modular reactor. In continuous use, the enzyme activities all fell to stable limits which corresponded to 35% of the original activity for β-D-fructofuranosidase and lysosyme, 25% in the case of urease and 15% for glucose oxidase. Collagen-enzyme (protein-protein) interaction was ascribed to multiple salt linkages, hydrogen bonds and Van der Waals interaction.

Entrapment

Polyacrylamide xerogels are widely used for the immobilisation of biological macromolecules by entrapment. In early work, difficulty arose owing to mechanical instability of the enzyme conjugates (4). Later, these difficulties were resolved by optimising the monomer to cross-linker ratio and polymerisation volume (5). The potential of included glucose oxidase and lactate dehydrogenase as analytical reagents was then demonstrated. Subsequently, more enzymes have been immobilised by this method than by any other single technique (14,40).

Other entrapment methods have been less successful. For example, practical application of cholinesterase entrapped in starch gel was only possible when the conjugate was dispersed in polyurethane foam to confer mechanical stability (41). A general disadvantage of the entrapment method, particularly acute in the case of starch gels, is the tendency of conjugate particles to "leak" biological activity into the surrounding solution. This occurs even after rigorous washing procedures. On the credit side, the inclusion method has the advantage of not involving chemical reaction with the biological molecule and also of simplicity.

In order to ensure that the maximum pore size of the constraining matrix is small enough for molecular entrapment, it is essential that the matrix density is relatively high. The xerogel character of the matrix means that the immobilised species is effectively immersed in an aqueous solution of polymer. This may effect denaturation of the biological molecule. For example, xerogel entrapment of urease in an aqueous polyacryloylmorpholine gel, prepared by free radical polymerisation at 0°, led to conjugates in which most of the enzyme was denatured (42). Entrapment in polyacrylamide xerogels under identical conditions resulted in conjugates of high activity. Clearly, there are only a limited number of matrices suitable for the entrapment process.

Immobilisation by entrapment may be effected by matrices other than aqueous xerogels. For example, inclusion of proteolytic enzymes in silastic resins leads to materials in which biological activity is confined to the surface (43). Consequently the conjugates are active against both small and large substrates.

A related technique to inclusion is to enclose whole droplets of enzyme solution within semi-permeable nylon microcapsules (44). This is achieved by dispersing an enzyme solution containing hexamethylene diamine in a chlorocarbon-hydrocarbon phase containing adipoyl chloride and a surfactant.

Covalent Coupling

Immobilisation of biological molecules by covalent coupling must necessarily not involve reaction with functional groups at sites responsible for biological activity (14). In the case of proteins, such as anti-bodies and enzymes, the amino acid residues responsible for activity or specificity will be duplicated, in non-essential form, several times in each molecule. Thus random coupling will denature often only a proportion of the bound molecules.

Unfortunately it arises often that amino acid residues responsible for enzyme activity are many times more reactive than those in the rest of the molecule. For example, the active site of chymotrypsin contains a very reactive serine residue which undergoes reaction rapidly and selectively with iodoacetate. When the enzyme is immobilised by reaction with ω-iodo-N-alkyl polymethacrylate, most of the enzyme molecules are inactivated (45). One approach to this problem is to protect the active site during the coupling reaction by addition of a competitive inhibitor (46). Active site cysteine residues may be protected by the use of either mercury or p-chloromercuribenzoate (47). In general, any information available on the effect of reagents on the enzyme's activity or on the amino acids involved in the active site, will assist in selecting a suitable coupling procedure.

In spite of the problem of chemical denaturation, covalent coupling procedures are very popular. Functional sites on proteins most often involved in coupling are the side-chain amino groups of arginine and lysine, the amino terminals of the protein chains, and the π-excessive aromatic nuclei of histidine, tryptophan and tyrosine. Less frequently the hydroxyl groups of serine, the thiol groups of cysteine and carboxyl groups such as those in the side-chains of aspartic and glutamic acid and at the carboxyl terminals of the protein chains may be utilised. Coupling must necessarily take place in aqueous solution, and under conditions of pH and temperature carefully selected to avoid denaturation.

Coupling of free amino groups is commonly effected by acylation or arylation with suitably active groups pendant on the carrier.

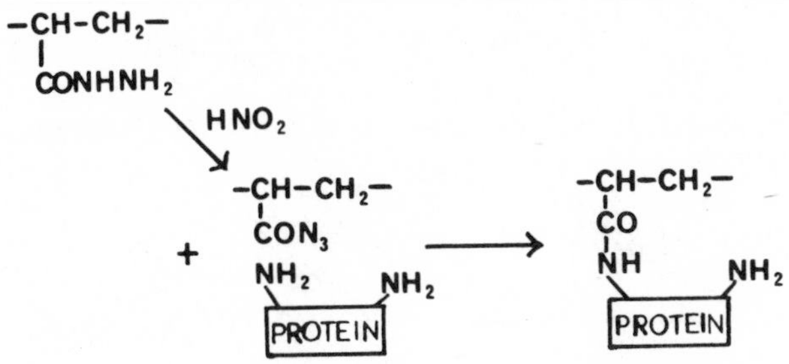

Fig. 2. Binding proteins by the acid azide technique.

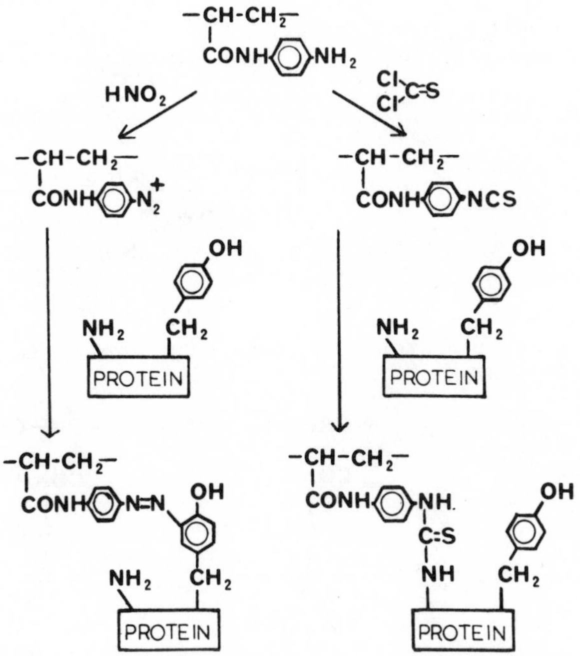

Fig. 3. Binding via the aryl diazo and isothiocyanato techniques Such versatility is important in the case of enzymes.

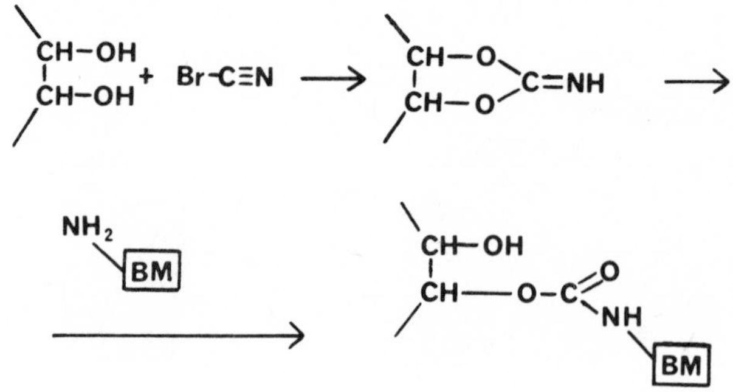

Fig. 4. Iminocarbonate binding of bio-organic molecules, BM.

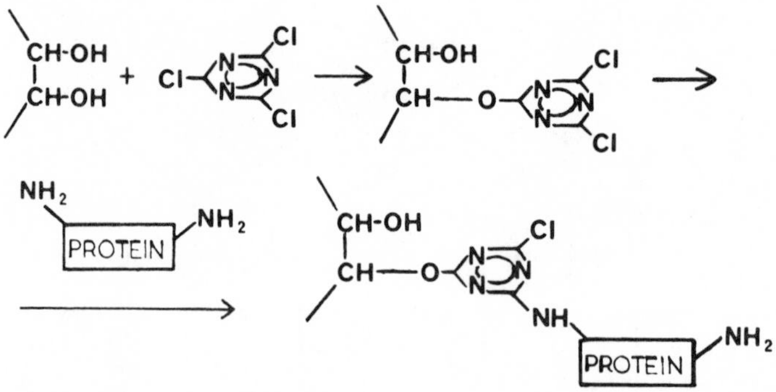

Fig. 5. Binding proteins to polysaccharides by the trichloro-sym-triazinyl method.

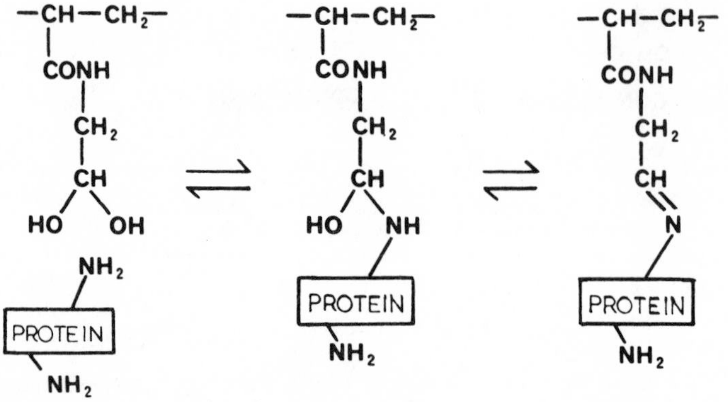

Fig. 6. Probable aldehydrol-protein binding reactions.

Acylation is achieved by such functional groups as acid azide (18), acid anhydride (6) and isothiocyanato (48), while arylation involves 2,4-dinitroaryl- (15) or sym-triazinyl- (49) halides. Coupling via the π-excessive aromatic nuclei is exclusively by means of diazo groups on the carrier (25). Acidic groups on the protein may be activated with water soluble diimides and coupled to amino groups on the carrier (50). Some of these methods are outlined in Figs. 2 - 6. Purpose synthesised carriers are commercially available for the application of most of these methods

Acid azide coupling (Fig. 2) is a popular method because it involves only aliphatic groups on the primary carrier. The aryl amino group has a deleterious effect on hydrophilicity but possesses the advantage of being amenable to activation with either nitrous acid, to give aryl diazo groups, or thiophosgene, to give aryl isothiocyanato groups (Fig. 3). Aryl diazo and isothiocyanato groups differ in their mode of coupling proteins. Such versatility is most desirable since one or the other coupling procedure may lead to loss of biological activity.

In cases where it is desired to avoid aromatic residues on the carrier, the isothiocyanato group may be derived from aliphatic amino groups. Following preliminary reaction with γ-aminopropyl triethoxysilane, porous glass, the important aerogel matrix, may be activated by this procedure (10). Alternatively the aliphatic amino groups may be reacted with p-nitrobenzoyl chloride, the nitro groups reduced and the resulting glass bound aryl amino groups diazotised for protein coupling (10). The complementary diazo and isothiocyanato procedures are therefore applicable to all the important classes of carrier.

Iminocarbonate formation (Fig. 4) may be carried out by reacting polysaccharides with cyanogen bromide at low temperature. This method is of vital importance as a preliminary step in the binding of biological molecules to preformed agarose gels of defined porosity (11). Agarose gels are destroyed by even moderate heat which precludes application of most other derivatization methods. The main use of agarose iminocarbonate is as a starting point for the synthesis of column packings for affinity chromatography (12). Analagous coupling procedures using cyclic carbonates of polysaccharides and polyallyl alcohol have been reported (51). Unfortunately cyclic carbonates are not so readily prepared as cyclic iminocarbonates.

The aldehydrol binding procedure (Fig.6) involves attack on the amino groups of biological molecules to give aminol and possibly azomethine linkages (52). Although, being a particularly mild binding procedure, the method is attractive, it has not been so widely applied as some other methods. This is probably because many synthetic polyaldehydes undergo spontaneous acetal and hemi-acetal formation in water. For example, spontaneous condensation of adjacent aldehydrol groups in polyacrolein leads to an extended polyacetal structure pendant on the hydrocarbon backbone. This results in loss of binding activity and gel character.

Suitable xerogel forming polymers for aldehydrol binding must be designed so that acetal and hemi-acetal formation is unfavourable for steric reasons. Cross-linked poly(acryloylaminoacetaldehyde) is such a polymer and it has been used to immobilise a variety of enzymes (52). A number of polyaldehydrol compounds suitable for enzyme coupling have been prepared by periodate oxidation of such polysaccharides as starch, cellulose and dextran (53,54).

Covalent cross-linking

Covalent cross-linking of biological molecules as a method of immobilisation is a rather limited procedure unless it is carried out in the presence of a suitable support. Although several cross-linking agents have been applied (14,40) only two have found widespread use. These are *bis* diazobenzidine-2,2'-disulphonic acid (29) and "glutaraldehyde" (55). Perhaps the most important application of *bis* diazobenzidine-2,2'-disulphonic acid has been the immobilisation of enzymes within hydrated nitrocellulose (collodion) membranes. "Glutaraldehyde" has found application in linking enzymes to aminoethyl cellulose (56) and to partially hydrolysed 6-nylon (57). Cross-linking with this reagent is unlikely to be a simple procedure involving solely aldehydrol groups. Aqueous "glutaraldehyde" contains appreciable amounts of unsaturated oligomeric condensation products. Primary amino groups of enzymes almost certainly undergo Michael addition across the conjugated double bonds in addition to reaction with the aldehydrol groups (58).

BIO-SPECIFIC ADSORBENTS

Selective adsorbents which exploit biological specificity are much more efficient than those depending on purely physico-chemical processes. Typical bio-specific adsorbents are prepared by covalently coupling an enzyme inhibitor or an antibody to a suitable synthetic or semi-synthetic carrier. Immobilised inhibitors are usually applicable only to the adsorption of a specific enzyme or group of enzymes. However, since antibodies may be raised, not only to proteins, but to a variety of other biological molecules, immunoadsorbents probably have the greater potential. Recently, the interrelated applications of affinity chromatography and immunoadsorbents have been critically catalogued and discussed in a number of excellent reviews (14,59-61).

Consequently, bio-specific adsorbents will be considered only insofar as to illustrate the principles involved in their preparation and design.

Immunoadsorbents

Early bio-specific adsorbents were based almost exclusively on cellulose and polystyrene matrices. Such materials were applied in the main to the preparation of immunoadsorbents and immobilised enzymes; the latter to serve as bio-specific adsorbents for their specific inhibitors. Since the preparation of immunoadsorbents involves first the chemical immobilisation of an antigen followed by interaction of the resulting conjugate with a macromolecular antibody, accessibility was a serious problem. Immunoadsorbents are based almost exclusively on organic gels of extreme pore size. Agarose and, to a lesser extent, the more porous xerogels such as cross-linked dextrans and polyacrylamides are the most popular. Porous glass has received comparatively little attention. This material has the advantages of being mechanically stable and resistant to microbial attack. However, non specific adsorption of proteins to glass surfaces is a serious problem.

Typical examples of the application of agarose immunoadsorbents are in the isolation of antibodies to insulin (62) and angiotensin (63). In the inverse procedure antibodies may be chemically coupled to agarose to isolate their specific antigens. For example the antibody to β-D-galactosidase has been coupled to agarose and used to isolate mutants of the enzyme in E. coli (64). Highly purified antibody is not essential for this technique providing the resulting adsorbent is biologically specific. Generally, dissociation of antigen-antibody complexes may be brought about by changes in pH or increase in ionic strength.

Recently, immunoadsorbents based on cross-linked dextran (finely divided Sephadex G200) have become commercially available (65). These are intended for application in radio-immuno assays using standard isotope dilution techniques. For example, Sephadex bound insulin antibody may be used to harvest insulin in serum to which radioactive insulin has been added. Following, determination of the amount of radioactive insulin bound, the concentration of natural insulin in the serum sample may be calculated.

Affinity Chromatography

In affinity chromatography the bio-specific adsorbent is used as a column packing. The main application is in the isolation of enzymes using immobilised competitive inhibitors. Although inhibitors are usually small molecules, accessibility problems are frequently acute. Not only must a gel of very high pore size be used but often the insolubilised ligand must be separated from the solvated matrix backbone by a spacer arm in order to allow unimpeded access of the enzyme. Studies directed towards the affinity purification of chymotrypsin gave a classical demonstration of this requirement. A bio-specific adsorbent prepared by coupling the inhibitor D-tryptophan directly to agarose was relatively ineffective compared to an adsorbent in which an ε-aminocaproic acid residue was interposed between the ligand and the agarose backbone (66).

Spacer arms are particularly important in the case of large enzymes in which the binding site may be buried deep in the molecule or when the ligand-enzyme interaction is particularly weak. However, the current trend towards very long spacer arms has been criticised recently on the grounds that unnecessarily long arms may introduce more chance of non-specific interference (67). Negative interference includes both localised steric hindrance and also "curl-back" of the hydrophobic ligand on the overlong hydrophobic arm. Positive interference infers non-specific adsorption of the enzyme as in hydrophobic chromatography (68). For example, a non-specific agarose gel containing N-phenylglycine pendant on a hydrophobic arm was shown to retain bacterial β-D-galactosidase in much the same way as an agarose gel carrying similar chains to which an inhibitor, a β-D-thiogalactoside, was attached (67).

Agarose iminocarbonate (Fig.4) is used generally for the preparation of bio-specific adsorbents. The optimum length of spacer arms for the attachment of enzyme inhibitors is probably between 0.5 and 1.5 nm for most enzymes. Suitable arms may be synthesised by coupling agarose iminocarbonate with an aliphatic primary diamine, reacting the resulting amino chain ends with the N-hydroxysuccinimide ester of bromoacetic acid, and using the bromoacyl group chain terminals to couple the specific ligand (69). Two agarose derivatives, AH Sepharose 4B and CH Sepharose 4B which carry $-(CH_2)_6NH_2$ and $-(CH_2)_5CO_2H$ groups respectively, are commercially available. These materials are intended as starting points for the synthesis of affinity supports. To date, only hydrophobic spacer arms appear to have been coupled to agarose. There is a real need for hydrophilic spacer arms for the immobilisation of hydrophobic ligands without the complication of "curl-back" (67).

IMMOBILISED ENZYMES

From a practical viewpoint, immobilisation of an enzyme has obvious manipulative advantages which facilitate its application, recovery and re-use. However, reactor technology is outside the scope of this lecture which is concerned more with the properties of the constrained enzyme within the semi-synthetic or carrier phase. In this context factors such as the physico-chemical modification of the enzyme during the binding procedure and the nature of the alien micro-environment within which the enzyme is constrained are important since it is these which lead to characteristic changes in stability and kinetic behaviour.

Stability and Denaturation

The conformation of a native enzyme is such that the more hydrophilic groups tend to be exposed to the external solution. Amino acids with the more hydrophobic side chains are grouped towards mainly the centre of the molecule. This results in extensive hydrophobic bonding which is important in preserving tertiary structure. In general, denaturation of an immobilised enzyme molecule will be promoted if functional groups on the constraining matrix can compete successfully for groups in the enzyme involved in hydrophobic bonding. Hydrophobic carriers or carriers which, though hydrophilic overall, have prominent hydrophobic features will promote denaturation (14).

Hydrophobic groups on the carrier matrix may effect perturbation of the enzyme protein, and thereby denaturation or, alternatively, compete for hydrophobic groups in an enzyme molecule which has already been perturbed by some other process. Low temperature denaturation of enzyme in hydrated conjugates probably takes place by the first mechanism. The second is most likely to be encountered during thermal perturbation and is detrimental to the heat stability of the enzyme. Heat stability is related to the probability of recovery of the native enzyme conformation following thermal perturbation (70).

Most carriers which are hydrophilic overall confer enhanced storage stability as compared to the free enzyme in solution. However, increased thermal stability is less common and usually results only when extremely hydrophilic carriers such as porous glass or aqueous xerogels are employed. Even then diminution rather than increase in heat stability may result.

The stability of enzyme conjugates to low temperature storage is probably a result of the binding interaction causing minor resistance to the conformational changes implicit in denaturation. Increased stability to heat denaturation is more difficult to explain but this may be a similar effect. Thus the extreme hydrophilicity of glass and its consequent enzyme compatibility together with its ability to non-specifically adsorb protein may produce a "cushioning effect" protecting the enzyme against disruptive perturbation. In the case of stabilisation by xerogels the total constrained dissolution of the carrier matrix precludes simple adsorption. Nevertheless other sorptive processes and consequent chain clustering around

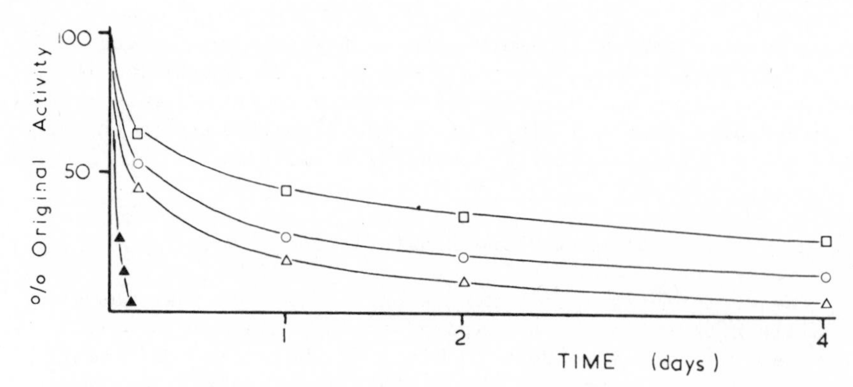

Fig. 7. Thermal denaturation of diazo coupled Enzacryl AA-α-amylase (△), isothiocyanato coupled Enzacryl AA-α amylase (○), azide coupled Enzacryl AH-α amylase (□) and soluble α amylase (▲) at 45°.

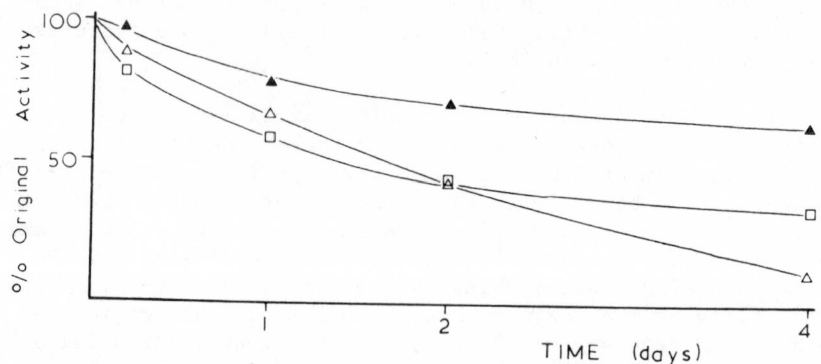

Fig.8. Thermal denaturation of diazo coupled Enzacryl AA-β amylase (△), isothiocyanato coupled Enzacryl AA-β amylase (○) and soluble β amylase (▲) at 45°.

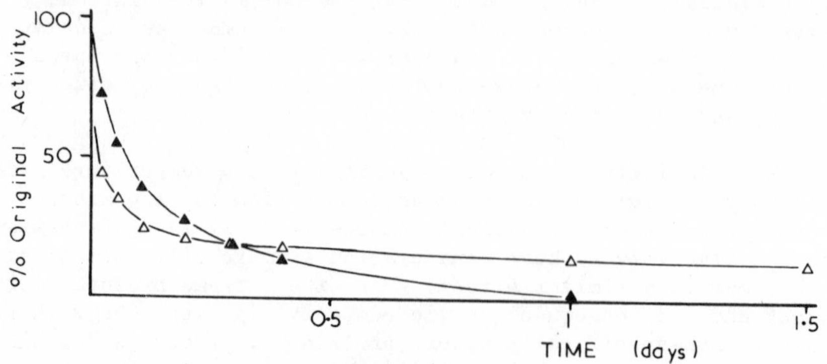

Fig. 9. Thermal denaturation of Enzacryl CHO-dextranase (△) and soluble dextranase (▲) at 50°.

the constrained enzyme molecule may contribute to stabilising tertiary structure.

Some typical heat denaturation profiles are presented in Figs. 7 - 9. The stabilisation of α-amylase following immobilisation by covalent binding to commercial polyacrylamides (Enzacryls AA and AH) is characteristic of this enzyme in conjugation with any neutral hydrophilic carrier (22,2). However, the conjugate prepared by coupling to Enzacryl AH by the acid azide procedure had considerably superior heat stability to conjugates prepared by aryl diazo and aryl isothiocyanato coupling with Enzacryl AA. Enzacryl AA, though hydrophilic overall, contains hydrophobic aromatic groups. A similar order of stabilities was observed on storage. The enzyme β-amylase, immobilised by coupling to Enzacryl AA had inferior heat stability (Fig. 8) when compared to the free enzyme in solution (2).

Dextranase, immobilised by coupling to commercial cross-linked poly(acryloylaminoacetaldehyde) (Enzacryl-CHO), was initially heat denatured (Fig.9) more rapidly than the soluble enzyme (71). However, a proportion of the immobilised dextranase molecules appear to be stabilised. This phenomenon is often encountered in immobilised enzyme technology. One explanation is that variation in the number of enzyme-to-carrier linkages or different orientations with respect to the carrier may result in differential stability. Alternatively if active enzyme molecules of different conformation and hence stability are present in the original coupling mixture, different rates of coupling may lead to variation in the proportion of the conformers in the immobilised as compared to the soluble enzyme preparation. This would be reflected in the heat denaturation curves.

Recent work with the thermostable proteolytic enzyme thermolysin, immobilised by coupling to cross-linked polyacrylamides by either the acid azide or aryl diazo coupling procedures, led to conjugates which, at 80°, exhibited considerably enhanced heat stability as compared to the native enzyme in solution (72). However, in this instance, the least heat stable conjugate (prepared by diazo coupling at pH 10.0) proved to have better storage properties. These observations are consistent with the hypothesis of a mode of enzyme binding being particularly effective in preventing perturbation of the tertiary structure within certain limits. This stabilises the enzyme at storage temperature. However, once perturbation occurs, as in heat denaturation, return to a catalytically active conformation will be impeded (70).

Proteolytic enzymes which undergo autodigestion in free solution are dramatically stabilised by immobilisation because this isolates the individual enzyme molecules. For example, trypsin immobilised by chemical coupling to cross-linked poly(acryloylaminoacetaldehyde) was much more stable at pH 7.5 and 50° than the native enzyme in solution under comparable conditions (73). At the same temperature but pH 3.5, at which trypsin is inactive, the immobilised and native enzyme both exhibited similar excellent stability. At 75° and pH 3.5 the immobilised enzyme was the least stable.

Kinetic Behaviour

The kinetic behaviour of an immobilised enzyme may be moderated by a number of effects. These include:

(a) micro-environmental changes in the vicinity of its active or effector sites
(b) steric interaction between enzyme, substrate and gel matrix
(c) diffusional limitations on the rate of substrate penetration of the system
(d) structural changes or constraints imposed on the enzyme molecule as a result of binding.

In practice, it is difficult to design kinetic experiments to differentiate between these effects. Any one may cause real or apparent changes in Michaelis constant (K_m) and all, except diffusional limitations, may result in modification of the enzymes maximum catalytic activity.

Microenvironmental changes encountered by an enzyme on coupling to a poly-ionic support are the easiest to relate to kinetic behaviour. If the substrate is of opposing charge to the constraining gel then, at low ionic strength an apparent decrease in K_m results (increased reaction rate at low substrate concentration). Consequently saturation kinetics are observed at much lower overall substrate concentration than in the case of the free enzyme in solution. For example, in the hydrolysis of benzoyl-L-arginamide by an anionic ethylene-maleic acid conjugate of trypsin (EMA-trypsin), K_m was lower by an order of magnitude compared to the free enzyme in solution (74). Similar effects have been observed for CM-cellulose conjugates of ficin (75) and bromelain (76) using benzoyl-L-arginine ethyl ester as substrate. In these examples, the opposing charge of the gel and substrate leads to an increased concentration of the latter in the vicinity of the immobilised enzymes. This effect, which disappears at high ionic strength, may be regarded as favourable micropartition of the substrate between the solvated, poly-ionic regions of the matrix and the bulk solution.

As well as producing apparent changes in K_m, enzyme immobilisation with poly-ionic carriers may result in apparent shifts in the pH-activity profile at low ionic strength. For example, when a poly-anionic ethylene-maleic acid conjugate of chymotrypsin was used to hydrolyse acetyl-L-tyrosine ethyl ester, the pH-activity profile was found to be displaced to alkaline pH when compared to that of the native enzyme (74). In the case of a poly-cationic polyornithine conjugate of the enzyme, the reverse effect was found to operate (77). As with apparent changes in K_m these discrepancies disappear at high ionic strength, the pH-activity profiles reverting to that of the native enzyme. Other examples of shifts in the pH-activity profile include shifts to alkaline pH with CM-cellulose conjugates of ficin and shifts to acid pH with DEAE-conjugates of invertase (78) and aminoacylase (38).

With poly-anionic carriers the hydrogen ion concentration within the conjugate at low ionic strength will be greater than in the external solution. The pH of the system as a whole must be raised to compensate and this results in an overall shift towards alkaline pH. The reverse will apply for cationic carriers.

The ability of a charged carrier to attract or repel charged substrates and thereby influence the catalytic activity of an immobilised enzyme can result in significant changes in specificity. This is apparent from the action pattern observed in the hydrolytic degradation of substrates with several potential sites for enzyme attack. For example, peptide mapping of tryptic hydrolysates of pepsinogen and carboxymethylated pepsinogen revealed that, whereas the native enzyme hydrolysed all fifteen lysyl bonds, EMA-trypsin hydrolysed only ten (79). The narrower specificity of EMA-trypsin as compared to the native enzyme has been exploited in the cleavage of the muscle protein, myosin, into heavy and light meromyosins. EMA-trypsin cleavage gave a preparation of light meromyosin which formed a lattice structure at low ionic strength. Light meromyosin prepared by the action of native trypsin formed fibrous aggregates (80). The improved specificity of EMA-trypsin compared to the native enzyme was ascribed to charge interactions between the immobilised enzyme and different charged regions of the protein substrate.

Steric interaction between a constrained enzyme molecule, its substrate and the gel matrix may result in either total loss or decrease in catalytic centre activity. If, as is likely in a population of immobilised molecules, individuals are affected to different extents, loss and decrease in activity will be difficult to differentiate since the sole effect of both is to decrease the maximum catalytic activity attainable by the conjugate. In this respect, steric interaction is also indistinguishable from denaturation.

The most serious steric effects are encountered when the enzyme substrate is very large. Such effects are the easiest to identify. In the case of a proteolytic enzyme, the percentage of the free solution activity retained on immobilisation may be determined for both large and small substrates. Low activity retention for the macromolecular substrate as compared to the small one may be ascribed to steric effects. For example, CM-cellulose conjugates of ficin exhibited only a fraction of the activity towards casein which might have been anticipated from the activity retained by the immobilised enzyme towards benzoyl-L-arginine ethyl ester (75). Other examples of synthetic conjugates exhibiting this type of differential activity include CM-cellulose derivatives of bromelain (76) and polyamino acid conjugates of trypsin (81).

In the case of enzymes immobilised within xerogel matrices, for example chymotrypsin in cross-linked dextran, it is arguable that poor activity retention towards large substrates is a result of the inability of the macromolecular substrate to penetrate the gel matrix (48), a situation similar to that of an enzyme immobilised by entrapment. Differential activity comparisons are open to criticism on the grounds that catalytic activity in the case of macromolecular substrates is rarely independent of substrate concentration. Quantitative comparison of observed activities may be of dubious validity owing to different kinetic characteristics of the free and bound enzyme.

Diffusional limitations on the rate of approach of reactants towards catalytically active sites generally have a significant effect on the kinetics of a heterogeneous reaction. All reactions involving immobilised enzymes are subject to diffusion control inasmuch as the conjugates are always separated from the bulk substrate solution by an unstirred layer of solvent. In the case of batch reactors the thickness of this layer depends on rate of stirring whereas, in a packed bed, flow rate and microturbulance between conjugate particles are the important parameters.

If one considers an idealised situation in which the enzyme molecules are attached to the surface of a smooth sphere, it is possible to envisage reaction taking place with an equilibrium concentration gradient of substrate set up across the unstirred layer. In reality, few systems approximate to ideality and it is clear that the unstirred layer must extend to enzyme molecules within the gel phase. Nevertheless it is possible to define an effective, if not real, thickness for the diffusion layer and relationships have been derived (82,83) relating both diffusion and electrical parameters with the Michaelis constants of the native and immobilised enzymes. Hornby et al (82) have derived the equation:

$$K_m^1 = \left(K_m + \frac{xV}{D}\right)\left(\frac{RT}{RT - ZxF\,\mathrm{grad}\,\psi}\right)$$

where: D = substrate diffusion constant
x = effective thickness of diffusion layer
V = maximum reaction velocity for the free enzyme
K_m = true Michaelis constant
K^1_m = apparent Michaelis constant
Z = electrochemical valence of substrate
F = Faraday constant
grad ψ = gradient of electrical potential
R = gas constant
T = absolute temperature of system

The diffusion term xV/D is most important for enzymes with high maximum velocities and in situations where the effective thickness of the diffusion layer is large; that is, for low stirring rates and large gelatinous particles. However, in most instances it is possible to choose operating conditions such that no more than a two or three fold increase in Michaelis constant results. This generalisation is most likely to break down in the case of enzyme conjugates based on xerogels.

An elegant demonstration of the effect of diffusional limitations on the observed Michaelis constant has been given with chymotrypsin immobilised by coupling with a cross-linked dextran (Sephadex G200) pre-activated by the cyanogen bromide procedure (84). On immobilisation of the enzyme an approximately tenfold increase in Michaelis constant was observed using acetyl-L-tyrosine ethyl ester as substrate. On dextranase catalysed hydrolysis of the xerogel matrix the enzyme was then released into solution when the Michaelis constant returned to its original value. The theoretical maximum velocity of the enzyme, as calculated from the Michaelis-Menten equation, did not change on immobilisation. This indicates that the activity of the immobilised enzyme was not moderated by steric effects.

Diffusional limitations are accentuated in the case of membrane bound enzymes. Not only is access of the substrate to the enzyme impeded but localisation of reaction products may be serious enough to cause enzyme inhibition or adverse changes in micro-environment. For example, in a study of benzoyl-L-arginine ethyl ester by papain immobilised in nitrocellulose membranes it was found that accumulation of benzoyl-L-arginine led to a local increase in acidity which could only be offset by raising the external pH by several units (85,86).

Structural changes or constraints which result in direct modification of the kinetic behaviour of an immobilised enzyme are comparatively rare. Shifts in pH-activity profiles, observed on immobilising papain, polytyrosyl-trypsin and subtilopeptidase A by diazo coupling to neutral carriers such as aminoaryl derivatives of starch dialdehyde, p-aminobenzyl cellulose and leucine p-aminophenylalanine copolymers, have been ascribed to short range interactions in the region of the active site of enzymes (87). These shifts are independent of the ionic strength of the medium indicating that they do not depend on modification of overall charge of the bound enzymes.

One possible approach to modifying the properties of an enzyme is to immobilise it in such a way that it is firmly supported at several points. This mode of attachment might have a marked influence on the mode of action of an enzyme with properties moderated by allosteric effects. In order to investigate this possibility, the enzyme urease has been immobilised by covalent coupling to an expanded xerogel, cross-linked poly(acryloylaminoacetaldehyde) Enzacryl CHO (88). The hydrophilicity of this carrier is due to aldehydrol groups which also constitute the sites functionally active in enzyme binding. The high concentration of aldehydrol groups favours multiple linkage. Expanded Enzacryl CHO-urease was found to differ from the native enzyme in that the conjugate exhibits saturation kinetics over a wide range of substrate concentrations and was much less susceptible to substrate inhibition. Saturation kinetics are barely obtained with native urease (89). These results are explicable if substrate inhibition involves an allosteric process operating from a site remote from the active centre of the enzyme.

Modification of enzymic properties by indiscriminate coupling reactions has also been observed in the case of polyacrylamide conjugates of the thermostable protease, thermolysin (88). Depending on the covalent binding procedure used, the Michaelis constant measured with furacryloylglycyl-L-leucinamide was only 0.5 to 4% of that of the native enzyme in solution. Unfortunately, the theoretical maximum catalytic activity of the bound enzyme also decreased. Perturbation of enzyme structure without denaturation is a distinct possibility in the case of thermolysin. The enzyme is known to exist in at least two active conformations (90). Cystine residues and, consequently, -S-S- bonds are absent. Tertiary structure is maintained solely by hydrophobic and ionic bonds the "length" of which are not necessarily defined exactly.

CONCLUDING REMARKS

Numerous methods have been published for the immobilisation of biological molecules. These methods are based on a small number of synthetic and semi-synthetic carriers. The best of these materials are effective because they constrain the biological molecule within an aqueous gel phase which confers a microenvironment approximately to that experienced in aqueous solution. Cross-linked dextran, cross-linked polyacrylamide, agarose and possibly cellulose are the most widely used carriers which fulfil these criteria. There is much scope for the purpose design and synthesis of alternative supports. The application of porous glass and of macroreticular forms of hydrophilic polymers such as polystyrene sulphonic acids and poly(hydroxyethyl methacrylates) must increase. Doubtless such materials will be systematically exploited in the future.

REFERENCES

(1) R. Axen and J. Porath, Nature (London), 210, 367 (1966).
(2) S. A. Barker, P. J. Somers, R. Epton, and J. V. McLaren, Carbohyd. Res., 14, 287 (1970).
(3) J. K. Inman and H. M. Dintzis, Biochemistry, 8, 4074 (1969).
(4) P. Bernfield and J. Wan, Science, 142, 678 (1963).
(5) G. P. Hicks and S. J. Updike, Anal. Chem., 38, 726 (1966).
(6) Y. Levin, M. Pecht, L. Goldstein, and E. Katchalski, Biochemistry 3, 1905 (1964).
(7) M. Le Page, R. Beau, and D. Jaques, French Patent 1,473,240; Chem. Abst., 68, 99044f (1968).
(8) W. J. Haller, J. Chem. Phys. 42, 686 (1965).
(9) H. H. Weetall and L. S. Hersh, Biochim. Biophys. Acta., 185, 464 (1969).
(10) H. H. Weetall, Science, 166, 615 (1969).
(11) J. Porath, R. Axen, and S. Ernback, Nature (London) 215, 1491 (1967).
(12) P. Cuatrecasas, M. Wilchek, and C. R. Anfingsen, Proc. Nat. Acad. Sci. U.S.A., 64, 636 (1968).
(13) G. Manecke, Naturwiss., 51, 25 (1964).
(14) I. H. Silman and E. Katchalski, Ann. Rev. Biochem., 35, 873 (1966).
(15) G. Manecke and H. J. Forster, Makromol. Chem., 91, 136 (1966).
(16) J. Coupek, M. Krivakova, and S. Pokorny, IUPAC International Symposium on Macromolecules, Helsinki, 1972.
(17) J. Turkova, O. Hubalkova, M. Krivakova, and J. Coupek, Biochim. Biophys. Acta. in press.
(18) F. Micheel and J. Ewers, Makromol. Chem., 3, 200 1949.
(19) M. A. Mitz and L. J. Summaria, Nature, 189, 576 (1961).
(20) D. M. Jones, Advan. Carbohyd. Chem., 19, 219 (1964).
(21) S. A. Barker, P. J. Somers, and R. Epton, Carbohyd. Res., 14, 323 (1970).
(22) S. A. Barker, P. J. Somers, and R. Epton, Carbohyd. Res., 8, 491 (1968).
(23) S. A. Barker, P. J. Somers, and R. Epton, Carbohyd. Res., 9, 257 (1969).
(24) S. S. Wang and W. R. Vieth, Biotech. Bioeng., 15, 93 (1973).
(25) J. J. Cebra, D. Givol, H. I. Silman, and E. Katchalski, J. Biol, Chem., 236, 1720 (1961).
(26) M. Gutman and A. Rimon, Can. J. Biochem., 42, 1339 (1964).
(27) P. V. Sundaram and E. M. Crook, Proc.7th Int. Cong. Biochem., Tokio, (1967).

(28) J. M. Nelson and E. G. Griffin, J. Amer. Chem. Soc., 38, 1109 (1916).
(29) R. Goldman, I. H. Silman, S. R. Caplan, O. Kedem, and E. Katchalski, Science, 150, 758 (1965).
(30) J. M. Nelson and E. G. Griffin, J. Amer. Chem. Soc., 38, 722 (1916).
(31) E. S. Vorobeva and O. M. Poltorak, Vestn. Mosk. Univ., Ser.II, 21, 17 (1966); Chem. Abst. 66, 62194 (1967).
(32) S. A. Barker and J. G. Fleetwood, J. Chem. Soc., 4857 (1957).
(33) E. Steers, Jr., P. Cuatrecasas, and H. B. Pollard, J. Biol. Chem., 246, 196 (1971).
(34) M. A. Mitz and R. J. Schlueter, J. Amer. Chem. Soc., 81, 4024 (1969).
(35) I. Langmuir and V. J. Schaefer, J. Amer. Chem. Soc., 60, 1351 (1968).
(36) L. H. Barnett and H. B. Bull, Biochim. Biophys. Acta., 36, 244 (1959).
(37) T. Tosa, T. Mori, N. Fuse, and I. Chibata, Enzymologia, 31, 214 (1966).
(38) T. Tosa, T. Mori, N. Fuse, and I. Chibata, Enzymologia, 32, 153 (1967).
(39) K. L. Smiley, Biotech.Bioeng. 13, 309 (1971).
(40) G. J. H. Melrose, Rev. Pure and Appl. Chem., 21, 83 (1971).
(41) E. K. Baumann, L. H. Goodson, G. G. Guilbault, and D. N. Kramer, Anal. Chem., 37, 1378 (1965).
(42) R. Epton, C. Holloway, and J. V. McLaren, unpublished work.
(43) H. D. Brown, A.B. Patel, and S. K. Chattopadhyay, J. Biomed. Mater. Res. 2, 231 (1968).
(44) T. M. S. Chang, Science Tools, 16, 33 (1969).
(45) E. Brown and A. Racois, Bull. Soc. Chim Fr., 12, 4351 (1971).
(46) I. H. Silman, D. Wellner, and E. Katchalski, Isr. J. Chem. 1, 65 (1963).
(47) H. D. Brown, A. B. Patel, S. E. Chattopadyay, and S. Pennington, Enzymologia, 35, 215 (1968).
(48) R. Axen and J. Porath, Nature (London) 210, 237 (1966).
(49) G. Kay and E. M. Crook, Nature (London) 216, 514 (1967).
(50) H. H. Weetall and N. Weliky, Nature (London) 204, 896 (1964).
(51) S. A. Barker, J. F. Kennedy, and A. Rosevear, J. Chem. Soc. (C), 2726 (1971).
(52) R. Epton, T. H. Thomas, and J. V. McLaren, Biochem. J. 123, 21P (1971).
(53) C. Flemming, A. Gabert, P. Roth, and M. Rudel, East German Patent 83,154, Chem. Abst. 79, 1077805 (1973).
(54) C. Flemming, A. Gabert, and P. Roth, Acta. Biol. Med. Ger. 30, 177 (1973).
(55) G. Broun, E. Selegny, S. Avrameas, and D. Thomas, Biochim. Biophys. Acta., 185, 260 (1969).
(56) M. L. Green and G. Crutchfield, Biochem. J., 115, 183 (1969).
(57) P. V. Sundaram and W. E. Hornby, FEBS Lett., 10, 325 (1970).

(58) F. M. Richards and J. R. Knowles, J. Mol. Biol., 37, 231 (1968).
(59) P. Cuatrecasas and C. B. Anfinsen, Ann. Rev. Biochem., 40, 259 (1971).
(60) P. Cuatrecasas and C. B. Anfinsen, Methods in Enzymology 22, 345 (1971).
(61) F. Friedberg, Chromatogr. Rev. 14, 121 (1971).
(62) P. Cuatrecasas, Biochem. Biophys. Res. Commun., 35,531 (1969).
(63) E. Hurwitz, F. M. Dietrich, and M. Sela, Eur. J. Biochem., 17, 267 (1970).
(64) F. Melchers and W. Messer, Eur. J. Biochem., 17,267 (1970).
(65) Phadbas Insulin Test - Pharmacia clinical and technical information booklet.
(66) P. Cuatrecasas,M. Wilchek, and C. B. Anfinsen, Proc. Nat. Acad. Sci. U.S.A., 61, 636 (1968).
(67) P. O'Carra, S. Barry, and T. Griffin, Biochemical Society (London) Transactions, 1, 289 (1973).
(68) R. J. Yon, Biochem. J. 126, 765 (1972).
(69) P. Cuatrecasas, J. Biol. Chem. 245, 3059 (1970).
(70) R. Goldman, L. Goldstein, and E. Katchalski, in Biochemical Aspects of Reactions on Solid Supports (ed. G. R. Stark), pp 1-78, Academic Press, New York.
(71) R. Epton, T. H. Thomas, and J. V. McLaren, Carbohyd. Res., 22, 301 (1972).
(72) R. Epton, T. H. Thomas, and J. V. McLaren, unpublished work.
(73) R. Epton, T. H. Thomas, and J. V. McLaren, unpublished work.
(74) L. Goldstein, Y. Levin, and E. Katchalski, Biochemistry 3, 1913 (1964).
(75) W. E. Hornby, M. D. Lilly, and E. M. Crook, Biochem. J. 107 669 (1966).
(76) C. M. Wharton, E. M. Crook, K. Brocklehurst, Eur. J. Biochem., 6, 565 (1968).
(77) L. Goldstein and E. Katchalski, Z. Anal. Chem. 243, 375 (1968).
(78) H. Suzuki, Y. Ozawa, H. Maeda, and O. Tanabe, Kogyo Gijutsuin, Hakko Henkyusho Kenkyu Hokoku, 31, 11 (1967); Chem. Abst., 70, 522 (1969).
(79) E. B. Ong, Y. Tsang, and G. E. Perlmann, J. Biol. Chem., 241, 5661 (1966).
(80) S. Lowey, L. Goldstein, C. Cohen, and S. M. Luck, J. Mol. Biol. 23, 287 (1967).
(81) A. Bar-Eli and E. Katchalski, J. Biol. Chem., 238, 1690 (1963).
(82) W. E. Hornby, M. D. Lilly, and E. M. Crook, Biochem. J. 107, 669 (1968).
(83) R. Goldman, O. Kedem, and E. Katchalski, Biochemistry., 7, 4158 (1968).
(84) R. Axen, P. A. Myrin, and J. C. Jonson, Biopolymers, 9, 401 (1970).
(85) R. Goldman, O. Kedem, I. H. Silman, S. R. Caplan, and E. Katchalski, Biochemistry, 7, 486 (1968).
(86) R. Goldman, O. Kedem, I. H. Silman, S. R. Caplan, and E. Katchalski, Science, 150, 758 (1965).

(87) L. Goldstein, M. Pecht, S. Blumberg, D. Atlas, and Y. Levin Biochemistry, 19, 2322 1970.
(88) R. Epton, T. H. Thomas, and J. V. McLaren, unpublished work.
(89) K. J. Laidler and J. P. Hoare, J. Amer. Chem. Soc., 71, 2599 (1949).
(90) Y. Ohta, J. Biol. Chem. 242, 509 (1967).

DISCUSSION SESSION

Discussion Leader - W. H. Daly

Goethals: I would like to propose a series of polymers for use in this area. We have reacted starch or cellulose (better cellulose) with propane sultone to form the sulfopropyl derivative which is soluble in water but can be crosslinked with epichlorohydrin to give a nice gel.

Cho: There is a technique for entrapping enzymes inside hollow fibers for use as fixed-bed reactors. How do you compare your insolubilized enzymes with these?

Epton: They are all part of the same picture. I didn't attempt to discuss hollow fibers because of time limitations.

Corett: Did I understand you to say that enzyme entrapment in Silastic resins was possible and that the enzymes are reactive?

Epton: I've no practical experience on this point,but there are 6-7 papers in the literature on that point. It is reported to be a surface-active system.

Corett: We once tried to reproduce one of those papers. It reported cholinesterase entrapped in silastic and we could't get anything except for the enzyme which was leaching into the solvent. The entrapped enzyme did not work at all.

Epton: It seems contrary to all theory that it should work, but there have been these several reports of it which I have mentioned.

Lewin: We have investigated the elementary structure of cellulose particularly with regard to the size and distribution of the "defects" between the elementary fibrils, presumably where binding of enzymes may occur. In this regard we have found that mild acid treatment followed by mild base dissolves about 40% of the original weight of cellulose and it is therefore much more porous and may be better for binding. We can also control the size of these openings by, e.g. drying.

Daly: We are growing micro-organisms on cellulose. We have carried out alkaline treatments on fibers and studied what happens when these organisms attack the treated cellulose. The <u>Cellulomonas</u> we use produces a cellulase which remains attached to the cell walls, and does appear to be extracellular. The fiber is treated with base and growth is allowed to take place. The fiber is removed from the culture medium and an electronmicrograph of the surface is taken. You see a cluster of organisms along the surface of the cellulose fiber. This cluster can not be easily disrupted by mechanical shear. We postulate that the enzymes penetrate the cellulose and that the actual cellulytic activity occurs within the fiber. It is in the amorphous region and the micro-organism must be attached to the cellulose fiber.

Parikh: Dr. Epton, you descibed the polyaldehyde resins which could be used to bind enzymes by the formation of Schiff bases. Does the enzyme slowly leak from the resin because of the instability of the Schiff base linkage?

Epton: Under the conditions used, if we get binding at all there is no reason to expect the reaction to suddenly reverse itself unless conditions are drastically changed. It would, in some instances, be desirable to be able to regenerate the carrier at will.

Daly: O'Driscoll has published on the entrapment of enzymes in poly(hydroxyethylacrylate) or Hydrogels. Your aldehyde system is so closely related to these that I wonder if you have tried to copolymerize hydroxyethylacrylate with acryloylaminoacetaldehyde so you would have a hydrogel system containing binding sites?

Epton: No, we haven't tried that.

Parikh: All these non-specific binding reactions (diazotization, azide or cyanogen bromide) cause reaction at a multitude of sites on the enzyme creating as inhomogeneous preparation. Would the bound enzymes be more "natural" in their behavior if they were attached to the resin by only one bond?

Epton: I would expect it to be very much the same or inactive. It is frequently an all or nothing process. There are very few cases where you get a real modification of the properties of an enzyme. You either immobilize it successfully or you destroy it.

Daly: Can you increase the activity if you carry out the immobilization in the presence of some substrate which will saturate the active site?

Epton: There have been a fairly large number of papers which use competitive inhibitors to do this,in fact, you can improve the activity in this way. In some very early work the enzyme in question was used as an antigen and then the immobilized enzyme antibody complex was used as the final preparation.

Corett: The problem with this method is that you can not predict what kind of an antibody you will get and sometimes you will get an antibody to an enzyme which binds near the active site and thereby decreases the reactivity or stops it altogether. In other cases you may get antibodies which may bind far from the active site and you get no change in activity. In fact, in some cases, perhaps due to allosteric effects you get an increase in re - activity.

Daly: I would like to describe briefly some work we are doing in the enzyme binding area. Although the choice of an inert support is governed by the presence of a suitable binding site, a much wider variety of polymers could be considered as inert supports for enzymes. Ideally, a support with good mechanical properties as well as high chemical stability and good physiological compatability should be selected. Since the compatability of an enzyme polymer system is not predictable, we are developing a system which will allow maximum flexibility in the selection of support polymers and the methods of binding proteins. The key to this approach is the preparation of a polyfunctional binding agent which is capable of reacting with a variety of functionalized polymers and introducing a protein binding site.

Our work with arylsulfonyl isocyanates [1] has led to the synthesis of several potential binding agents of the following type, $OCN\text{-}SO_2\text{-}C_6H_4\text{-}X$, where X is a functional group which is known to react with proteins. The high reactivity of arylsulfonyl isocyanates with substrates containing active hydrogens, i.e., alcohols, phenols, amines, and mercaptans, as well as substituted olefins,[2] gives us a wide selection of polymer supports. In the case of arylsulfonyl carbamates (3) the adduct may be methylated with diazomethane to produce an N-methylated derivative which can be hydrolysed under very mild conditions. Thus it would be possible to reverse the binding process if necessary.

Our initial attention was directed toward the use of poly(hydroxy ether) (1) derived from bisphenol-A and epichlorohydrin as the enzyme support. The polyether is a neutral substrate containing secondary hydroxyl groups as reactive sites; it can be cast into films with good mechanical properties and exhibits excellent chemical stability. The hydrophobic character of this support can be modified by grafting ethylene oxide to the corresponding polyalcoholate. The short chain polyether grafts terminate with primary hydroxyls which are more reactive and more accessible than the secondary hydroxyls attached directly to the polymer backbone. The hydrophilicity of the support is definitely increased but there is also a significant change in the mechanical properties; the graft copolymers become rubbery materials. Thus the polyether-ethylene oxide system will provide a range of mechanical and chemical properties without markedly changing the chemical composition.

p-Nitrobenzenesulfonyl isocyanate was selected as the binding agent since the nitro substituent can be reduced to an amine which can be modified to bind proteins. The addition of arylsulfonyl isocyanates to alcohols produces acidic sulfonyl carbamate linkages which serve as ion-exchange sites and help activate the enzymes. The reaction scheme illustrating the synthesis of the reagents and steps involved in coupling the enzyme is shown in Scheme 1.

SCHEME I

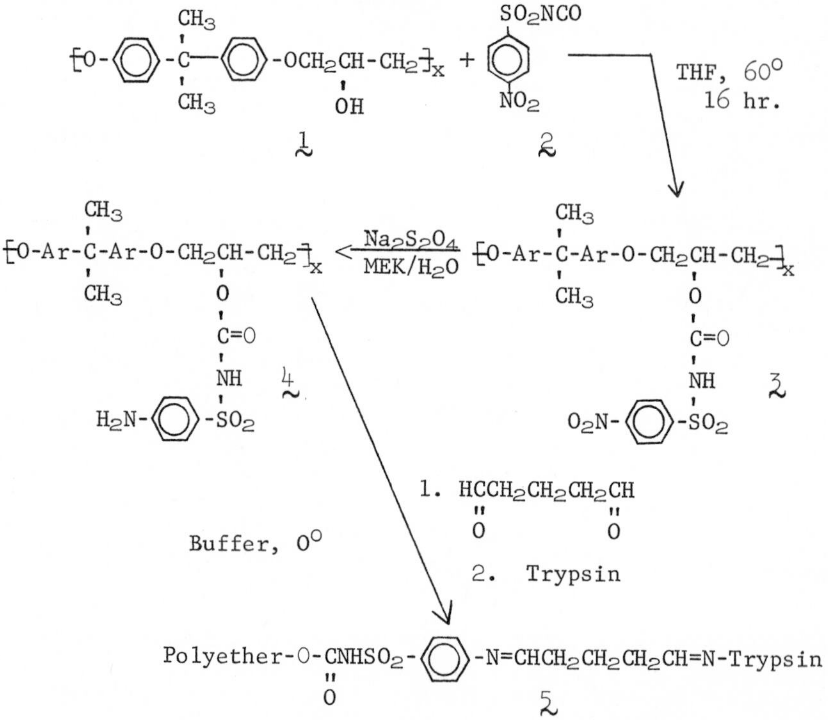

$$\left[\text{O-}C_6H_4\text{-}C(CH_3)_2\text{-}C_6H_4\text{-OCH}_2\text{CH(OH)-CH}_2\right]_x \; (\mathbf{1}) + \text{O}_2\text{N-}C_6H_4\text{-SO}_2\text{NCO} \; (\mathbf{2}) \xrightarrow{\text{THF, } 60^\circ,\ 16\text{ hr.}}$$

$$\left[\text{O-Ar-C(CH}_3)_2\text{-Ar-O-CH}_2\text{CH(O-C(=O)-NH-SO}_2\text{-}C_6H_4\text{-NO}_2)\text{-CH}_2\right]_x \; (\mathbf{3})$$

$$\xrightarrow{\text{Na}_2\text{S}_2\text{O}_4,\ \text{MEK/H}_2\text{O}} \left[\text{O-Ar-C(CH}_3)_2\text{-Ar-O-CH}_2\text{CH(O-C(=O)-NH-SO}_2\text{-}C_6H_4\text{-NH}_2)\text{-CH}_2\right]_x \; (\mathbf{4})$$

$$\xrightarrow[\text{Buffer, } 0^\circ]{1.\ \text{HC(=O)CH}_2\text{CH}_2\text{CH}_2\text{CH(=O)};\ 2.\ \text{Trypsin}} \text{Polyether-O-C(=O)NHSO}_2\text{-}C_6H_4\text{-N=CHCH}_2\text{CH}_2\text{CH}_2\text{CH=N-Trypsin} \; (\mathbf{5})$$

The reduction of the nitro group in **3** proved to be unexpectedly difficult. Stannous chloride in HCl cleaved the carbamate linkage; similar results were obtained using phenylhydrazine in DMSO. Sodium dithionite failed to effect the reduction under the conditions reported in the literature,[3] but utilization of a two-phase system (MEK/H_2O) where **3** was dissolved in MEK was effective. Polyamine **4** is very hydrophobic and can not be protonated in 4N HCl; it probably exists as a zwitterion between the amine and the sulfonyl carbamate groups. It was the original intent to diazotize the aryl amine and use the diazonium salt to couple the enzyme. Treatment of **4** with i-amyl nitrite in chloroform and trifluoroacetic acid[4] produced a gummy reddish-brown polymer which formed a bright red derivative with β-naphthol. However, attempts to bind the diazotized support to trypsin in both aqueous and mixed solvent systems (DMF/H_2O) were not successful. The enzyme appeared to be adsorbed only on the surface of the support; several washings destroyed the esterolytic activity of the system. Glutaraldehyde had been successfully used to immobilize enzymes by Habeeb[5] and it seemed to increase the hydrophillic properties of the substrate so we selected this reagent to bind trypsin to **4**. The binding process appeared to be quite efficient; amino acid analysis indicated that ca. 30% of the total weight of **5** is bound enzyme.

The esterase activity (V_{max}) of bound trypsin was found to be approximately 1/4 that of free tryspin while Km is about 4 times larger. A slight reduction in the pH required for optimum activity of immobilized enzyme was observed but this enzyme remained relatively reactive at high pH ranges, which normally deactivate trypsin. We have also bound thrombin under similar conditions and have shown that the bound enzyme exhibits about 1/6 of the esterolytic activity of native thrombin. The thrombin-polyether system is also effective in catalyzing blood clotting.

REFERENCES

1. R. W. Rousseau, C. D. Callihan, and W. H. Daly, Macromolecules, **2**, 502 (1969).

2. H. Ulrich and A. A. R. Sayigh, Angew. Chem. Int. Ed., **5**, 704 (1966).

3. (a) P. Cuatrecasas, J. Biol. Chem., **245**, 3059 (1970); (b) J. K. Inman and H. W. Dintzis, Biochemistry, **8**, 4070 (1969); (c) R. Axen and J. Porath, Nature, 367 (1966).

4. Z. B. Papanastassion, A. McMillan, U. T. Czebotar, T. J. Bardow, J. Am. Chem. Soc., **81**, 6056 (1959).

5. A. F. S. A. Hebeeb, Archives of Biochem. and Biophys., **119**, 264 (1967).

SYNTHESIS AND PROPERTIES OF POLYMERS WITH SULFUR-CONTAINING FUNCTIONAL GROUPS

Eric J. Goethals

Laboratory of Organic Chemistry
Rijksuniversiteit Gent
Gent, Belgium

INTRODUCTION

Owing to the fact that sulfur can exist in different oxidation states and that in each oxidation state it can form different functions, the number of different organic sulfur-containing compounds is very large. The compounds can be highly reactive or chemically inert. This article gives a review of the synthesis of polymers which contain the most common sulfur containing functional groups which are moderately or highly reactive. It also discusses the chemical reactions which have been carried out with these polymers or gives some suggestions of potential uses.

A. Thiols

1. Synthesis. Polythiols have been sunthesized by two general methods: polymerization of thiol-containing monomers and introduction of the thiol group into an already existing macromolecule. In the first method, the thiol group must be masked because it is an inhibitor for ionic polymerization and a powerful chain-transfer agent for radical polymerizations.

In the following paragraph some examples of monomers with masked thiol groups which have been used to synthesize polymeric thiols are discussed.

For the preparation of poly(vinylmercaptan) (3), several monomers have been used but the best ones seem to be vinyl thiolacetate (1-3) (1) and S-vinyl-O-t-butyl thiolcarbonate (3,4) (2).

$$CH_2{=}CH(SAc)\ (\underline{1}) \xrightarrow{AIBN} \ldots{-}CH_2{-}CH(SAc){-}\ldots \xrightarrow{1)\ NaOH\ \ 2)\ HCl} {-}CH_2{-}CH(SH){-}\ (\underline{3})$$

$$CH_2{=}CH(S{-}COOt.Bu)\ (\underline{2}) \xrightarrow{AIBN} \ldots{-}CH_2{-}CH(SCOOt.Bu){-}\ldots \xrightarrow{\text{dry HBr or heat}} {-}CH_2{-}CH(SH){-}\ (\underline{3})$$

Poly (p.mercaptostyrene) (5) is also obtained via the acetylated monomer (4) (5).

$$CH_2{=}CH{-}C_6H_4{-}SAc\ (\underline{4}) \xrightarrow{AIBN} \ldots{-}CH_2{-}CH(C_6H_4{-}SAc){-}\ldots \xrightarrow{KOH} \ldots{-}CH_2{-}CH(C_6H_4{-}SH){-}\ldots\ (\underline{5})$$

Water-soluble polymers can be obtained by copolymerization of (3) with N-vinylpyrrolidone or N-isopropylacrylamide (4). The water-solubility decreased markedly when the tert-butyloxy-carbonyl group was removed.

Cationic ring-opening polymerization of 3,3-bis(mercapto-methyl)oxetane diacetate (6) with boron trifluoride etherate leads to a polyether which, after hydrolysis, produces a polythiol (7) with a polyether backbone (6).

$$\text{O}(CH_2)_2C(CH_2SAc)_2\ (\underline{6}) \qquad \ldots{-}CH_2{-}C(CH_2SH)_2{-}CH_2{-}O{-}\ldots\ (\underline{7})$$

A thiol-containing polyamide (9) has been synthesized by Overberger and coworkers by condensing αα'-dimercapto adipoyl chloride (with thiol functions protected by carbobenzoxy groups) (8) with hexamethylene diamine.

After polycondensation, the carbobenzoxy groups were removed with dry hydrogen bromide in trifluoroacetic acid (7).

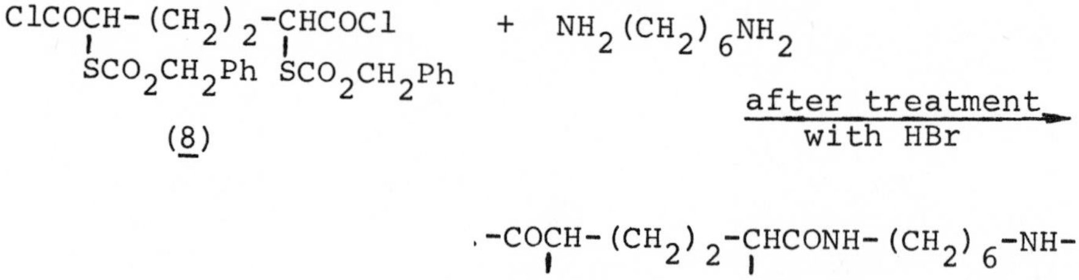

$$\underset{(\underline{8})}{\underset{|\qquad\qquad\qquad|}{ClCOCH-(CH_2)_2-CHCOCl}}\ \ \begin{matrix}\\ SCO_2CH_2Ph\ \ SCO_2CH_2Ph\end{matrix} + NH_2(CH_2)_6NH_2 \xrightarrow[\text{with HBr}]{\text{after treatment}}$$

$$-\underset{SH}{COCH}-(CH_2)_2-\underset{SH}{CHCONH}-(CH_2)_6-NH- \quad (\underline{9})$$

The same authors reported the preparation of a polyurethane, containing thiol side chains (8).

Dijkstra and Smith described the polymerization of different methacroyloxyalkylisothiouronium salts ($\underline{10}$) and their copolymerization with several acrylates or methacrylates (9). On treatment with two equivalents of base, the polymers were converted into the

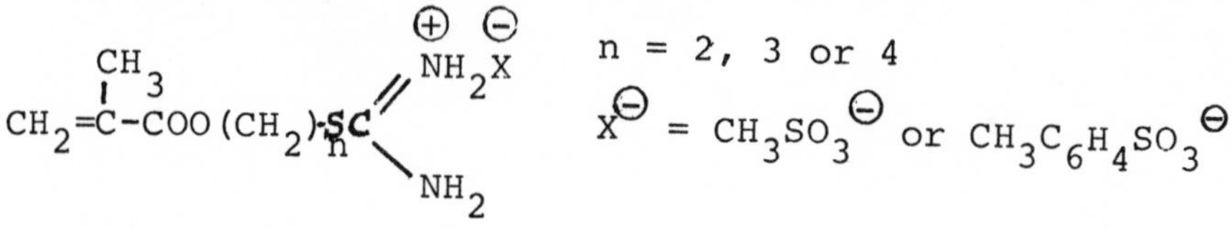

$$CH_2=\overset{CH_3}{\overset{|}{C}}-COO(CH_2)_n-SC\begin{matrix}\overset{\oplus}{=}NH_2X^{\ominus}\\ \diagdown NH_2\end{matrix} \qquad (\underline{10})$$

$n = 2, 3$ or 4

$X^{\ominus} = CH_3SO_3^{\ominus}$ or $CH_3C_6H_4SO_3^{\ominus}$

corresponding polythiols. Mercaptide ion formation was always markedly slower in the polymers than in the corresponding monomers or model substances.

Introduction of thiol groups into an already existing polymer has mostly been achieved with polystyrene or styrene-divinylbenzene copolymers. The following reaction schemes give two examples, the first one for the synthesis of poly (p-mercaptostyrene) (10) and the second one for the synthesis of poly (p-mercaptomethylstyrene) (11,12).

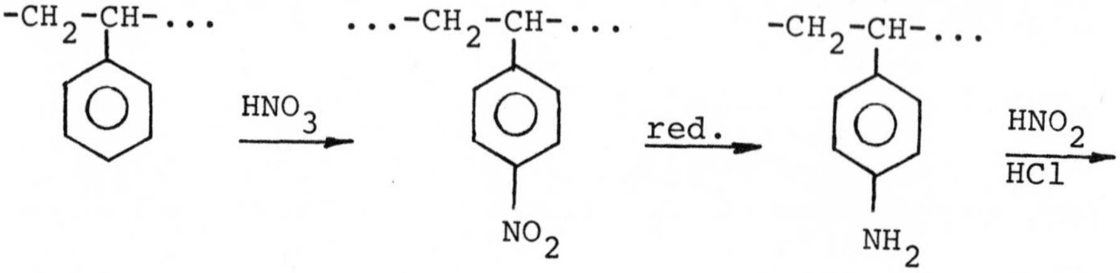

$$-CH_2-\underset{C_6H_5}{CH}-\ldots \xrightarrow{HNO_3} \ldots-CH_2-\underset{C_6H_4-NO_2}{CH}-\ldots \xrightarrow{red.} -CH_2-\underset{C_6H_4-NH_2}{CH}-\ldots \xrightarrow[HCl]{HNO_2}$$

$$-CH_2-CH-\ldots(C_6H_4)N_2Cl \xrightarrow{C_2H_5OCS_2K} \ldots-CH_2-CH-\ldots(C_6H_4)S-\underset{\|\atop S}{C}OC_2H_5 \xrightarrow[H_2O]{KOH} \ldots-CH_2-CH-(C_6H_4)SH$$

$$CH_2-CH-\ldots(C_6H_5) \xrightarrow{CH_2O\ +\ HCl} \ldots-CH_2-CH-\ldots(C_6H_4)CH_2Cl \xrightarrow{(NH_2)_2C=S}$$

$$\ldots-CH_2-CH-\ldots(C_6H_4)CH_2SC(=NH_2Cl)NH_2 \xrightarrow{H_2O,\ HCl} \ldots-CH_2-CH-\ldots(C_6H_4)CH_2SH$$

2. Reactions. Thiols are reducing agents according to the reaction

$$2\ RSH \underset{+2e}{\overset{-2e}{\rightleftarrows}} RSSR + 2\ H^+$$

With strong reducing agents, thiols can be recovered from disulfides. Consequently, insoluble macromolecular thiols can be used as redox resins or electron exchanger. Synthetic polymeric thiols are also interesting in relation to the important role of the thiol function in some enzymes, and they have been studied because of their antiradiation properties. Finally, polymeric thiols could be used as specific ion exchangers.

Most of the studies on reactions of polythiols concern the redox reactions. The first reports on this subject are from Cassidy (13). Kun (14) described the preparation of redox resins using conventional and macroreticular chloromethylated styrene-divinylbenzene copolymers. The macroreticular resins have the advantage to retain the spherical form during the synthesis and also after an oxidation-reduction cycle. The oxidized form of the resin, i.e., the disulfide form was reduced with 10% aqueous sodium bisulfite.

A number of papers especially by Overberger and coworkers have described the oxidation of soluble polythiols to disulfides by means of different oxidizing agents (15-18). The purpose of these studies was to compare the reactivity of the polymers with low molecular weight model substances.

The results depend on the kind of oxidation reaction used to follow the disappearance of thiol groups. A technique for following the rate of oxidation of thiols by 2,6-dichlorophenolindophenol spectrophotometrically was used in pH 10 aqueous buffer solution and in organic solvents (15). In the pH 10 buffer systems, a proximity effect was illustrated by the enhanced oxidation rates of 2,4-di(p-mercaptophenyl)pentane and poly(p-mercaptostyrene) over p-thiocresol (16). Only a slight increase in the rate of oxidation of polymers in comparison with their model compounds was observed. When the oxidation was carried out in dimethylformamide, the proximity effect of neighboring thiol substituents became more pronounced. The results obtained for poly(vinylmercaptan) and a series of dithiols are summarized in Table I.

Table I: (15) Relative oxidation rates of thiols in DMF

Thiol	Observed oxidation rates, μmole. ml^{-1} min^{-1} × 10^5 (a)	Relative rates
2,6-heptanedithiol	0.328	1
2,5-hexanedithiol	3.90	11
2,4-pentanedithiol	9.70	28.7
2,3-butanedithiol	20.1	61.2
poly(vinylmercaptan)	91.4	280

[a] On the basis of one SH group oxidized.

Oxidation by molecular oxygen in dimethyl sulfoxide with ferric sulfate as a catalyst (17) is proposed to occur via the rapid formation of a ferric-thiol complex, a slow rate-determining reduction of the complex by mercaptan and finally rapid regeneration of the iron (III) complex by oxidation with the oxygen. As shown in Table II, under these conditions also the ease of oxidation increases as the distance between sulfhydryl groups decreases. Although poly(vinyl mercaptan) oxidizes faster than 2,4-pentane dithiol a polymer effect of the magnitude predicted from the oxidations by 2,6-dichlorophenolindophenol was not observed.

In addition to a slight rate enhancement over the dithiol

model compound, poly(vinylmercaptan) exhibits other characteristic differences suggesting that the actual concentration of mercapto groups capable of being oxidized is less than the mercaptan titer of the polymer solution. This was attributed to the conformation of the chain (stabilized by a few disulfide linkages) which would render sequences of sulfhydryl groups inaccessible.

Another aspect of the polymer effect is the mobility of the sulfhydryl groups attached to the polymer chain. The polymer does not precipitate during the initial 30% of reaction indicating that intramolecular disulfide formation is the predominating mode of oxidation.

Table II: Relative oxidation rates of thiols in dimethyl sulfoxide (17).

	K (min^{-1})	Relative rates
2,6-heptanedithiol	1.64	1
2,5-hexanedithiol	3.55	2.2
2,4-pentanedithiol	25.8	15.1
poly(vinylmercaptan)	50.6	30.8

The relative oxidation rates of poly(vinylmercaptan), poly-[3,3-bis(mercaptomethyl)oxetane] and a number of model compounds of the latter in aqueous 0.5 N sodium hydroxide solution with molecular oxygen in the presence of different metal salts as catalyst, have been measured (18). Cullis, Swan and Trimm (19) who made an extensive study of the oxidation of mercaptans under these conditions, found that depending on the metal salt used as catalyst, these oxidations were of pseudo zero order or of first order with respect to thiol. The same dependence was found for the polymeric mercaptans. Table III gives a summary of the results.

Under these conditions poly(vinylmercaptan) is also oxidized at a higher rate than low molecular weight compounds if the reaction is of first order with respect to thiol concentration (for catalysis with Co^{2+}). When the rate is independent of the thiol concentration (catalysis with Zn^{2+} or Cu^{2+}) the low molecular weight compounds are oxidized at a higher rate than poly(vinylmercaptan). That a polymer effect is absent is plausible since it cannot be expected that the presence of neighboring thiolgroups or the existence of micro regions with very high thiolconcentrations could have an influence, if the rate determining step of the reaction is independent of thiol concentration. In these cases the rate determining step probably is the formation of a complex between the metal salt, mercapto groups, solvent and oxygen.

probable that these complexes must meet certain steric requirements and the lower rate observed for poly(vinylmercaptan) could be ascribed to a greater difficulty for the polymer to form these complexes.

Poly[bis(mercaptomethyl)oxetane] always reacts at a lower rate. This could be ascribed to the fact that in this case, the polymer chain is much more rigid than in poly(vinylmercaptan) so that the formation of the complex is always the, or one of the rate determining steps.

Table III (18): Relative oxidation rates of several dithiols and polythiols with oxygen in aqueous 0.5N sodium hydroxide under the influence of metal salts (10^{-3} mol l^{-1}).

Thiol	Zn^{2+} (a)	Cu^{2+} (a)	Co^{2+} (b)
$...-CH_2-C(CH_2SH)_2-CH_2-O-...$	1	1	1
$O(CH_2)_2C(CH_2SH)_2$ (oxetane ring)	2.9	3.6	8.0
$CH_3-C(CH_2SH)_2-CH_3$	6.7	5.4	18
$CH_3OCH_2C(CH_2SH)_2-CH_2OCH_3$	6.9	4.8	14
$...-CH_2-CH(SH)-...$	2.6	2.25	34

(a) Reactions are pseudo zero order in thiol

(b) Reactions are first order in thiol.

B. Sulfides

1. Synthesis. Polymers in which the sulfide group is part of the polymer chain can be obtained by various polycondensation reactions or by ring-opening polymerization of cyclic sulfides. Examples of polycondensation reactions are:

$$NaS\text{-}CH_2CH_2\text{-}SNa + Br(CH_2)_nBr \rightarrow \ldots\text{-}S\text{-}CH_2CH_2\text{-}S\text{-}(CH_2)_n\text{-}\ldots$$

$$HS(CH_2)_6SH + CH_2{=}CH\text{-}CH_2CH_2CH{=}CH_2 \rightarrow \ldots\text{-}S\text{-}(CH_2)_6\text{-}S\text{-}(CH_2)_6\text{-}\ldots$$

$$Cl\text{-}C_6H_4\text{-}SNa \longrightarrow \ldots\text{-}C_6H_4\text{-}S\text{-}\ldots$$

Ring-opening polymerization of 3- and 4-membered cyclic sulfides can be of the anionic, cationic or coordination type of mechanism.

$$CH_2\text{—}CH(CH_3) \text{ (bridged by S)} \longrightarrow \ldots\text{-}CH_2\text{-}CH(CH_3)\text{-}S\text{-}\ldots$$

Polymers in which the sulfide group is substituted on the chain are obtained by free radical or cationic polymerization of alkyl vinyl sulfides.

$$CH_2{=}CH(SR) \longrightarrow \ldots\text{-}CH_2\text{-}CH(SR)\text{-}\ldots$$

2. Reactions. Although polysulfides have been studied intensively in many laboratories in the last decades, they are of minor importance in the present context because of their relative chemical inertness.

Sulfides form complexes with heavy metals and with halogens (20). In the presence of water, the halogen-complexes react with breaking of a sulfur-carbon bond and the formation of sulfonyl halides. These properties have not yet been studied with polymeric sulfides.

C. Sulfoxides

1. Synthesis. Only a few polymers with sulfoxide groups have been synthesized. The main reason is that the sulfoxide function retards or even inhibits radical polymerization. So methyl sulfoxide does not polymerize and copolymerization with styrene or with methyl methacrylate yields only small amounts of polymer containing very little methyl vinyl sulfoxide units (21).

Other routes for the preparation of polysulfoxides are polycondensation of sulfoxide-containing bifunctional compounds and controlled oxidation of polythioethers. So phosgene and bis(4-hydroxyphenyl)sulfoxide yield a sulfoxide-containing polycarbonate with melting point 230-250° (22).

$$HO-C_6H_4-S(=O)-C_6H_4-OH + COCl_2 \longrightarrow -O-C_6H_4-S(=O)-C_6H_4-OC(=O)-$$

Several polyphenylene sulfoxides have been obtained by oxidation of the corresponding polyphenylene sulfides with nitric acid (23).

Oxidation of polythietane to the corresponding polysulfoxide was achieved with 1 equivalent of hydrogen peroxide (24).

$$..-S-CH_2CH_2CH_2-... \xrightarrow{H_2O_2} ...-S(=O)-CH_2CH_2CH_2-...$$

2. Reactions. Reactions of polymeric sulfoxides have not yet been described. It is well known that sulfoxides are good reagents for oxidation reactions (25). Dimethyl sulfoxide or tetramethylene sulfoxide are mostly used. The reactions include transformation of thiols into disulfides, phosphines into phosphine oxides, and alcohols, esters, chloroformiates, halides into aldehydes or ketones. The reactions are often characterized by high yields and are possible also at highly sterically hindered substrates. The last series of reactions proceed via the corresponding alkoxysulfonium salts (11).

$$R_2S{=}O + R'_2CHX \longrightarrow R_2\overset{\oplus}{S}-O-CHR'_2\ X^{\ominus} \xrightarrow{base} R_2S + O{=}CR'_2$$

(11)

D. Sulfones

1. Synthesis. Polymers with sulfone functions in the polymer chain are obtained by radical copolymerization of an alkene with sulfur dioxide or by polycondensation of bifunctional monomers containing a sulfone function

$$CH_2{=}\underset{\displaystyle R}{\underset{|}{C}}H + SO_2 \longrightarrow \ldots{-}CH_2{-}\underset{\displaystyle R}{\underset{|}{C}}H{-}SO_2{-}\ldots$$

$$Cl{-}C_6H_4{-}SO_2{-}C_6H_4{-}Cl + NaO{-}C_6H_4{-}C(CH_3)_2{-}C_6H_4{-}ONa \longrightarrow$$

$$\ldots{-}C_6H_4{-}SO_2{-}C_6H_4{-}O{-}C_6H_4{-}C(CH_3)_2{-}C_6H_4{-}O{-}\ldots$$

Vinyl sulfones have been polymerized by anionic initiators (26). Radical homopolymerization was difficult but copolymers with other vinyl monomers were readily obtained (27).

2. Reactions. The sulfone group is a chemically inert function and is of little interest in the context of this review.

E. Sulfonium Salts

1. Synthesis. Bailey and Lacombe described the synthesis, polymerization and copolymerization of a series of acrylic and methacrylic sulfonium monomers (12) (28). The monomers were prepared just prior to conversion to polymer by reacting a thioether, e.g., S-methyl thioethylacrylate, with an alkylating agent such as dimethylsulfate.

$$CH_2{=}CH{-}COOCH_2CH_2SCH_3 \xrightarrow{(CH_3)_2SO_4} CH_2{=}CH{-}COOCH_2CH_2\overset{\oplus}{S}(CH_3)_2 \quad CH_3OSO_3^{\ominus}$$

(12)

The monomers readily homopolymerize by a free radical mechanism to yield high molecular weight, water soluble polymers which demonstrate typical polyelectrolyte behavior. Copolymers with acrylamide, N-vinylpyrrolidone, styrene, etc., were also water-soluble (29).

Hatch and coworkers (30) described the synthesis of a series of vinylbenzyl sulfonium monomers and polymers by reacting a mixture of ortho and para vinylbenzyl chloride with various dialkyl sulfides. The monomers polymerized spontaneously at room temperature in aqueous solution forming highly crosslinked gels. The crosslinking is ascribed to an interchange of the substituent groups of the sulfonium monomer during or after its polymerization which would form some divinyl derivative (13) or form cross-links in the polymer.

$$CH_2{=}CH{-}C_6H_4{-}CH_2\overset{\oplus}{S}R_2Cl^{\ominus} \rightleftharpoons CH_2{=}CH{-}C_6H_4{-}CH_2SR + RCl$$

$$CH_2{=}CH{-}C_6H_4{-}CH_2SR + CH_2{=}CH{-}C_6H_4{-}CH_2Cl \rightleftharpoons (CH_2{=}CH{-}C_6H_4{-}CH_2)_2\overset{\oplus}{S}R\ Cl^{\ominus}$$

(13)

2. Reactions. Polymeric sulfonium salts have properties which are similar to those of the quaternary ammonium salt polymers. In a recent review, Hoover (31) has described the (potential) uses of this class of polymers. They are especially useful as coagulants and flocculants, have been used as anion exchange resins and may find applications as permselective membranes, sensitizers for photographic films, etc. One disadvantage of the sulfonium salts over the ammonium salts is that they are more readily attacked by nucleophilic reagents (which can also be the counter ion) to form the corresponding thioethers. When the nucleophile is a water soluble amine, this reaction can be used to prepare polymeric amines.

F. Sulfonic Acids

1. Synthesis. Polymers containing sulfonic acid groups in the form of the free acid or of a salt have been studied intensively. The two methods for the preparation of these polymers

are polymerization of sulfonated monomers and introduction of the sulfonic acid groups in an existing polymer.

a. Polymerization of sulfonated monomers. Ethenesulfonic acid polymerizes in aqueous solution (32-34), under the influence of hydrogen peroxide, potassium persulfate, or ultraviolet light. In the latter case the molecular weight can reach 250,000 (32). The polymer can be separated from unreacted monomer by adding barium chloride since the barium salt of poly(ethenesulfonic acid) is insoluble, that of ethenesulfonic acid, soluble in water.

Generally it is preferred to use an alkali salt of ethenesulfonic acid to prepare the polymer. These salts are easier to synthesize and can be stored without decomposition. Mostly the polymerization is carried out in aqueous solution under the influence of ultraviolet light, peroxides or redox systems (35).

Copolymerization of ethenesulfonic acid or other sulfonated vinylmonomers and their salts with different vinyl monomers has been studied intensively because that way it is possible to introduce a reactive function, in this case the sulfonic acid group, into a polymer. Ethenesulfonic acid and its sodium salt have been copolymerized with different vinyl monomers (35).

In the copolymerization with N-vinylpyrrolidone, the composition of the copolymer is about the same as that of the monomer mixture. In all other cases it was necessary to take a large excess of ethenesulfonate when a high content of sulfonate groups in the copolymer is desired. A patent however claims that the copolymerization of sodium ethenesulfonate with acrylic acid or acrylamide or a mixture of both gives polymers with a high sulfonate content when the copolymerization is carried out in alcohol medium (36).

Wiley and coworkers (37-39) studied the polymerization and copolymerization of potassium styrene sulfonate in aqueous solution, and in dimethylformamide. In both cases water-soluble high viscosity polymers showing polyelectrolyte properties were obtained.

b. Polycondensation of sulfonic acid-containing compounds. A great number of papers and patents describe the polycondensation of sulfonated phenol with formaldehyde (40), yielding a sulfonated polymer which, depending on the reaction conditions may be either soluble or insoluble in water. Sulfonation of phenol gives predominantly p-phenolsulfonic acid, and it can normally be expected that polycondensation of this compound with formaldehyde leads to a linear and therefore soluble polymer (14).

$$HO-C_6H_4-SO_3H + CH_2O \longrightarrow \cdots-C_6H_2(OH)(SO_3H)-CH_2-C_6H_2(OH)(SO_3H)-CH_2-\cdots$$

(14)

Nevertheless, insoluble products are obtained by allowing sufficient time for the polycondensation. Probably some sulfonic acid groups are split off by hydrolysis whereby the reactive para position of the phenol becomes free, leading to crosslinked products (41). This hypothesis is supported by the observation that crosslinking becomes more important with increasing water concentration. When the condensation is carried out in alkaline medium, soluble polymers are obtained. When the condensation is carried out in alkaline medium, soluble polymers are obtained. When insoluble products are desired, it is necessary to add unsubstituted phenol to the reaction mixture.

c. Introduction of sulfonic acid groups in polymers. Direct sulfonation of polymers by means of concentrated sulfuric acid, chlorosulfonic acid, or sulfur trioxide, as such or in the form of complexes, is certainly the most important method of obtaining sulfonic acid-containing polymers. Of all polymers, polystyrene is the most used in this way.

Two types of sulfonated polystyrene are of commercial interest, one completely water-soluble, prepared from styrene homopolymer, the other insoluble in water and made from styrene-divinylbenzene copolymer.

Sulfonation of styrene homopolymer by sulfuric acid to form soluble derivatives is difficult because of sulfone formation, which leads to crosslinked and therefore insoluble products. When sulfur trioxide is used, the production of soluble derivatives is favored by low concentration of the reagents, efficient agitation, low temperatures, pure solvents, concurrent reactant feeding, use of a small reaction vessel, low excess of sulfonating agent, use of vinyltoluene-containing polymers, and rapid handling of the finished product (42). Concentrated sulfuric acid largely yields colored insoluble materials. Under special conditions, however, crosslinking can be avoided. Thus, soluble derivatives were obtained (43) by treating finely ground polystyrene of molecular weight below 250,000 with 95% sulfuric acid at 100°C. Soluble derivatives are also obtained when ethylene chloride is used as solvent (44) or when the sulfonation is done in the presence of silver sulfate as a catalyst (45). By using this catalyst, it is possible to sulfonate polystyrene without change in the molecular-weight distribution (46). When sulfur trioxide is the sulfonating agent, the reaction must be carried out in a solvent to avoid carbonization. When the concentration of the reactants is kept low (maximum 10%) and the temperature is between -20 and 45°C, colorless water-soluble products are obtained (47). Preferred solvents are chlorinated hydrocarbons such as methylene chloride, ethylene chloride, or a mixture of sulfur dioxide and carbon tetrachloride.

Sulfur trioxide adducts largely yield water-soluble poly(styrenesulfonic acid) (48).

Comparison of the infrared spectra of sulfonated polystyrene and of poly(p-styrenesulfonic acid) obtained by polymerization of the vinyl monomer indicates that sulfonation of polystyrene occurs practically exclusively at the para position of the benzene nucleus (49).

The properties of soluble sulfonated polystyrenes as a function of the degree of sulfonation are discussed by Signer and Demagistri (50). The polymers are soluble when 25% of the benzene rings are sulfonated and they behave as typical polyelectrolytes in solutions.

Sulfonation of crosslinked polystyrene gives insoluble products which are commercially available as ion-exchange resins. The most common crosslinking agent is divinylbenzene. Besides styrene-divinylbenzene copolymers, a few other copolymers of styrene have been sulfonated. Those with butadiene (51) and with divinyl-acetylene (52) are crosslinked, and their sulfonated derivatives can also be used as ion-exchange resins. A patent describes the sulfonation of poly(2,6-dimethylphenylene oxide) to form cation-exchange resins (53).

Sulfoalkylation occurs when a mixture of formaldehyde and sodium bisulfite forms sodium hydroxymethanesulfonate; this mixture can be used to introduce the sulfomethyl group into a molecule carrying an active hydrogen atom.

$$CH_2O + NaHSO_3 \rightarrow HOCH_2SO_3Na$$

$$R\text{-}H + HOCH_2SO_3Na \rightarrow R\text{-}CH_2SO_3Na + H_2O$$

This reaction has been used to sulfonate water-soluble phenol-formaldehyde resins, yielding a tanning agent (54). Commonly, such sulfonated phenol-formaldehyde resins are formed in one step by reacting a mixture of aqueous sodium sulfite, a phenol, and 30% formaldehyde in alkaline medium (55). When excess formaldehyde is used, water-insoluble products are obtained (56). These can be used as ion-exchange resins.

Sulfomethylcellulose has been made by reaction of sodium chloromethanesulfonate, easily obtained from methylene chloride and sodium sulfite, with alkali cellulose in isopropyl alcohol (57) or in water (58). In an analogous manner, sulfoethylcellulose has been made from alkali cellulose and sodium 2-chloroethanesulfonate (59).

$$R_{cell}OH\text{-}NaOH + ClCH_2CH_2SO_3Na \rightarrow R_{cell}OCH_2CH_2SO_3Na + NcCl + H_2O$$

Sulfoethylcellulose containing more than 0.1 sulfoethyl group for each anhydroglucose unit is soluble in alkali; with more than 0.2 sulfoethyl group, it is soluble in water (60).

Starch has been transformed into sulfoethylstarch by reaction with sodium 2-chloroethanesulfonate in alcoholic medium (61).

Sulfopropyl derivatives of various polymers have been made by reaction with propanesultone. Cellulose (62) and starch (62) react best in a water-miscible solvent such as isopropyl alcohol or dioxane after addition of an aqueous sodium hydroxide solution. Poly(vinyl alcohol) reacts with propanesultone in dimethyl sulfoxide solution in the presence of potassium carbonate (63).

Poly(ethylenimine) reacts with propanesultone in methanol or in benzene to form an amphoteric polymer (15). This is insoluble in water but soluble in strong acids or strong bases (64). Reaction of 4-vinylpyridine with 1,4-butanesultone yields an inner salt which has been polymerized to a polyampholyte (16). An analogous polymer is obtained when poly(2-vinylpyridine) is treated

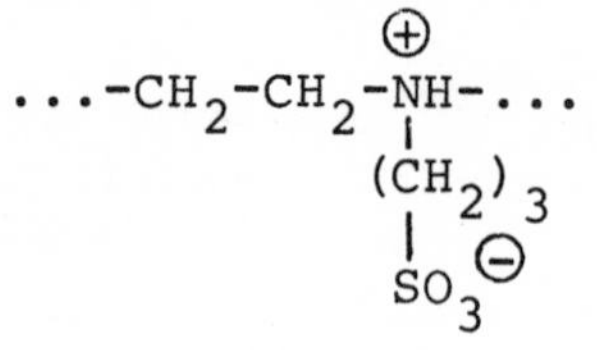

$$\ldots\text{-}CH_2\text{-}CH_2\text{-}\overset{\oplus}{N}H\text{-}\ldots \quad (CH_2)_3SO_3^{\ominus} \text{ on N}$$

(15)

$$\ldots\text{-}CH_2\text{-}CH\text{-}\ldots \text{ (4-pyridinium, } N^{\oplus}\text{-}(CH_2)_4SO_3^{\ominus})$$

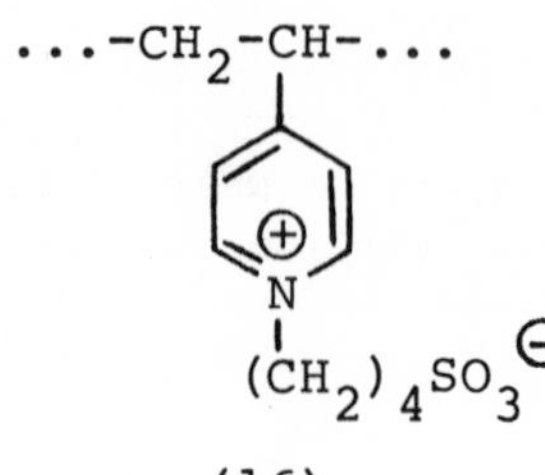

(16)

with 1,4-butanesultone. In this case the product is not completely substituted, since it precipitates before 100% is reached.

2. Reactions. Reactions with polymeric sulfonic acids have mostly been related to the catalytic activity of the strongly

acidic sulfonic acid group. The reactions can be divided into two classes depending on the polymer being soluble or insoluble in the reaction medium. In the first case ethenesulfonic acid polymers or copolymers and soluble sulfonated polystyrene are mostly used. In the second case the polymers are the classical and commercially available ion exchange resins.

a. Reactions with soluble polysulfonic acids. In a solution of a polymeric acid, the hydrogen ions, formed by dissociation, tend to congregate in the neighborhood of the polyanion as a counterion layer. Their distribution is thus non-uniform and Morawetz and Westhead (66) have treated mathematically the effect of this circumstance on the rates of acid-catalyzed reactions.

For the benzidine rearrangement they pointed out that a marked increase in rate is to be expected when a monobasic acid is replaced by the equivalent concentration of a polymeric acid.

The rate of this rearrangement is known to be proportional to the square of acid concentration which implies that it is the diprotonated species (18) which rearranges.

$$C_6H_5\text{-}NH\text{-}NH\text{-}C_6H_5 \underset{}{\overset{+H^{\oplus}}{\rightleftharpoons}} C_6H_5\text{-}\overset{\oplus}{N}H_2\text{-}NH\text{-}C_6H_5 \ (17) \underset{}{\overset{+H^{\oplus}}{\rightleftharpoons}}$$

$$C_6H_5\text{-}\overset{\oplus}{N}H_2\text{-}\overset{\oplus}{N}H_2\text{-}C_6H_5 \ (18) \longrightarrow \overset{\oplus}{N}H_3\text{-}C_6H_4\text{-}C_6H_4\text{-}\overset{\oplus}{N}H_3$$

This phenomenon has been studied by Arcus and coworkers (67). As shown in Table IV the rate of the rearrangement under the influence of poly(styrenesulfonic acid) is indeed much higher than for the non-polymeric acids.

Table IV (67): The benzidine rearrangement in ethanol (96%) at 0°C.

Hydrazobenzene : 0.010 M	Acid : 0.05 N
Acid	10^3 K (min^{-1})
Hydrogen chloride	0.24
Benzenesulfonic Acid	0.27
Poly(styrenesulfonic acid)	32

It is assumed that the monoprotonated species (17) associates electrostatically with the polyanion, whence its concentration is augmented in the region of increased hydrogen ion concentration. The observed 120-fold acceleration of the rearrangement is ascribed to the combination of the two non-uniformities in distribution, the increase of hydrogen concentration in the volume enveloping the polyanion and the increase in the concentration of (17) in the same region, leading to an accelerated rate of formation of the rearranging species (18).

Sakurada and his collaborators (68) have studied the homogeneous hydrolysis of low molecular weight esters with polymeric sulfonic acids such as poly(ethenesulfonic acid), copolymers of ethenesulfonic acid and styrene, sulfonated polystyrenes, etc. As shown in Table V, when longer alkyl esters such as butyl acetate are hydrolyzed with partially sulfonated polystyrenes, the rates of hydrolysis are 10 times greater than those of hydrolysis with hydrogen chloride.

Table V (68): Hydrolysis of methyl acetate and n-butyl acetate with polystyrene sulfonic acids in water at 40°C.

Catalyst	a) Methyl acetate Ester concn. : $2.50 \cdot 10^{-3}$M catalyst concn. : $5 \cdot 10^{-3}$N		b) Butyl acetate Ester concn. : $2.85 \cdot 10^{-2}$M catalyst concn. : $5 \cdot 10^{-3}$N	
	$k \cdot 10^2$ (l mole^{-1} min^{-1})	$r^{(b)}$	$k \cdot 10^2$ (l mole^{-1} min^{-1})	$r^{(b)}$
HCl	2.31	1.00	2.30	1.00
PS-S(23)[a)]	5.06	2.19	23.5	10.22
PS-S(33)	4.86	2.10	18.3	7.96
PS-S(40)	3.63	1.57	14.5	6.31
PS-S(52)	3.53	1.53	10.53	4.50
PS-S(65)	3.25	1.41	6.94	3.02
PS-S(77)	2.79	1.21	5.91	2.57
PS-S	2.39	1.03	4.42	1.92

a) Figures in parentheses mean the degree of sulfonation in mole %.

b) Ratio of the rate constant for the polymer catalyst to that for HCl : $r = k/k_{HCl}$

This is ascribed to the occurrence of a hydrophobic interaction between a benzene ring of the polymer and the alkyl group which results in a higher concentration of ester in the neighborhood of the polymer molecule. The catalytic efficiency of poly(styrenesulfonic acid) increased with decreasing degree of sulfonation which means that an unsulfonated monomer unit is more effective for the hydrophobic binding of the ester molecule to the polymer catalyst, than a sulfonated one. As shown in Fig. 1, the effect is more pronounced for n butyl acetate than for t butylacetate. Esters which contain an amino group are hydrolyzed with poly(styrenesulfonic acid) with a much higher rate than with hydrogen chloride due to the electrostatic interaction with the polymeric acid. As shown in Fig. 2, in this case the rate-enhancement becomes more pronounced with increasing degree of sulfonation. The third example of reactions catalyzed by soluble polymeric sulfonic acids, is the case of poly(styrenefulfonic acid) in the controlled fragmentation of labile polysaccharides, described by Painter (69).

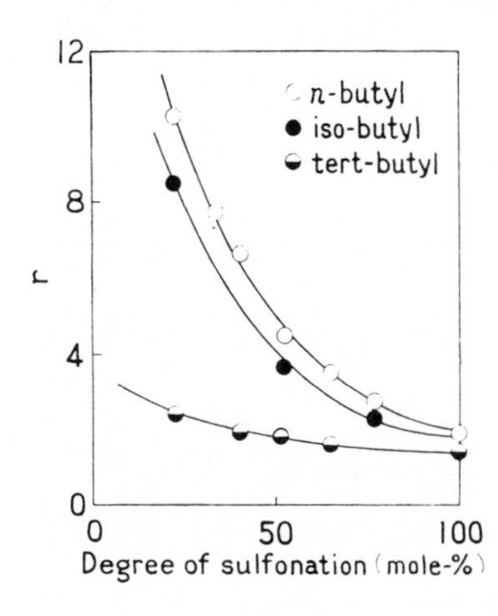

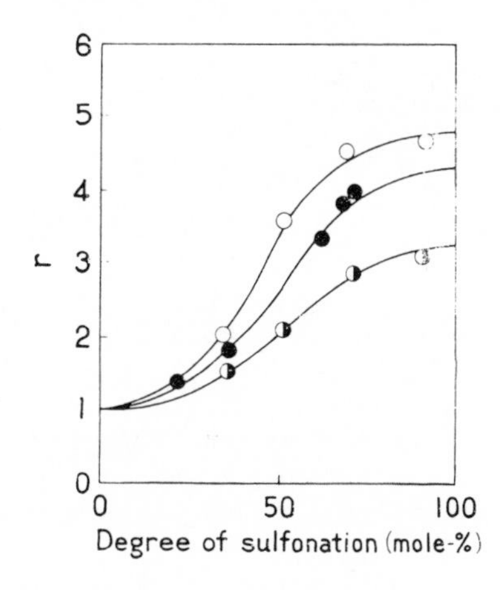

Fig. 1. Hydrolysis of butyl acetates with polystyrenesulfonic acids of various degree of sulfonation in water (Ester conc.: $2.85 \cdot 10^{-2}$M; catalyst conc.: $5.00 \cdot 10^{-3}$N; temp.: 40°C).

Fig. 2. Hydrolysis of ethyl p-aminobenzoate with polystyrenesulfonic acids of various degrees of sulfonation in 50% aqueous acetone at 80°C. (Catalyst conc.: 0.05 N; ester concn.: upper curve 0.08 M, middle curve 0.10 M, lower curve 0.18 M).

In the study of polysaccharide structure by partial hydrolysis, it is important to recover as much as possible of the original polysaccharide as oligosaccharides rather than as monosaccharides, or as fragments which are too large to separate and identify. The yield of oligosaccharides can be improved if the desired fragments are removed continuously from the reaction mixture by dialysis. To achieve this, a non-dialysable catalyst must be used.

The water-insoluble cation exchange resins are non-dialysable but they are rather unsatisfactory for the hydrolysis of polysaccharides probably owing to the difficulty with which large molecules penetrate the resin particles (70). Therefore Painter (69) used poly(styrenesulfonic acid) to hydrolyse the polysaccharide inulin. When the hydrolysis was carried out in a specially designed apparatus which provided 16 cm^2 of membrane per ml of reaction mixture, the yield of higher oligosaccharides was substantially increased at the expense of fructose and the lower oligosaccharides (see Table VI). The usefulness of the technique however was limited by the poor selectivity of the cellophane membranes which allowed too much high molecular-weight material to pass, thereby reducing the yield of the more useful fragments.

Table VI (69): Hydrolysis of polysaccharide by poly(styrenesulfonic acid) using dialysis.

Area of membrane cm^2/ml	Saccharides obtained (% of total)								
	mono	di	tri	tetra	penta	hexa	hepta	octa	higher
0	51.2	8.9	8.5	6.6	5.3	3.9	3.5	3.2	8.9
4	26.4	9.9	9.8	10.4	9.8	10.1	7.0	3.8	12.8
16	17.6	6.1	7.7	9.5	9.7	9.8	11.0	10.1	18.5

b. Reactions with insoluble polysulfonic acids. Many chemical reactions can be catalyzed by ion exchangers, and a great number of papers and patents have been published on this subject (71).

The use of solid ion exchangers has several advantages over homogeneous catalysis by dissolved electrolytes: (1) The catalyst is readily removed from the reaction products by say, decantation or filtration. (2) Continuous operation in columns is possible. (3) Often the purity of the product is superior since side reactions are less significant. (4) In the case of acid catalysis, corrosion is less important with the solid resins.

On the other hand, the applicability is restricted by the chemical and thermal stability of the ion exchangers. Also, ion exchange catalysts are more expensive than dissolved electrolytes.

In many cases, ion exchangers were found to produce other results than a dissolved electrolyte. As a rule, the ion exchanger is more selective, i.e., it distinguishes more sharply between large and small molecules. Small molecules have free access to the interior of the resin where the catalytically active counter ions are located, while large molecules are excluded by sieve action and thus have little contact with the catalytically active species. For example, and ion exchanger resin, when added to a mixture of monomeric and a polymeric ester (e.g., ethylacetate and polyvinylacetate) hydrolyzes the monomer without significantly attacking the polymer (72). Also, maltose is hydrolyzed completely in 10 hrs. at 98° whereas glycogen is stable under the same conditions when the catalyst is Amberlite IR-120 in the H form (72). A further application of resin catalysis is in the isolation of a reaction intermediate. For example, when acetone is refluxed through an intimate mixture of mannitol and resin, contained in a soxhlet thimble, a high yield of a mixture of mono and di-isopropylidene mannitols was obtained, the former predominating. The reaction is largely prevented from giving the fully substituted triisopropylidene mannitol by the first rapid removal, from the presence of the catalyst of the soluble, partly substituted derivatives (70).

G. Sulfonyl Halides

1. Synthesis. Ethenesulfonyl chloride can be polymerized but the yields are low (73). Better results are obtained with ethenesulfonyl fluoride (74), which can be transformed to a polymer with a degree of polymerization of 370 under the influence of AIBN.

$$ClCH_2CH_2SO_2Cl \xrightarrow{3\ KF} CH_2{=}CH(SO_2F) \xrightarrow{AIBN} \ldots{-}CH_2{-}CH(SO_2F){-}\ldots$$

Studies of the copolymerization of m and p styrenesulfonyl fluoride with styrene and methyl acrylate showed that the introduction of a sulfonyl fluoride group enhanced the overall reactivity of the monomer (75). The reaction of aminobenzenesulfonyl fluorides (19) with acrylic or methacrylic acid chloride or with vinyl isocyanate yields substituted acrylamides (20) or vinylureas (21) which can be polymerized or copolymerized with other vinyl monomers (76).

$$H_2N{-}C_6H_4{-}SO_2F\ (19) + CH_2{=}CHCOCl \longrightarrow CH_2{=}CHCONH{-}C_6H_4{-}SO_2F\ (20)$$

$$H_2N{-}C_6H_4{-}SO_2F\ (19) + CH_2{=}CH{-}NCO \longrightarrow CH_2{=}CHNHCONH{-}C_6H_4{-}SO_2F\ (21)$$

Another method of preparing polymers containing arylsulfonyl fluoride groups is the reaction of aminobenzenesulfonyl fluoride with polymers containing reactive groups such as poly(acrylic anhydride) and copolymers of maleic anhydride (76).

2. Reactions. Fluorosulfonyl-substituted polymers and copolymers are readily hydrolyzed to the corresponding sulfonic acids in slightly alkaline medium (74). Poly(ethenesulfonyl fluoride) has been reacted with ammonia, aliphatic and aromatic amines to yield the corresponding sulfonamides in 80 mole % conversion (74). The amines must be used in high excess to avoid crosslinking.

Other possibilities are the formation of sulfonate esters by reaction with alcohols. Sulfonate esters in turn are good alkylating agents.

H. Sulfonates

1. Synthesis. Overberger and coworkers studied the copolymerization of n-butyl ethenesulfonate with various vinyl monomers (77). Propyl ethenesulfonate and allyl ethenesulfonate have been polymerized with AIBN. In the last case a soluble polymer that contains 64-88% sultone rings (22), depending on monomer concentration, is obtained (78).

$$CH_2{=}CH{-}SO_2{-}O{-}CH_2{-}CH{=}CH_2 \xrightarrow{AIBN} \cdots{-}CH_2{-}CH{-}CH_2{-}CH{-}\cdots \text{ (six-membered ring: } CH{-}SO_2{-}O{-}CH_2{-}CH{-}CH_2\text{)}$$

(22)

The same monomer has been copolymerized with styrene, methylacrylate and vinyl acetate. Polymers in which the sulfonate is part of the polymer chain have been made by polycondensation of disulfonyl halides and diols. Reactions with these polymers result in degradation of the macromolecules and therefore this kind of polysulfonate falls beyond the scope of this review.

2. Reactions. Esters of sulfonic acids are known to be good alkylating atents. In the reaction, one molecule of sulfonic acid (eventually as a salt) is formed. This reaction can occur in two directions: the alkyl group introduced can be the polymer itself thereby splitting of a low molecular sulfonic acid, or the polymer can act as an alkylating agent forming a polymeric sulfonic acid. For example with an amine:

$$\sim\!\sim\!\underset{SO_2OR}{|}\!\sim\!\sim + R'NH_2 \longrightarrow \sim\!\sim\!\underset{SO_2OH}{|}\!\sim\!\sim + R'NHR$$

$$\sim\!\sim\!\underset{OSO_2R}{|}\!\sim\!\sim + R'NH_2 \longrightarrow \sim\!\sim\!\underset{NHR'}{|}\!\sim\!\sim + RSO_3H$$

The cyclopolymer obtained from allyl ethenesulfonate (22) combines the two reactions:

$$...-CH_2-CH(SO_2-O-CH_2-)CH-... \;(\text{ring: } CH-CH_2-CH, \; CH-SO_2-O-CH_2-CH) + R'NH_2 \longrightarrow ...-CH_2\underset{SO_3H}{CH}-CH_2\underset{CH_2NHR'}{CH}-...$$

(22)

This reaction has been carried out for example with methyl orange which leads to colored polymers which were soluble in water because of the presence of the sulfonic acid groups on the polymer chain (79).

I. Sulfonamides

1. Synthesis. Ethenesulfonamide polymerizes under the influence of radical initiators in aqueous or alcohol solution. High yields are obtained only within a pH range 6.5-7.0 (80). The polymers have the viscosity characteristics of polyelectrolytes. When the polymerization of ethenesulfonamide is initiated with sodium, a hydrogentransfer type of polymerization occurs with the formation of a linear polysulfonamide (81).

$$CH_2=CH-SO_2NH_2 \xrightarrow{Na} (-CH_2-CH_2-SO_2NH-)_n$$

N-substituted ethenesulfonamides could be copolymerized with styrene or with acrylonitrile, but did not homopolymerize under the influence of radical initiators (74). However, a series of poly(N-substituted ethenesulfonamides) have been made by reaction of poly(ethenesulfonyl fluoride) with ammonia and with several aliphatic and aromatic amines (74). The route via the sulfonyl fluoride is preferred because ethenesulfonyl fluoride is more readily polymerized than ethenesulfonyl chloride.

Radical polymerization of p-styrenesulfonamide by bulk,

solution, and emulsion techniques has been studied (82). Base-catalyzed polymerization of this monomer yields polymers (23), consisting of two different recurring units: a normal vinyl unit and a linear sulfonamide unit formed by hydrogen-transfer polymerization (83). Infrared absorption studies showed that 60-80% of the polymer consisted of vinyl recurring units (84).

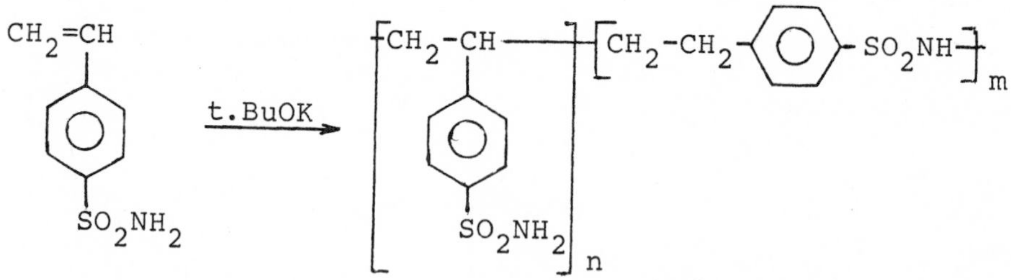

$$CH_2{=}CH{-}C_6H_4{-}SO_2NH_2 \xrightarrow{t.BuOK} \left[CH_2{-}CH(C_6H_4SO_2NH_2)\right]_n\left[CH_2{-}CH_2{-}C_6H_4{-}SO_2NH\right]_m$$

(23)

Polycondensation of disulfonyl dichlorides with diamines yields polysulfonamides. When the polycondensation is carried out in bulk or in solution, the polymers obtained generally are of low molecular weight. High-molecular-weight polymers are obtained by interfacial polycondensation. Daly (85) reported the synthesis of poly(arylsulfonimides) (24).

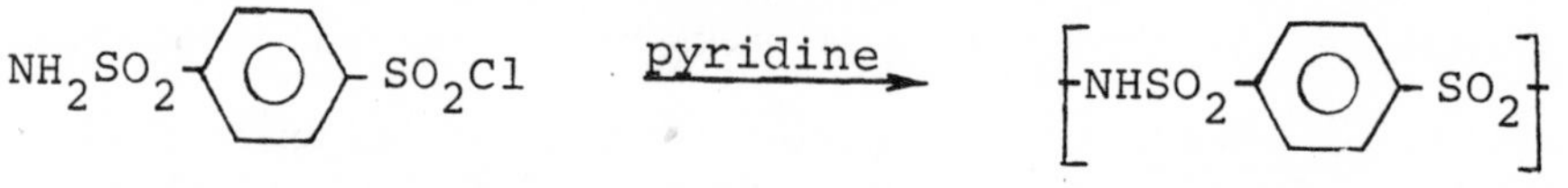

$$NH_2SO_2{-}C_6H_4{-}SO_2Cl \xrightarrow{pyridine} \left[NHSO_2{-}C_6H_4{-}SO_2\right]$$

(24)

An entirely different method for the preparation of a linear aliphatic polysulfonamide is the polymerization of propane sultam (25) under the influence of strong bases (86). The polymerization is also initiated by strong acids, but the yields are lower.

$$\text{(cyclic } CH_2CH_2CH_2SO_2NH) \xrightarrow{t.BuOK} (-CH_2CH_2CH_2SO_2NH-)_n$$

(25)

2. Reactions. Sulfonamides which carry a hydrogen atom on the nitrogen are weak acids and can be transformed into water-soluble salts. Also, polymeric sulfonamides can form salts and they can become soluble in aqueous sodium hydroxide (87).

Some papers and patents describe reactions on polysulfonamides. Murphy (88) chlorinated polysulfonamides to products showing properties similar to the well-known chlorinated monosulfonamides. They color starch potassium iodide paper blue and split off chlorine under the influence of heat or light. Analogous reactions have been reported by Schulz (89) recently. The polymers could be used as oxidizing agents which readily transformed secondary alcohols into ketones.

J. References

1) C. G. Overberger, J. J. Ferraro and F. W. Orttung, J. Org. Chem., 26, 3458 (1961).
2) G. Hardy, J. Vargu, K. Nyrtrai, I. Czajlik and L. Zubonyai, Vysokomolekul. Soedin., 6, 758 (1964).
3) C. G. Overberger and W. H. Daly, J. Am. Chem. Soc., 86, 3402 (1964).
4) W. H. Daly, C. D. S. Lee and C. G. Overberger, J. Polymer Sci., Part A-1, 1723 (1971).
5) C. G. Overberger and A. Lebovits; J. Am. Chem. Soc., 77, 3675 (1955).
6) E. J. Goethals and E. Du Prez, J. Polymer Sci., [A-1] 4, 2893 (1966).
7) C. G. Overberger and H. Aschkenasy, J. Org. Chem., 25, 1648 (1960).
8) C. G. Overberger and H. Aschkenasy, J. Am. Chem. Soc., 8, 4357 (1960).
9) T. K. Dykstra and D. A. Smith, Makromol. Chem., 134, 209 (1970).
10) H. P. Gregor, D. Dolar and G. K. Hoeshele, J. Am. Chem. Soc., 77, 3675 (1955).
11) J. R. Parrish, Chem. Ind. (London) 1956, 137.
12) C. L. Arcus and N. L. Salomons, J. Chem. Soc., 1963, 1174.
13) H. G. Cassidy, J. Am. Chem. Soc., 71, 402 (1949).
14) K. A. Kun, J. Polymer Sci., A-1 4, 847 (1966).
15) C. G. Overberger and J. J. Ferraro, J. Org. Chem., 27, 3539 (1962).
16) C. G. Overberger, J. J. Ferraro and F. W. Orttung, J. Org. Chem., 26, 3458 (1961).
17) C. G. Overberger, K. H. Burg and W. H. Daly, J. Am. Chem. Soc., 87, 4125 (1965).
18) E. Du Prez and E. J. Goethals, Makromolek. Chem., 146, 145 (1971).
19) C. J. Swan and D. L. Trimm, J. Appl. Chem., 18, 340 (1968).
20) Houben-Weyl, Methoden der Organischen Chemie, (E. Muller, ed.) Georg Thieme Verlag, Stuttgart, Teil IX, p. 143.
21) C. C. Price and R. D. Gilbert, J. Am. Chem. Soc., 74, 2073 (1952).
22) H. Schnell and H. Krimm (to Farbenfabriken Bayer A.G.) U.S. Pat. 2,999,846 (1961).

23) H. A. Smith (to Dow Chemical Co.) Belg. Pat. 644780 (1964).
24) C. C. Price and E. A. Blair, J. Polymer Sci., [A-1] 5, 171 (1967).
25) W. W. Epstein and F. W. Sweat, Chem. Revs., 247 (1967).
26) J. Boor, Jr. and A. M. T. Finch, J. Polymer Sci., A-1 9, 249 (1971).
27) C. C. Price and H. Morita, J. Am. Chem. Soc., 75, 4747 (1953).
28) F. E. Bailey, Jr. and E. M. La Combe, J. Macromol. Sci., A4, 1293 (1970).
29) E. M. La Combe and F. E. Bailey, Jr., (Union Carbide) U.S. Pat. 3,280,081 (Oct. 18, 1966).
30) M. J. Hatch, F. J. Meyer and W. O. Floyd, J. Appl. Polym. Sci., 13, 721 (1969).
31) M. F. Hoover, J. Macromol. Sci.-Chem., A-4, 1327 (1970).
32) D. S. Breslow and G. E. Hulse, J. Am. Chem. Soc., 76, 6399 (1954).
33) W. Kern and R. C. Schulz, Angew. Chem., 69, 153 (1967).
34) D. S. Breslow and A. Kutner, J. Polymer Sci., 27, 295 (1958).
35) A. Kutner and D. S. Breslow, Encyclopedia of Polymer Science and Technology, Interscience Publishers, Vol. 9, p. 455.
36) R. L. Baechtold (American Cyanamid Co.) U.S. Pat. 3,203,938 (1965).
37) R. H. Wiley, J. Polymer Sci., 28, 163 (1958).
38) R. H. Wiley, N. R. Smith and C. C. Ketteren, J. Am. Chem. Soc., 76, 720 (1954).
39) R. H. Wiley and S. F. Reed, J. Am. Chem. Soc., 78, 2171 (1956).
40) A. O. Jakubovic, Chem. Prod., 23, 257 (1960).
41) A. O. Jakubovic, J. Chem. Soc., 1960, 4820.
42) H. H. Roth, Ind. Eng. Chem., 49, 1820 (1957).
43) D. C. Ingles, Australian J. Chem., 12, 97 (1959).
44) Imp. Chem. Industries, Bit. Pat. 829,704 (1960).
45) M. Kato, T. Nakagawa and H. Akamatsu, Bull. Soc. Chim. Japan, 33, 322 (1960).
46) W. R. Carrol and H. Eisenberg, J. Polymer Sci. [A-2], 4, 599 (1966).
47) H. H. Roth, Ind. Eng. Chem., 46, 2435 (1954).
48) R. Signer, A. Demagistri and C. Müller, Makromol. Chem., 18/19, 139 (1956).
49) R. Hart and R. Janssen, Makromol. Chem., 43, 242 (1961).
50) R. Signer and A. Demagistri, J. Chim. Phys., 47, 704 (1950).
51) G. L. Starobinets, L. T. Sevast'yanova and G. N. Bulatskaya, Zh. Prikl. Khim., 33, 107 (1960).
52) G. F. d'Alelio (Koppers Co.) U.S. Pat. 2,631,127 (1953).
53) D. W. Fox and P. Skenian (General Electric Co.) U.S. Pat. 3,259,592 (1966).
54) D. E. Nagy (Am. Cyanamid Co.) U.S. Pat. 2,621,164.
55) K. Faber and E. Kamareck, Ullman's Encyclopädie der Technischen Chemie, vol. 11 (W. Foerst, ed.), p. 595 (1960).
56) A. S. Nyguist (Am. Cyanamid Co.), U.S. Pat. 2,610,171 (1952).

57) F. Filbert and M. F. Fuller, U.S. Pat. 2,820,788 (1958).
58) J. O. Porath, U.S. Pat, 2,891,057 (1959).
59) T. Timell, Swed. Pat. 124,025 (1949).
60) T. Timell, Svensk Papperstid., 51, 254 (1948).
61) M. F. Fuller (du Pont de Nemours) U.S. Pat. 2,883,375 (1959).
62) G. Natus and E. J. Goethals, J. Macromol. Chem., A2, 489 (1968).
63) E. J. Goethals and G. Natus; Makromol. Chem. 116, 152 (1968).
64) G. Natus and E. J. Goethals, Makromol. Chem., 123, 130 (1969).
65) R. Hart and D. Timmerman, J. Polymer Sci., 28, 639 (1958).
66) H. Morawetz and E. W. Westhead, Jr., J. Polymer Sci., 16, 273 (1955).
67) C. L. Arcus, T. J. Howard and D. S. South, Chem. and Ind., 1964, 1756.
68) I. Sakurada, Y. Sakaguchi, T. Ono and T. Ueda, Makromol. Chem., 91, 243 (1966).
69) T. J. Painter, Chem. and Ind., 1960, 1214.
70) W. H. Wadman, J. Chem. Soc., 1952, 3051.
71) N. G. Polyanski, Usp. Khim., 39, 244 (1970).
72) H. Deuel, J. Solms, L. Anyas-Weisz and G. Huber, Helv. Chim. Act., 34, 1849 (1951).
73) W. Kern, R. C. Schulz and W. Schaefer; Angew. Chem., 49, 167 (1957).
74) W. Kern, R. C. Schulz and H. Schlesman, Makromol. Chem., 39, 1 (1960).
75) R. Hart, Makromol. Chem., 49, 33 (1961).
76) R. Hart and D. Timmerman, J. Polymer Sci. 48, 151 (1960).
77) C. G. Overberger, D. E. Baldwin and H. P. Gregor, J. Am. Chem. Soc., 72, 4864 (1950).
78) E. J. Goethals and E. De Witte, Makromol. Chem., 115, 234 (1968).
79) E. De Witte and E. J. Goethals, unpublished results (1969).
80) H. F. Park (Monsanto Chemical Co.), U.S. Pat. 2,754, 287 (1956).
81) D. S. Breslow, G. E. Hulse and A. S. Matlack, J. Am. Chem. Soc., 79, 3760 (1957).
82) R. H. Wiley and C. C. Ketterer, J. Am. Chem. Soc., 75, 4519 (1953).
83) N. Yoda and C. S. Marvel, J. Polymer Sci., A3, 2229 (1965).
84) A. Konishi, N. Yoda and C. S. Marvel, J. Polymer Sci. A 3, 3833 (1965).
85) W. H. Daly and H. J. Hoelle, J. Polymer Sci., B 10, 519 (1972).
86) A. D. Bliss, W. K. Cline, C. E. Hamilton and O. J. Sweeting, J. Org. Chem., 28, 3537 (1963).
87) R. A. Edington, P. Markly, R. J. B. Marsden, J. Mather and J. W. Stimpson (Imp. Chem. Industries Ltd.), Brit. Pat. 875,072 (1961).
88) W. A. Murphy (du Pont de Nemours Co.) U.S. Pat. 2,853,475 (1958).

89) R. C. Schulz, Pure and Appl. Chem. (Butterworths London) vol. 30, p. 239 (1972).

DISCUSSION SESSION

Discussion Leader - G. Challa

Epton: I would like to comment on a further important application of thiol-containing polymers. This is the reversible immobilization of enzymes. Prof. S.A. Baker, et al (in "The Birmingham University Chemical Engineer", 22, 74(1971) have descibed a suitable thiol-containing copolymer for this application. This copolymer, Enzacryl Polythiol, is commercially available from Aldrich Chemical Co. in the United States and Koch-Light Laboratories, Ltd. in Europe.
Enzacryl Polythiol is a cross-linked copolymer of acrylamide and N-acryloyl cystein. The polymer swells readily in water to form a gel. This will undergo oxidative coupling with molecules containing sulfhydryl groups. Since most enzymes contain either no free sulfhydryl groups or a very small number, prior enrichment with sulfhydryl groups is desirable . This is achieved by reaction with N-acetylhomocysteine thiolactone. Oxidative coupling is brought about by potassium ferricyanide. The conjugate may be dissociated, if desired, by treatment with cysteine or ethane thiol in aqueous solution. Both enzyme and copolymer are quantitatively recovered.
Another important aspect of this work is the conversion of Enzacryl Polythiol to the reactive Enzacryl Polythiolactone by diimide treatment. The thiolactone polymer is effective in the acylative, irreversible coupling of enzymes with concurrent generation of sulfhydryl groups. The enzyme molecules are therefore constrained within a reducing micro-environment. This situation could be valuable in the case of such enzymes as papain which require pre-activation by cystein and a generally reducing environment.

Rempp: I would like to refer to two research projects carried out at our Institut in Strasbourg on the preparation of sulfur-containing groups by a substitution reaction on a polymer chain. Substitution reactions may exhibit random character, or a sequential character owing to reactivity enhancements of the groups next to the substituted ones.
The work of Dr. Galin deals with nucleophillic substitution onto the ester function of poly (methyl methacrylate). The reactions were carried out in DMSO solution using dimsyl sodium as reagent:

$$\text{PMMA dyad: } (C{=}O)OCH_3,\ (C{=}O)OCH_3 \ + \ CH_3SOCH_2^- \rightarrow (C{=}O)CH_2SOCH_3,\ (C{=}O)OCH_3$$

Two molecules of dimsyl sodium are needed to substitute one ester function because of the markedly acidic character of the β- keto-sulfoxide functions (polymers substituted more than 20% are soluble in aqueous alkali). Below 30° C, and for degrees of substitution up to 30%,the reaction exhibits random character without any side reactions. At higher temperatures,cyclization may occur by attack at a neighboring ester funtion by the β-ketosulfoxide carbanion:

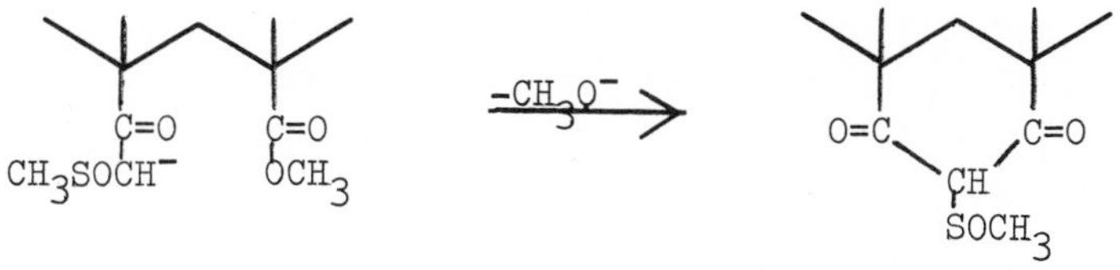

C=O C=O
CH_3SOCH^- OCH_3 $\xrightarrow{-CH_3O^-}$ O=C C=O
CH
$SOCH_3$

When excess dimsyl sodium and higher temperatures are used, another type of cyclic, disubstituted unit may some times be found in the polymer

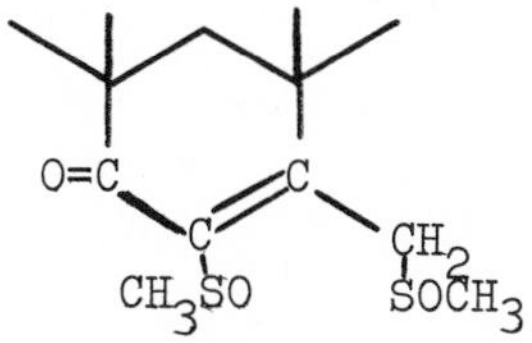

O=C C
C CH_2
CH_3SO $SOCH_3$

However, when the reaction is carried out under mild conditions the polymer is homogeneous in composition and the substitution can be considered to occur at random.

The project of Dr. Gramain deals with the substitution of poly(vinyl alcohol), PVA, with propane sultone. First PVA is treated with dimsyl sodium in DMSO solution. Metallation can involve up to 80% of the alcohol units. The solution has by then become a gel because of association of ONa units. When propane sultone is added, the expected reaction takes place up to about 40% conversion:

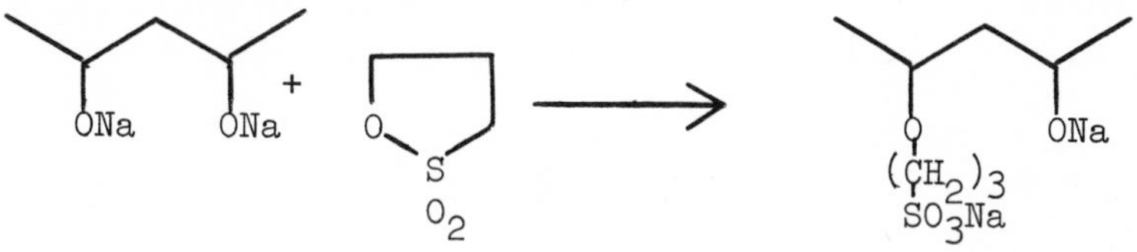

ONa ONa + O S O_2 ⟶ O ONa
$(CH_2)_3$
SO_3Na

Using an excess of dimsyl sodium, degrees of substitution of 65-70% were attained. This substitution does not occur at random since dehydration of the substituted polymer ($KHSO_4$) produces a black product. The UV-visible absorption spectrum reveals the presence of rather long (up to 15) sequences of conjugated double bonds. Such long sequences of conjugated double bonds point to the existence of rather long nonsubstituted sequences in the sulfonated polymer.

The same results were obtained by Gramain for Williamson's reaction applied to PVA:

ONa ONa + $C_{18}H_{37}Br$ ⟶ ONa O–$C_{18}H_{37}$

Though the maximum degree of substitution is much lower than in the preceding case, it appeared that this substitution also takes place sequentially and not at random. This was not only proved bv dehydration of the remaining PVA units but also by crystallographic data. The birefringence of the sample shows the occurrence of intra-molecular phase separation, even at low degrees of substitution. The structure is lamellar, as expected for crystallizable long hydrocarbon chains. The thickness of the lamellae does not vary with the degree of substitution. This definitely shows that the substituted units are very near each other along the chain.

It should be pointed out that the compositional heterogeneity of the substituted PVA is quite large in both cases. This as another argument in favor of sequential substitution of the PVA chains. So it seems that in both these cases the reactivity of units neighboring substituted ones is enhanced, whereas the reactivity might well be reduced in the case of substitution of PMMA by dimsylsodium. Studies like these are worthwhile as products with the same degree of substitution may have quite different properties depending on whether the substitution occurs randomly or sequentially.

Goethals: Propane sultone reacts with DMSO to form dimethyl (3 sulfopropoxy) sulfonuim salt which rearranges to dimethyl (3 sulfopropyl) oxosulfonium salt at higher temperatures (G.Natus and E. J. Goethals, Bull. Soc. Chim. Belg. 74, 450 (1965)).

$$(CH_3)_2S{=}O + \text{propane sultone (O–SO}_2\text{ ring)} \xrightarrow{30^\circ} (CH_3)_2\overset{+}{S}O(CH_2)_3SO_3^-$$

$$\xrightarrow{85^{\circ}} (CH_3)_2S^{+}(=O)(CH_2)_3SO_3^{-}$$

The first formed compound is still an excellent sulfopropylating agent and it is possible that the reaction partially or even predominantly proceeds via this intermediate in DMSO. This fact could perhaps explain the non-random nature of the sulfopropylation of PVA in DMSO. Since the reagent is now an ion, instead of neutral propane sultone, electrostatic attraction to the already substituted parts of the macromolecule could be controlling the reaction.

DeLaMare: In the metallation with R Li/TMED (tetramethylethylene diamine) do you expect a purely random process or some type of "ordering" as you found in the metalation of PVA?

Rempp: Probably a random distribution.

Carraher: There are many routes to include sulfur into polymers. Several routes not previously mentioned here are used in our laboratory.

1. 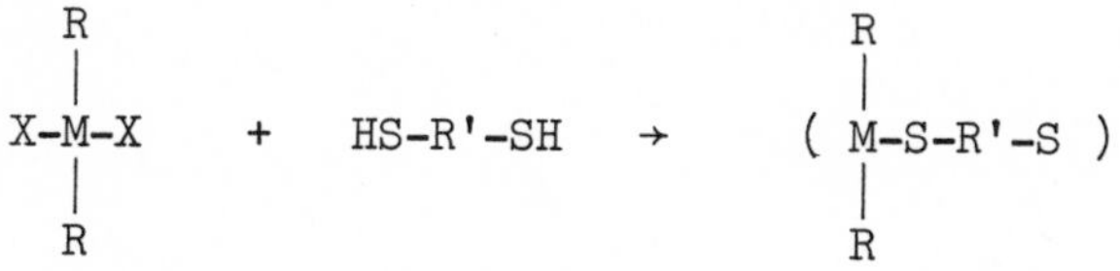 $$X\text{-}MR_2\text{-}X + HS\text{-}R'\text{-}SH \rightarrow (MR_2\text{-}S\text{-}R'\text{-}S)$$

X=halide, M=Sn,Ge,Si,Hf,Ti

2. 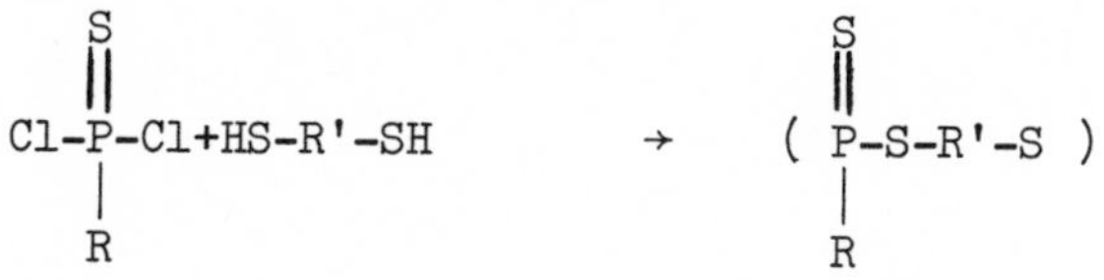 $$Cl\text{-}P(=S)(R)\text{-}Cl + HS\text{-}R'\text{-}SH \rightarrow (P(=S)(R)\text{-}S\text{-}R'\text{-}S)$$

OR may also be used instead of R.

3. 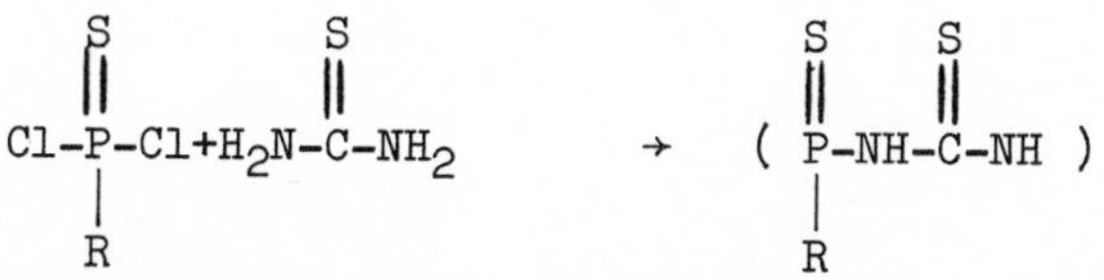 $$Cl\text{-}P(=S)(R)\text{-}Cl + H_2N\text{-}C(=S)\text{-}NH_2 \rightarrow (P(=S)(R)\text{-}NH\text{-}C(=S)\text{-}NH)$$

similary with HO-R'-Oh, H_2NH_2 etc.
$SO_2Cl_2 + H_2N\text{-}R\text{-}NH_2$, HO-R-OH, etc $\rightarrow$ corresponding products

Lewin: We have recently found that polymeric sulfates may be interesting flame-retarding agents. Can you comment on this?

Goethals: The formation of a poly (bis phenol sulfate) from sulfuryl chloride and bis phenol-A has been reported, but I don't know any details.

Harwood: We have recently prepared polymers containing mixed acetic-styrene sulfonic anhydride units by refluxing sulfonated polystyrene resins with acetyl chloride. Such resins are strong acylating agents and can be used as reagents for preparing esters of alcohols and phenols.

DeLaMare: Polydienes and EPDM terpolymers are easily metalated with RLi and TMED. Has this route been used to introduce sulfur functions along the backbone either by reaction with,e.g. propane sultone or directly with sulfur?

Daly; Such a process can be used with S_8.

APPLICATION OF BENDER'S SALTS TO THE SYNTHESIS OF MERCAPTAN - CONTAINING POLYMERS *

W. H. Daly and Chien-Da S. Lee

Department of Chemistry, Louisiana State University,

Baton Rouge, Louisiana 70803

Interest in the preparation of polymers containing free mercaptan groups for potential applications as reducing agents, ion exchange resins, heavy metal scavengers, radiation prophylatics and polymeric catalysts has prompted a thorough investigation of potential synthetic techniques.[1] In contrast to vinyl alcohol, vinyl mercaptan can be isolated in very low yield from the addition of hydrogen sulfide to acetylene in the presence of light.[2] However it is useless as a monomer because the sulfhydryl group is too labile, i.e., it is a very efficient chain transfer agent and undergoes trimerization *via* a tautomeric thioacetaldehyde intermediate.[3] Thus one is forced to resort to indirect synthetic techniques of the following general types: (a) synthesis and polymerization of monomers containing a "blocked" mercaptan group and (b) introduction of mercaptan groups into polymers with suitable reactive sites.

The application of several different blocking groups to polymercaptan synthesis has been thoroughly evaluated. In general, this technique enables one to prepare homopolymers and copolymers of vinyl mercaptan under conditions where the structure of the polymer can be assigned accurately. For example, vinylthioesters such as S-vinyl thioacetate or thiobenzoate can be prepared in low yields ($< 25\%$) and polymerized under free radical conditions.[4,5] However, these monomers also exhibit high chain transfer constants ($\sim 5\times10^{-2}$) so the molecular weights of their polymers do not exceed 2000.

* This article has been reproduced from a photocopy, as the original typed sheets were not available.

Subsequent hydrolysis of the thioester functional group requires alkaline conditions and is accompanied by oxidative crosslinking.

An alternate approach considered by H. Ringsdorf et al.[6] was to prepare S-vinyl mercaptals and polymerize them in the presence of azobisisobutyronitrile or boron trifluoride etherate. Only low molecular weight polymers were obtained regardless of the reaction conditions. Further, the polymeric mercaptal could not be hydrolysed to polymercaptan. This is unfortunate since it is possible to prepare a polymer containing dithiane rings (2) by cyclopolymerizing S,S'-divinyldithioformal (1).[7] We were able to alkalate these rings by preparing a lithio salt of the polymer with butyl lithium followed by treatment with an alkyl halide but all attempts to release the elaborated aldehydes (5) generated by this process failed.

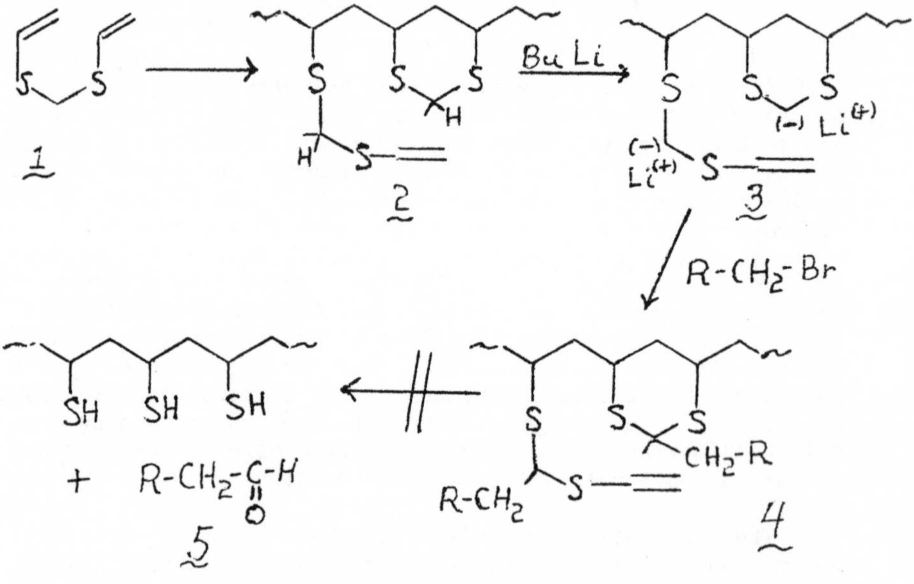

Recently a polymercaptan synthesis has been described which involves the preparation of methacryloxyalkyl isothiuronium salts.[8] Polymerization of these monomers proceeds very slowly even in polar solvents and high degrees of conversion were difficult to obtain. Alkaline hydrolysis of the isothiuronium salt was accompanied by ethylene sulfide evolution in some cases; this side reaction reduced the yield of mercaptide groups significantly.

In an effort to improve the mechanical properties of polymercaptans Geothals turned to cationic polymerization of 3,3-bis(methylenethioester)oxetanes.[9] Using dithiocarbonates or bis(thioacetate) derivatives, polymers with molecular weights as high as 21,000 could be obtained. Higher molecular weight materials were prepared by copolymerizing the sulfur containing monomers with 3,3-bis(chloromethyl)-oxetane but extensive cross-linking occurred during the deblocking step due to the nucleophilic attack of liberated mercaptide ions on residual chloromethyl substituents.

In conjunction with a study of the application of sulfur containing polymers as radiation prophylatics, Overberger et al. prepared a series of vinyl thiocarbamates, thiocarbonates and dithiocarbonates.[10] Several of these derivatives yielded reasonably high molecular weight polymers but the polymers proved difficult to hydrolyse to the desired polymercaptans. However, utilization of the labile carbo-t-butoxy blocking group in monomer synthesis by preparing S-vinyl-O-t-butylthiocarbonate alleviated the hydrolysis problems and produced an excellent monomeric precursor for polymercaptans.[11] The blocking group could be removed by treatment of the polymer with anhydrous HBr in 6:1 chloroform-tetrachloroethane or by thermolysis at 150°C in N-methylpyrrolidone.

S-vinyl-O-t-butylthiocarbonate (6) polymerizes readily in the presence of free radical initiators to yield polymers with a maximum molecular weight of 55,000 (DP ≅ 335). Although this is an order of magnitude higher than the molecular weight attainable with vinyl thioacetate under similar conditions, 6 exhibits a chain transfer constant of $3.9\ 10^{-3}$ which limits the molecular weight of both homopolymers and copolymers prepared under free radical conditions. The copolymerization characteristics of S-vinyl-O-t-butylthiocarbonate were typical of vinyl thiocarbonyl monomers; the Q and e copolymerization parameters are consistant with a resonance stabilized monomer and radical with a negatively polarized double bond.[12] Water soluble copolymers of 6 can be prepared by incorporating 60% vinyl pyrrolidone but the hydrophobility of free mercaptan groups renders these copolymers insoluble in water upon removal of the blocking group; a copolymer of vinyl mercaptan containing 95% vinyl pyrrolidone remains water soluble.

One of the problems encountered in working with S-vinyl-O-t-butylthiocarbonate was a tendency for this monomer to yield crosslinked polymer. The crosslinking was apparently due to the presence of 3-5% S,S'-divinyldithiocarbonate (10) formed by the reaction sequence shown in Scheme 1.

Scheme 1

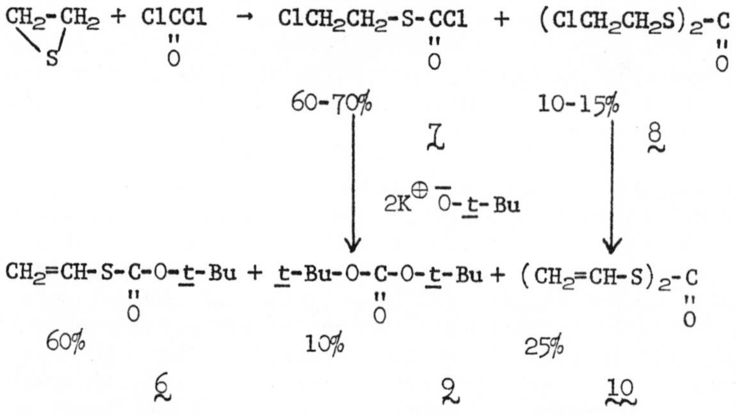

$$\underset{\text{S}}{CH_2\text{-}CH_2} + ClC(=O)Cl \rightarrow \underset{\mathbf{7}\ (60\text{-}70\%)}{ClCH_2CH_2\text{-}S\text{-}C(=O)Cl} + \underset{\mathbf{8}\ (10\text{-}15\%)}{(ClCH_2CH_2S)_2\text{-}C{=}O}$$

$$\xrightarrow{2K^{\oplus}\ {}^{-}O\text{-}t\text{-}Bu}$$

$$\underset{\mathbf{6}\ (60\%)}{CH_2{=}CH\text{-}S\text{-}C(=O)\text{-}O\text{-}t\text{-}Bu} + \underset{\mathbf{9}\ (10\%)}{t\text{-}Bu\text{-}O\text{-}C(=O)\text{-}O\text{-}t\text{-}Bu} + \underset{\mathbf{10}\ (25\%)}{(CH_2{=}CH\text{-}S)_2\text{-}C{=}O}$$

It was very difficult to prevent multiple addition of ethylenesulfide to phosgene, and the tendency of 7 and 8 to decompose during distillation made separation of the two products difficult. Therefore, we sought an alternate route to the preparation of 6.

Okawara et al.[13] reported that the addition of sodium N,N-dialkyldithiocarbamate (11) to 1,2-dichloroethane (Scheme 2) in DMF yielded a diadduct (12) which could be cracked to produce S-vinyl-N,N-dialkyldithiocarbamate (13). Monomer 13 could be polymerized under free radical conditions but it exhibited a high chain transfer constant so the molecular weight of the resultant polymer (14) was rather low. Furthermore, the thermal stability of the polymer was poor, which is not surprising since the polymer structure was very similar to diadduct 12 which could be cracked at relatively low temperatures. Polymercaptan could be generated from 14 by treatment with dimethylamine.

Scheme 2

$$ClCH_2CH_2Cl + Na^+ \, {}^-S\text{-}\underset{\underset{S}{\|}}{C}\text{-}NR_2 \xrightarrow[90\%]{DMF} R_2N\text{-}\underset{\underset{S}{\|}}{C}\text{-}S\text{-}CH_2CH_2S\text{-}\underset{\underset{S}{\|}}{C}\text{-}NR_2$$

11 12

$$\xrightarrow{\Delta, 230^\circ} R_2N\text{-}\underset{\underset{S}{\|}}{C}\text{-}S\text{-}CH{=}CH_2 + CS_2 + R_2NH$$

13

$$\xrightarrow{AIBN} \sim CH_2\underset{\underset{S\text{-}\underset{\underset{S}{\|}}{C}\text{-}NR_2}{|}}{CH} \sim$$

14

$$\xrightarrow{(CH_3)_2NH} \text{insol.} \sim CH_2\underset{\underset{SH}{|}}{CH} \sim + (CH_3)_2\text{-}N\text{-}\underset{\underset{S}{\|}}{C}\text{-}NR_2$$

Okawara has also conducted an extensive study on nucleophilic displacement of reactive functional groups on polymeric substrates with N,N-dialkyldithiocarbamates and xanthates.[14] Substituted polymers derived from polyvinyl chloride exhibited poor thermal stability and decomposed to yield polyenes. However derivatives of polymers which do not contain hydrogen atoms beta to the reactive site, i.e., chloromethylated styrene and chloromethyl acetals, could be converted to polymercaptans. The photochemical properties of these polymers were dependent upon the nature of the sulfur containing substituent. For example, photolysis of polymers containing dithiocarbamate groups (15) in the presence of vinyl monomers produced graft polymers which did not contain sulfur (Scheme 3). In contrast, polymers containing xanthate groups (16) produced polyvinylbenzylthiol radicals which lead to disulfide crosslinks upon irradiation. The potential application of these polymers as photoresist resins was considered.

Scheme 3

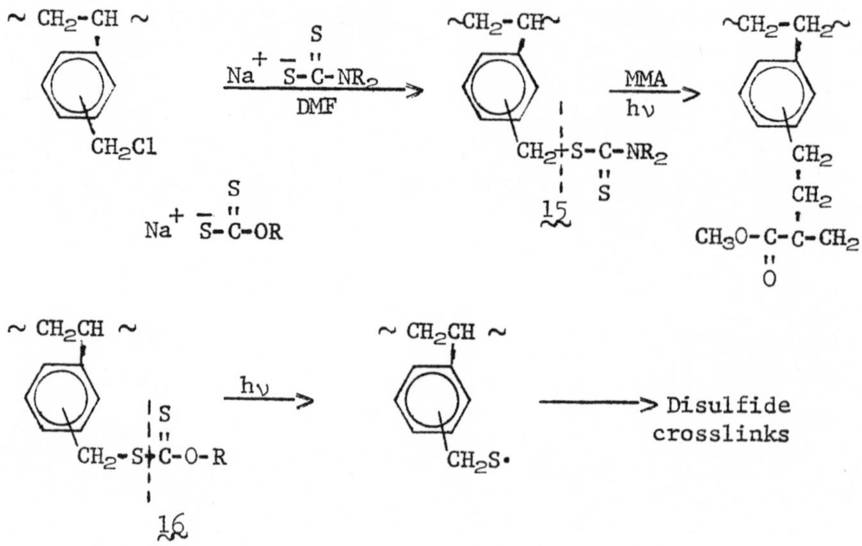

$\sim CH_2-CH \sim$ (ring, CH_2Cl) $\xrightarrow[DMF]{Na^+ \; ^-S-C(=S)-NR_2}$ $\sim CH_2-CH \sim$ (ring, $CH_2 \vdots S-C(=S)-NR_2$) **15** $\xrightarrow[h\nu]{MMA}$ $\sim CH_2-CH_2 \sim$ (ring, $CH_2-CH_2-C(CH_2)-C(=O)-OCH_3$)

$Na^+ \; ^-S-C(=S)-OR$

$\sim CH_2CH \sim$ (ring, $CH_2-S \vdots C(=S)-O-R$) **16** $\xrightarrow{h\nu}$ $\sim CH_2CH \sim$ (ring, $CH_2S\cdot$) $\longrightarrow$ Disulfide crosslinks

BENDER'S SALTS IN MONOMER SYNTHESIS

Application of similar nucleophilic displacement reactions to the preparation of S-vinyl-O-alkylthiocarbonates simply required the synthesis of appropriate Bender's salts. Bender's salts are stable crystalline reagents which can be prepared by treating potassium alcoholates with carbonyl sulfide in an anhydrous solvent mixture of alcohol and dimethylformamide (DMF).[15] In contrast to potassium xanthates, the salts are insoluble in most organic solvents but a mixture of 1-2% water in DMF is a good solvent system for homogeneous displacement reactions. In the dry form, the salts are quite stable and samples have been stored over a year in an argon atmosphere without noticeable loss of reactivity. Bender's salts also form colored complexes with transition metals[16] so the reaction mixtures may be dark colored initially, but the color disappears as the reaction nears completion. The reaction conditions for nucleophilic displacement reactions were established by optimizing the reaction of potassium O-*t*-butylthiocarbonate **17a** with ethyl bromide and benzyl bromide. The displacement was essentially quantitative in DMF-H_2O (98:2) after stirring the reagents for 6 hr at 45°.

As Scheme 4 illustrates, selective displacement of bromide in the presence of chloride can be effected in THF: t-butanol (4:1) at -10°. Subsequent dehydrohalogenation of the S-(β-chloroethyl)-O-t-butylthiocarbonate (18a) produced monomer 6a in 58% overall yield. The monomer is still contaminated by 2-3%

Scheme 4

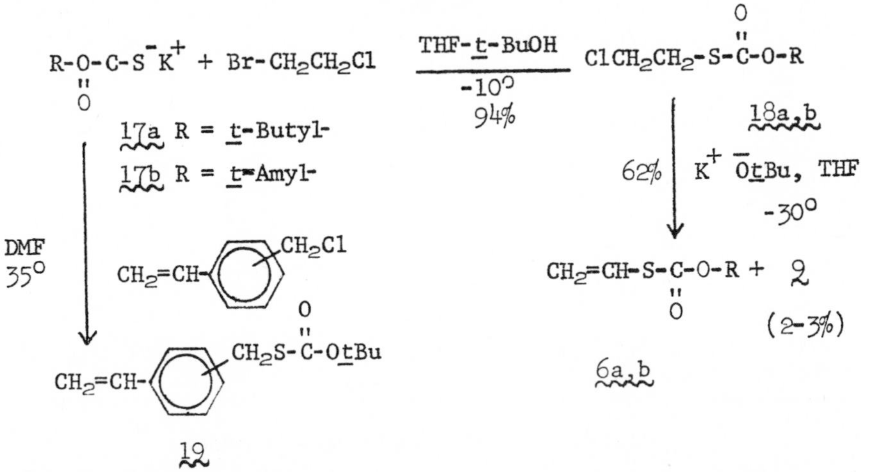

$$R-O-C(=O)-S^-K^+ + Br-CH_2CH_2Cl \xrightarrow[-10^\circ,\ 94\%]{THF-t-BuOH} ClCH_2CH_2-S-C(=O)-O-R$$

17a R = t-Butyl-
17b R = t-Amyl-

18a,b

$$\xrightarrow[-30^\circ]{62\%\ \ K^+\ ^-OtBu,\ THF} CH_2=CH-S-C(=O)-O-R + 2\ (2\text{-}3\%)$$

6a,b

17a + $CH_2=CH-C_6H_4-CH_2Cl$ $\xrightarrow{DMF,\ 35^\circ}$ $CH_2=CH-C_6H_4-CH_2S-C(=O)-OtBu$

19

di-t-butylcarbonate but this appears to act as an inert diluent in the polymerizations. Using potassium O-t-amylthiocarbonate (17b) as the nucleophile, S-vinyl-O-t-amylthiocarbonate (6b) can be prepared via Scheme 4 in 51% yield.

An alternative approach to a monomer synthesis utilizes a commercially available mixture of 60% meta and 40% para-vinylbenzyl chloride as the substrate. Treatment of the mixture with 17a in DMF produces S-vinylbenzyl-O-t-butylthiocarbonate (19) in quantitative yield. Polymerization of 19 under free radical conditions yields a mixture of benzene soluble and insoluble crosslinked polymer (20). Thermolysis of the soluble fraction in N-methylpyrrolidone yields a soluble poly(vinylbenzylmercaptan).

TABLE I

SUBSTITUTION EFFICIENCY

Substrate		Product			
Structure	Initial Cl content,%	Cl Content,%	Extent of subst.	S Content,% Found	Calcd.
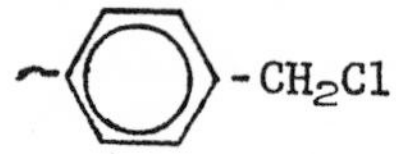 ~C_6H_4-CH_2Cl	21.46	1.17	92	11.65	12.17
~CH_2-CH(Cl)~	56.69	13.93	54.6	18.96	20.0
~CH_2-CH(CH_2Cl)-O~	38.24	2.02	95	23.65	16.84
~CH_2-CH(CH_2Cl)-O~	25.18	7.23	67.6	14.20	13.42

Copolymer with ethylene oxide

MODIFICATION OF PREFORMED POLYMERS

Although the synthesis of monomeric mercaptan precursors has many advantages it is still very difficult to convert these to polymers with sufficient molecular weight to exhibit good mechanical properties. Therefore, we have applied the nucleophilic properties of Bender's salts to the displacement of reactive halides on polymer substrates. Polymers containing halomethyl functional groups such as: chloromethylated polystyrene, poly(epichlorohydrin) (Herchlor H), copoly(epichlorohydrin-ethylene oxide) (Herclor-C), poly(2,6-bis-bromomethylphenylene oxide) and poly(vinyl chloride) were utilized as substrates with varying degrees of success.

All of the polymer substrates employed dissolve in DMF so the displacement reactions occurs initially in a homogeneous medium. The mixture soon became heterogeneous as both potassium chloride and substituted polymer precipitated from solution. Based upon residual chlorine content, the substitution is nearly quantitative in most cases (Table I). However, sulfur analyses revealed that the substitution is accompanied by side reactions which increase the sulfur content of the products when the reactive sites are attached to a flexible backbone such as poly(oxyethylene). These derivatives are no longer soluble, but they swell rapidly in benzene indicating that crosslinking is one of the side reactions. Careful exclusion of oxygen and light fails to inhibit the crosslinking process. Further, the reaction conditions are too mild to effect thermal deblocking of the carbo-t-butoxy substituent so the crosslinking process probably involves a series of nucleophilic reactions. The cross-linking mechanism postulated in Scheme 5 is consistant with the experimental observation that di-t-butylcarbonate and t-butanol is a by-product of the reaction. The dithiocarbonate structure (21) has a higher sulfur composition than the S-alkyl-O-t-butyl-thiocarbonate substituents which would account for the high

Scheme 5

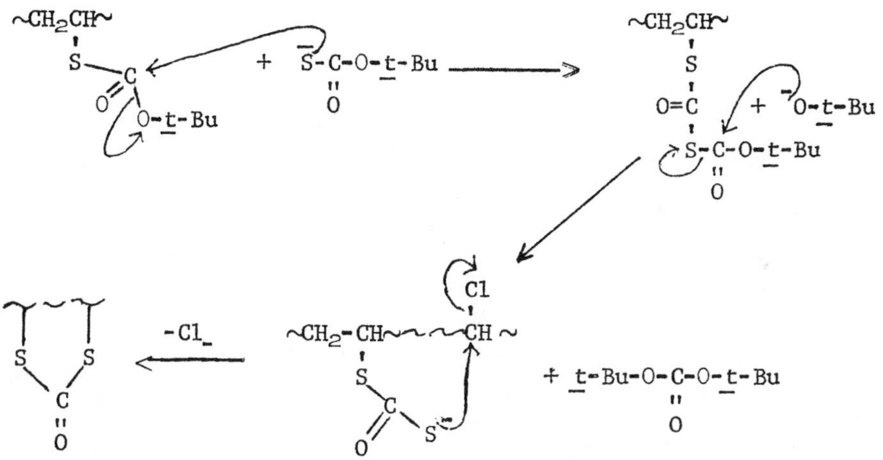

~CH2CH~ S–C(=O)–O-t-Bu + ⁻S-C(=O)-O-t-Bu ⟶ ~CH2CH~ S–C(=O)–S-C(=O)-O-t-Bu + ⁻O-t-Bu

~CH2-CH~~~CH(Cl)~ with S–C(=O)–S⁻ + t-Bu-O-C(=O)-O-t-Bu (9)

–Cl⁻ ⟵ cyclic S–C(=O)–S crosslink (21)

sulfur content of the substituted polymers. However precise ellucidation of the crosslink structure will require further investigation.

It is possible to reduce the extent of crosslink formation by using lower mole ratios of Bender's salts to reactive functional groups in the substitution reactions. Since quantitative conversion is impossible under these conditions the resultant copolymers still contain active functional groups which will react with sulfhydryl substituents as soon as they are liberated. Thus polymercaptans derived from flexible substrates will be crosslinked inevitably and high free mercaptan contents can not be attained with this technique.

THERMAL ANALYSIS OF POLYMERCAPTAN PRECURSORS

The extent of O-t-butylthiocarbonate substitution could be determined quantitatively by thermal gravimetric analysis (TGA). Differential thermal analysis (DTA) reveals that the deblocking process beings around 200° for all of the polymeric derivatives (Figure 1). A kinetic study of the thermolysis of poly(S-vinyl-

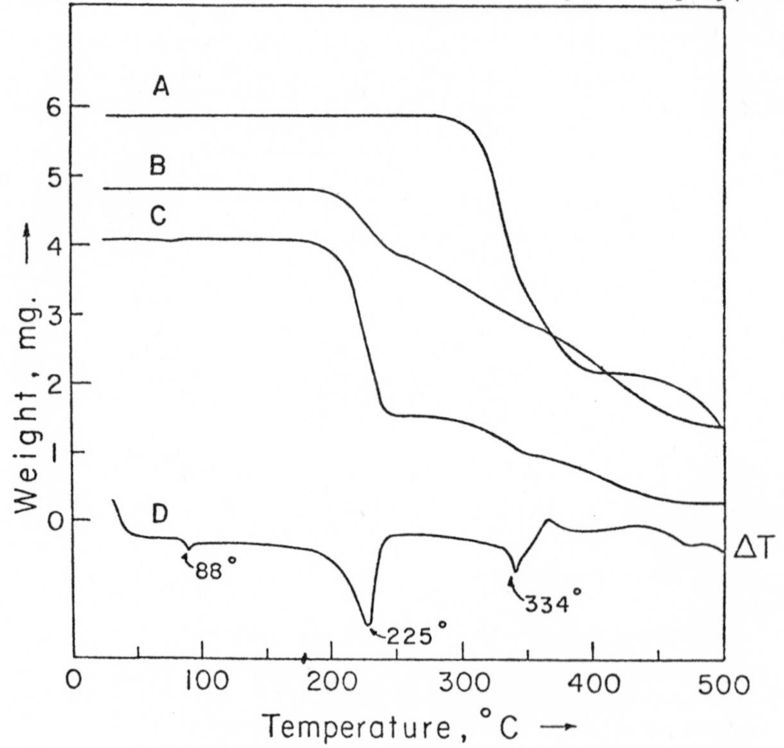

Figure 1. Thermogravimetric analysis of polythiocarbonates: A, TGA of polyvinyl chloride (PVC); B, TGA of thiocarbonate derivative prepared from PVC; C, TGA of poly-(S-vinyl-O-t-butylthiocarbonate) (22); D, DTA of 22.

O-*t*-butylthiocarbonate) by isothermal TGA showed that the reaction is first order and exhibits a linear Arrhenius temperature dependence in the range from 170-250° (E_a = 31.22 kcal/mol, ΔS‡ = -2.42 e.u.). The low activation energy and negative entropy of activation are consistent with the cyclic transition state usually proposed for Chugaev reactions.[17]

The total weight loss averaged over ten isothermal runs was 62.60 ± 0.1% (calc'd 62.5%). Mass spectral analysis of the gas mixture evolved showed that it was composed of CO_2 and isobutylene only, which is consistent with the deblocking mechanism. Thus, the deblocking reaction is quantitative for polymeric substrates containing only O-*t*-butylthiocarbonate substituents. This fact coupled with the low chlorine content of the substituted polymers gives some insight into the nature of the crosslinking reaction (Table II). Chloromethylated

TABLE II

THERMOLYSIS EFFICIENCY

Substrate	Decomp. temp., ^{o}C	Weight loss,%	Calcd.wt loss,%[a]	Calcd.SH content,%[b]
Poly(S-vinyl-O-*t*-butylthiocarbonate)	215	62.60	62.5	100
Poly(S-vinyl-O-*t*-amyl thiocarbonate	222	64.60	65.51	98.6
Poly(vinylbenzyl thiocarbonate)	222	35.5	40.0	89
Derivative of:				
Chloromethylated polystyrene	223	35.23	38.02	92.5
Poly(vinyl chloride)	202	19.5	47.17	41.3
Herchlor-H	214	20.0	51.3	39.0
Herchlor-C	220	4.07	32.69	12.4

a. Based upon the residual chlorine content.
b. Based upon the actual weight loss.

polystyrene, although slightly crosslinked, has undergone the expected substitution reaction since the chloromethyl groups are practically isolated from each other. The TGA curve of 20, which was obtained from a monomeric precursor, was essentially identical to that of chloromethylpolystyrene which had been treated with 19a; they exhibited a sharp break at 220° and approximately 90% of the theoretical weight loss as CO_2 and isobutylene by-products was recorded. The mercaptan content of the residual polymer was confirmed by titration. The more flexible polymer substrates with neighboring chloride leaving groups appear to have incorporated more cyclic structures. These polymers exhibited smaller initial breaks in the TGA and continued to degrade slowly over a temperature range from 250-475°C. The cyclic structures are more stable to thermolysis and do not degrade to polymercaptans.

CONCLUSIONS

The application of Bender's salts to the synthesis of monomeric precursors for polymercaptans is the most efficient method for preparing low molecular weight, well defined homopolymers and copolymers. The technique may also be employed to convert reactive polymeric substrates to polymercaptans if the substrate is a relatively rigid polymer containing isolated leaving groups. However, quantitative conversion of functional groups on flexible polymeric substrates is not possible due to the propensity for S-alkyl-O-*t*-alkylthiocarbonates to undergo side reactions leading to crosslinked polymers which can not be converted to polymercaptans.

REFERENCES

1. E. J. Goethals, J. Macromol. Sci. Rev., 1, 74 (1968).

2. H. E. Gunning, U. S. Patent 3,474,016; C.A 72, 54733w (1970).

3. O. P. Strausz, T. Hikida and H. E. Gunning, Can. J. Chem., 43, 717 (1965).

4. K. Kinoshita, T. Irie and M. Imoto, Kogyo Kagako Zasshi, 72, 1210 (1969).

5. G. Hardy et.al., Vysokomol. Soedin., **6**, 758 (1964).

6. H. Ringsdorf et.al., Makromol. Chem., **121**, 227, 240 (1969); **92**, 122 (1966).

7. H. Ringsdorf and C. G. Overberger, J. Polymer Sci., **61**, 511 (1962).

8. T. K. Dykstra and D. A. Smith, Makromol. Chem., **134**, 209 (1970).

9. E. Duprez and E. J. Goethals, Makromol. Chem., **146**, 145 (1971); J. Polymer Sci. Part A-1, **4** 2893 (1966).

10. C. G. Overberger, H. Ringsdorf and N. Weinshenker, J. Org. Chem., **27**, 4331 (1962); Makromol. Chem., **64**, 126 (1963).

11. C. G. Overberger and W. H. Daly, J. Amer. Chem. Soc., **86**, 3402 (1964).

12. W. H. Daly, Chien-Da S. Lee and C. G. Overberger, J. Polymer Sci. Part A-1, **9**, 1723 (1971.

13. T. Nakai, K. Shioya and M. Okawara, Makromol. Chem., **108**, 95 (1967).

14. (a) M. Okawara, H. Yamashina, K. Ishiyama and E. Imoto, Kogyo Kagaku Zasshi, **66**, 1383 (1963); (b) M. Okawara, T. Nakai and E. Imoto, ibid., **68**, 582 (1965); (c) M. Okawara and T. Nakai, Bull. Tokyo Inst. Technol., **78**, 1 (1966).

15. Potassium O-ethylthiocarbonate was originally synthesized by C. Bender in 1868, and the entire class of compounds with the general formula, $R\text{-}O\text{-}C(=O)\text{-}S^{-}Me^{+}$, have been named in his honor. C. Bender, Ann, **148**, 137 (1868); Bull. Soc. Chem. (2), **12**, 256 (1869).

16. R. Fischer and G. Fessler, Pharmazie, **10**, 349 (1955).

17. R. F. W. Bader and A. W. Bourns, Can. J. Chem., **39**, 348 (1961).

DISCUSSION SESSION

Discussion Leader - G. Challa

St. Pierre: Have you tried to react polyvinyl chloride with hydrazine or azides?

Daly: The reaction of azides with PVC has been reported, but we have not repeated this work. We have done the displacement reaction on Herclor (poly(epichorohydrin)) and the reaction is essentially quantitative in DMF. We have not tried to use hydrazine but I would guess that it would produce a crosslinked polymer under S_N2 conditions.

Rivin: It should be possible to prepare random mercapto-substituted polymers having good physical properties by reacting a diene homo- or co-polymer with a sulfur vulcanizing system to produce a crosslinked intermediate. Subsequent reduction of inter- and intra- molecular polysulfides with triphenyl phosphine and a metal hydride (e.g. $LiAlH_4$) would give the corresponding mercaptan in high yield. This process would be most convenient with low molecular weight polymers (~ 10,000) and should be free of significant chain scission and rearrangement side reactions.

Daly: Yes, this would be a good approach. The major disadvantage would be the need to conduct a series of reactions on the polymer, particularly if a crosslinked matrix is formed. Braun has prepared poly(p-mercapto styrene) by treatment of poly (p- lithiostyrene) with sulfur. A polymercaptide which could be converted to a polymercaptan with acid. Your proposal is an analogous system and it should be a good method for producing polymers containing randomly distributed mercaptan groups.

PARAOXON ANALOGS AS HAPTENS

R. Corett, A. Kalir, E. Katz, R. Shahar and M. Torten

Israel Institute for Biological Research

Ness Ziona, Israel

INTRODUCTION

The use of parathion and other similar insecticides is known to be hazardous to crop-dusters and other personnel exposed to them. Accidental poisoning was, until now, treated by pharmaceutical means only. We have decided to check whether prophylactic protection can be achieved in a completely different manner, i.e., by the use of parathion analogs as immunogens.

Parathion (I) is known to act in the body after oxidation to its more toxic analog, paraoxon (II).

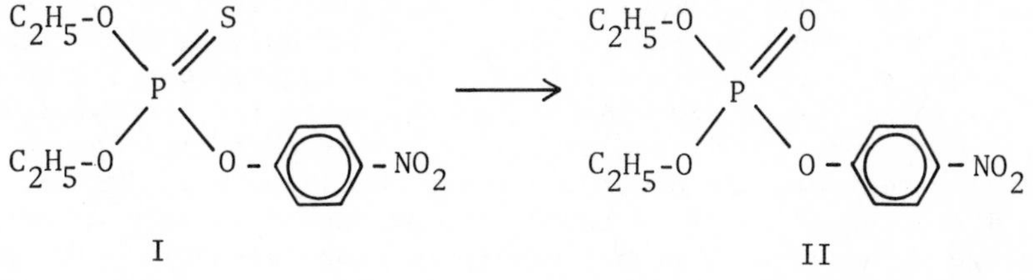

$$(C_2H_5\text{-}O)_2P(=S)\text{-}O\text{-}C_6H_4\text{-}NO_2 \longrightarrow (C_2H_5\text{-}O)_2P(=O)\text{-}O\text{-}C_6H_4\text{-}NO_2$$

I   II

Therefore, in our work, we concentrated on the last-mentioned compound.

It is well known that molecules the size of paraoxon are too small to stimulate antibody production. Antibody production may, however, be stimulated by attaching analogs to the small molecule (haptens) to a suitable high molecular weight carrier. In immunological terminology a carrier with haptens coupled to it is known as a conjugate. Upon injection to test animals, a conjugate can cause the body to produce antibodies to either the carrier, the carrier-hapten linkage and vicinity or to the hapten itself.

PREPARATION AND PROPERTIES OF CONJUGATES

Paraoxon was synthesized from diethylchlorophosphate (III) and paranitrophenol. It was reduced catalytically to give O-(p-aminophenyl) diethyl phosphate (IV).

Bovine serum albumin (BSA), a globular protein of molecular weight 69,000, was used as the carrier molecule. This protein contains, among others, 56 lysine residues, 19 tyrosine residues, 18 histidine residues, 53 aspartic and 26 glutaric acid residues. Its isoelectric point is at about pH 4.8. For serological studies, conjugates with rabbit serum albumin (RSA) were also prepared.

Two different haptens were chosen: 1) O-(p-aminophenyl) diethyl phosphate (IV), which contains almost the whole paraoxon entity and 2) diethylchlorophosphate (III), comprising only that part of the molecule which binds to esterolytic enzymes in the body.

Diethyl chlorophosphate (III) was coupled to the protein directly. A dioxane solution of III was slowly added to an aqueous solution of the appropriate protein. pH was kept constant at 8.5 by simultaneous addition of NaOH. The resulting conjugates will be referred to as BSA-Cl and RSA-Cl. The haptens are linked to the ε-amino groups of lysine residues.

O-(p-aminophenyl) diethyl phosphate (IV) was coupled to protein in two different ways: (a) IV was diazotized in the usual way. A buffered solution (pH 9.5) of the resulting diazonium salt was added to the protein solution. These conjugates will be referred to as BSA-DA and RSA-DA. According to the literature, diazo compounds are linked primarily to the phenolic rings of tyrosine residues, to histidine and probably also to lysine.
(b) The amino group of IV could also be caused to form an amide linkage with carboxyl groups of aspartic and glutaric acid residues. This was achieved by the use of N-cyclohexyl-N'[2-(4-morpholinyl) ethyl] carbodiimide-methyl-p-toluenesulphonate, a water soluble coupling agent used in peptide synthesis. These conjugates will be referred to as BSA-CD and RSA-CD.

After the completion of each of the three coupling reactions, the solutions were rinsed with distilled water in a ultrafiltration cell, then concentrated in the same cell to a convenient volume.

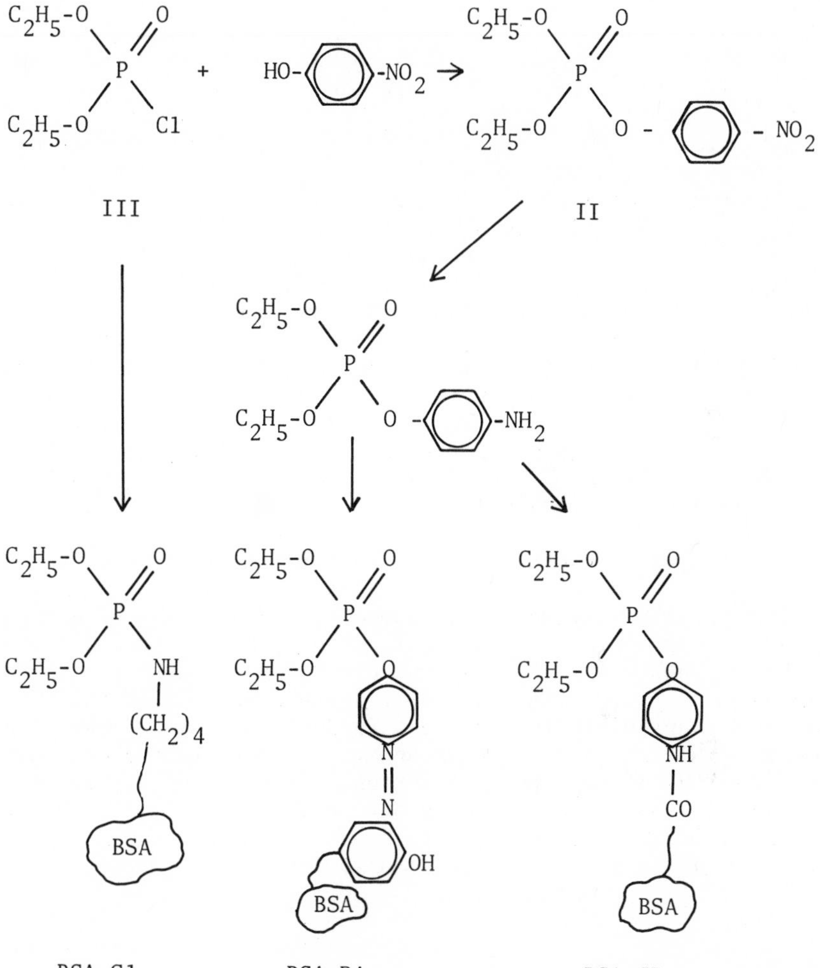

III II

BSA-Cl BSA-DA BSA-CD

Protein and phosphorus analysis was carried out by the conventional methods, and the number of haptens per protein molecule calculated.

In the case of -DA conjugates, it was found that the number of haptens per molecule could be calculated from the U.V. spectra of the conjugates, using the formula

$$\frac{O.D._{360}}{O.D._{280}} = 0.2 \frac{[hapten]}{[protein]}$$

which was found to hold true for molecules containing up to four haptens.

Solubility in water of the conjugates was studied in pH range 2.5 - 11. BSA is soluble through the entire range tested. BSA-DA, with 17 haptens per protein molecule, was also soluble in this range. BSA-C1, which contained 37 or 51 haptens per molecule, was insoluble between pH 3.85 and 4.95. It seems plausible that this difference in behavior is due to the difference in the nature of the chain connecting the phosphorus atom with the protein backbone: in BSA-DA this chain consists of two phenyl rings linked by an azo group - a rather stiff structure, which sticks out and holds the bulky diethyl phosphate group well away from the protein surface. In BSA-C1, the flexible aliphatic connecting chain enables the diethyl phosphate group to come close to the protein surface and increase its hydrophobic nature.

The number of haptens per conjugate molecule could be varied by changing the pH of the coupling reaction and the relative concentration of the reagents.

A conjugate containing less than the maximum possible number of haptens could be either a homogeneous compound or a mixture of molecules with varying hapten numbers. In the case of a homogeneous compound, the coupling reaction proceeds in a strictly predetermined way, i.e., haptens are being bound to the various residues on the carrier in a well defined order. In the case of a mixture, hapten binding proceeds randomly and the number of haptens per molecule which we find is an average for all the molecules present. By attempting to fractionate the conjugates by electrophoresis, isoelectric focusing or careful precipitation by stepwise changing of the solution pH, we hope to find out which of these two possible mechanisms actually occurs.

IMMUNOLOGICAL DATA

Rabbits were given three 1 ml consecutive intramuscular injections, on days 1, 10 and 30, of either BSA-C1 or BSA-DA. Each injection contained 10 mg protein per ml. The first injection was given with complete Freund's adjuvant, the second and third with incomplete adjuvant. Six days after the third injection the animals were bled, the serum separated and frozen at -20°C until used.

Ring tests for identification of specific antibody-antigen reactions were performed in the usual manner. Precipitation reactions in agar were done by the modified Ouchterlony method on glass slides. The results of the ring tests are summarized in Table 1 and those of the precipitation tests, in Figure 1.

TABLE I

SPECIFIC ANTIBODY-ANTIGEN REACTIONS AS DETECTED BY RING TESTS

Antigen			Antibody			
	Protein mg/ml	Haptens per Molecule	Anti BSA-C1	Anti BSA-DA R-184	Anti BSA-DA R-186	Normal Rabbit Serum (Control)
BSA	10	-	++	+	+	-
BSA-C1	3.5	37	+++			-
BSA-DA	11	6.5		++	+	-
RSA	10	-	-	-	-	-
RSA-C1	31	50	+	-	++	-
RSA-DA-11	7.5	8.9	++	+	+++	-
RSA-DA-6	61	3.4	++	+	++	-
RSA-CD-22	14	1.1	+	±	+	-
RSA-CD-23	8.2	2.2	++	+	+	-

The results presented in Table 1 and Figure 1 clearly show that both BSA-C1 and BSA-DA can stimulate antibody production against the carrier BSA as well as against the haptens themselves. Our results point to the fact that the immunocompetent system is able to recognize the diethyl phosphate moiety, whether or not it is linked to a phenyl ring.

Further research in progress is aimed at studying the relative affinity of the antibodies against the diethyl phosphate moiety alone versus the whole paraoxon molecule. Such studied should give us a clue as to which immunogen might best be used for practical purposes.

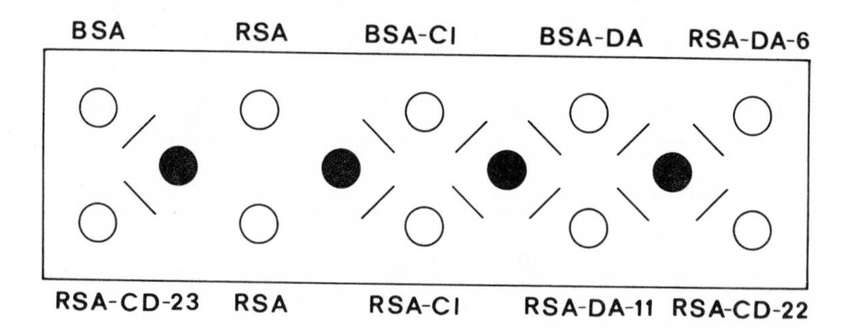

Figure 1. Specific antibody-antigen reactions as detected by precipitation in agar.

● - antisera: anti BSA-Cl or anti BSA-DA.

DISCUSSION SESSION

Discussion Leader - W. H. Daly

Moore: Do these indications of antibody formation show that these materials will in fact induce immune responses in vivo?

Corett: We hope so, but we don't know yet. There might also be other uses, e.g. it certainly can be used as an analytical tool. We have also not checked if these antibodies would respond to parathion derivatives. I'm not at all sure, but if so this would be a very sensitive analytical method to detect traces of parathion or paraoxon.

Stewart: Are these paraoxon -BSA conjugates toxic?

Corett: No, they are not. At least none of those that we made.

Williams: In the preparation of the antigen, could not the response that is seen in the rabbit be caused by conformational changes in the rabbit serum albumin?

Corett: I don't think so because we get antibodies which are active against unreacted serum albumin. That means that any changes which occurred were not large enough to affect the immunogenic site of BSA, whatever that may be.

CHEMICAL REACTIONS OF MACROMOLECULES IN THE SOLID STATE

G. Smets

Laboratory of Macromolecular Chemistry
Katholieke Universiteit Leuven.Louvain.Belgium

When chemical reactions are carried out in solid polymeric systems, different cases have to be considered depending the final purposes of these modifications. Chemical transformations of solid high polymers are very often a part of the synthesis of a polymer with desirable final properties. Such type of reactions can be illustrated for example by the second condensation step in the synthesis of ladder-type thermostable polymers, in which the final aromatization is achieved by heating the film or fiber at high temperature, usually with secundary ring cyclization and evolution of low molecular compound. Photoresist materials is another important example of chemical transformation; they are photosensitive materials which applied as thin coatings on substrates provide a chemically resistant coating in the light-struck regions. Their photosensitivity results from a photoinitiated polymerization, a photo crosslinking reaction or a combination thereof. In the case of polyvinylcinnamate photocyclodimerization with cyclobutane ring formation occurs together with photo crosslinking initiated by free radicals formed during irradiation.
While the degree of completion of such reactions can be very high, in other cases, e.g. the heterogeneous grafting of vinyl monomers on fibers or films the conversion is intentionally kept very low, because only surface modifications are intended.
Solid cross linked macromolecules can also be used as heterogeneous catalysts for the synthesis of compounds at their surface. The analogies existing between such

heterogeneous systems and some enzymatic processes are very attractive and suggest many questions concerning the molecular dispersion and molecular aggregation of dissolved macromolecules, the structuration of their solution in function of their concentration, temperature etc. Reactions carried out in mesomorphic systems (smectic, nematic) are of course connected with the same problems.
In the present paper we will concentrate our attention mainly to reversible organic reactions taking place in a solid polymer matrix. Here also different questions arise. We will therefore consider which is the influence of the chemical structure and of the physical properties of the matrix on the reaction rates as compared to the rates in liquid systems. Inversely, in how far can such reactions be used as a probe for the investigation of polymer orientation and polymer properties, and which are the possible applications offered by some of the systems.

As a general statement, it can be foreseen that, when compounds are bound to a polymeric substrate or are incorporated in a rigid medium strong restrictions are imposed to their translation and rotation possibilities. On account of differences of chain segment mobility these restrictions will be most pronounced in the rigid glassy state. In the rubbery region, above the glass-transition domain, the matrix can be compared to a high viscous medium, and the transition and rotation possibilities will strongly increase and approaches progressively those observed in solution. On the basis of such considerations, a drastic change in reaction rate has to be expected in the neighbourhoud of the glass transition domain.

With these assumptions in mind, we will discuss successively reversible photochromic phenomena, particularly the cis-trans isomerization of aromatic azo-derivatives, the ring opening/closure of spirobenzopyranes derivatives and some related phenomena.

A. CIS-TRANS ISOMERIZATION OF AROMATIC AZO-COMPOUNDS

When aromatic azo-derivatives are exposed to irradiation of an adequate wavelength, trans-cis and cis-trans isomerizations occur; in addition a thermal cis-trans isomerization takes place, the trans isomer being more stable than its cis-isomer (10 Kcal/mol). Due to lack of coplanarity of the cis-compound, this isomer is strongly hypsochromic with respect to the trans-form. Whether the rate of isomerization around the azo-group is influenced by a polymer matrix, was firstly examined by Kamogawa and al. (1). These authors compared the cis-trans isomerization rates of the model compounds (in solution, and embedded in a polystyrene matrix) with the corresponding copolymers of vinyl amino azobenzene and styrene (1/9), in solution and in bulk. Table 1 summarizes recovery halftimes of the systems as given by these authors, and which makes the comparison solution/bulk behavior possible.

Table 1 : Halftime values of copolymers of vinylamino-azobenzenes with styrene (1/9) and model compounds.

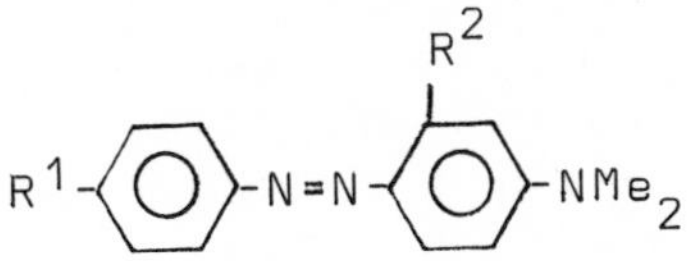

R^1-C6H4-N=N-C6H3(R^2)-NMe_2

	R^1	R^2	physical state	1/2 time in min.
a	$-CH_2-CH-(ST)_9$	H	S	30
			F	150
a'	CH_3	H	S	90
			F*	240
b	H	$-CH_2-CH-(ST)_9$	S	15
			F	7
b'	H	H	S	30
			F*	8
c	CH_3	$-CH_2-CH-(ST)_9$	S	12
			F	10
d	NH_2	H	S	10
			F*	5
e	p.$NH_2-C_6H_4$	H	S	10
			F*	15

S=solution, F=film of copolymer, F*=embedded in polystyrene film.

From this table, it can be seen that the copolymers b and c behave quite differently from a, on account of the site of attachement (ortho or para) of the polymeric back-bone with respect to the azo-link. For b and c, where high steric hindrance exists, the decoloration rates is even much higher in the film than in solution. Similarly for the model compounds no apparent regularity in relative solution/bulk behavior can be found. In order to account for the results the authors suggest the existence of two opposite effects : i) the well known aggregation of azobenzenes in solution and film, which can be prevented in a copolymer where styrene constitutes the main component (9 ST/1 azo) and therefore isolates the azo-group from each other; ii) the steric hindrance existing in polymeric systems. These two effects may compensate each other, the final behavior resulting from the balance between them.
The influence of the glass transition of the matrix in which embedding was achieved, as well as its hydrogen bonding properties, has also been considered by these authors. It can be illustrated with para-acetamino-azobenzene : p.CH_3-CO-NH-C_6H_4-N=N-C_6H_5. The halftimes/minutes were 8 minutes in benzene solution, 30 minutes in a polystyrene matrix, 10 minutes in butylacrylate-MMA film, and only 1 minute in a terpolymer N-hydroxy-methylacrylamide-butylacrylate-MMA.
Priest and Sifain (2) examined the photochemical and thermal trans-cis isomerization of 2.2'-azonaphthalene and 4-ethoxy-azobenzene in polystyrene-n.butyl benzene compositions. The cis content depends strongly on the matrix viscosity, and in the case of 2.2'-azonaphthalene, it increases from 45% in pure polystyrene (Tg ∿ 109°) to 58% for compositions with 44% butylbenzene (Tg ∿ 1°C) and reaches 78% in pure butylbenzene.
Above their respective glass temperature, all thermal isomerizations obey first-order kinetics. Contrarily, below Tg, the initial isomerization rates were abnormally fast, and decay to normal first-order process, only after an appreciable degree of conversion. The terminal linear portions of the first-order plots show that the isomerizations were slowest in unplasticized polystyrene.
A more quantitative study of the photochemical and thermal isomerization of azoaromatic residues in polymers in bulk was recently described by Paik and Morawetz (3). Copolymers of 4-(N-methacryl)amino azobenzene and of 4-(N-methacrylamino) -1,1'-azonaphthalene (1 mole %) with methylmethacrylate present in the bulk an isomerization quantum efficiency which is about 1/4 to 1/5 of that in solution. The photostationary states i.e. the

photoresponse of the system, were similar in bulk and in solution if the specimens were irradiated above Tg and cooled under irradiation; they were substantially lower below Tg (the photostationary fraction of cis isomer being 0.52 instead of 0.76). Thus, the photostationary state depends on the history of the specimen.
Above Tg first-order kinetics were found for the thermal cis-trans isomerization in the bulk, the rate constants being equivalent to those found in solution.
Contrarily, in the glassy state a portion (1/7) of azo-groups reacts anomalously rapidly about 8 to 12 times faster than the remainder groups which react with the normal rate constant as found for the rubbery state.
The authors assume that this anomaly reflects the non-equilibrium nature of the glassy state, where the free volume is unequally distributed : cis-isomers, trapped in a strained conformation, will more easily return to the trans form than the cis-isomer in an equilibrium environment. Indeed such anomaly was reduced by prolongation of the irradiation, or irradiation above Tg and cooling while irradiating.
In the case of glassy polymers containing azoaromatic groups in the back-bone, (polyamides and polyesters), the photostationary fraction of cis-isomer is much smaller than in glassy polymers with side azo-groups; f.e. with poly-bisphenol-A isophtalate, containing 1.4% p.p' azobenzene dicarbonyl units, this fraction is only 0.025 in bulk, while 0.45 in chloroform solution; the initial rates of thermal isomerization is however only slightly different.

B. PHOTOCHROMIC SPIRO-BENZOPYRANES

Spirobenzopyranderivatives undergo under ultraviolet irradiation a C-O-bond scission followed by cis-trans isomerization during which deeply coloured open ring merocyanines are formed.

CH_3 CH_3 N O $-NO_2$ $\underset{\Delta t}{\overset{h\nu}{\rightleftharpoons}}$ NO_2 $O^{\ominus}$ $N^{\oplus}$ CH_2 C_6H_5 CH_2 C_6H_5

N(benzyl)-6-nitro DIPS[*]

[*]The acronym DIPS is used for 3.3'dimethylspiro-[2H-1 benzopyran-2,2'-indoline].

These coloured species revert to the colourless compound by a thermal proces (4).
Gardlund (5) and Gardlund and Laverty (6) observed that the thermal decolouration follows first-order kinetics in solution. The reaction becomes biphasic when these compounds are incorporated in a glassy PMMA-matrix. In a rubbery matrix, e.g. polybutylmethacrylate above Tg, first-order kinetics were restored, the rate constant being nevertheless much lower than in solution. The existence of two or three isomers, each characterized by a different rate constant, is assumed in a glassy matrix; in solution or in the rubbery state these isomers interconvert and, the first-order rate constant corresponds only to the most stable isomer.
As will be seen later on, the data with spirobenzopyranes are much more coherent than for the azo compounds. It is the reason why we have concentrated most of our efforts on these derivatives. Our previous results have been reviewed recently and will only be summarized here (7,8). They will be completed with new data, as obtained recently. Only their behaviors in solid matrices and in bulk will be considered.

i) the rate of decolouration depends on the chemical nature of the matrix, on account of the negative solvatochromism of the spirobenzopyranes.

Table II : Influence of the polymer matrix on decolouration kinetics of N(β-isobutyroxyethyl)DIPS

	Polystyrene	PMMA
λ_{max} nm	609	585
k_{27} : $10^4 sec^{-1}$	17	2
E_a kcal/mole	17.6	24.4
$\Delta S^{\neq}$ e.u.	-14.6	+3.9

Not only the activation energies differ considerably, but the activation entropies have opposite signs in the two matrices. Similar results were found by Flannery (9) in the case of N(methyl)6-nitro DIPS in polar (ΔS positive) and apolar, non-hydrogen bonding (ΔS negative) solvents.

ii) the rate of decolouration of the photochromes

is strongly depending on Tg, as stated by Gardlund; moreover the activation energies have much higher values above Tg than below it. This increase is related with an additional activation energy for viscous flow.
This general statement is valid as well for photochromes dissolved in a polymeric matrix, than when they are bounded to the polymer chain.
For example, compositions containing five weight % N(benzyl)DIPS dissolved in polyester matrix are characterized by the following physical constants and activation parameters, as indicated in table III.
For a and d with high Tg only measurements were significant below Tg since the photochrome shows a pronounced thermochromism above 120°C. For b and c two different domains of temperatures above and below Tg were examined : the very important increase of apparent E_a is attributed to a contribution of the viscous flow activation energy.
Similarly, copolymer of n(β-methacrylamido-ethyl)DIPS (0.63 molar %) with isobutylmethacrylate (Tg 61°) and n.propylmethacrylate (Tg 53°) show a marked kinck in the Arrhenius plots around Tg. (fig. 1)
Figure 1 represents the comparison of the same DIPS-photochrome (2.7 weight %) when attached as side group of a polymer by copolymerization, and when dissolved in the same polymeric matrix N(β-isobutyramidoethyl) DIPS in poly isobutylmethacrylate). Tg of the mixture was around 30°C.

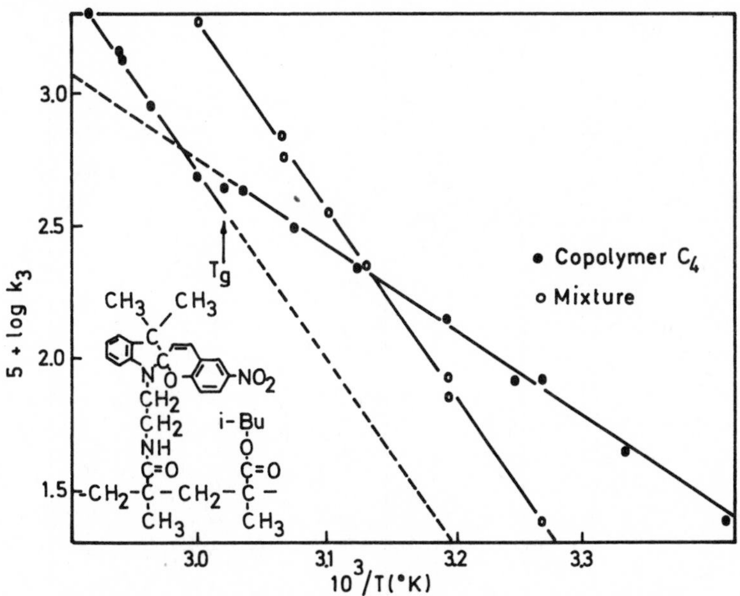

Fig.1: Arrhenius plot of decoloration reaction.

Table III : Decolouration kinetics of N(benzyl)6-nitro DIPS polyester matrices.

Polymer	T_{go} pure polymer	T_g mixture	Apparent activation energy kcal mole^{-1}		Molecular weight matrix	Temperature range of measurements
			below T_g	above T_g		
a. polybisphenol-A-carbonate	150	-	23.6	-	33.500	30.2-57.6°C
b. polybisphenol-A-adipate	70-75	53-55	20.4	61	∿44.000	22.4-63°C
c. polybisphenol-A-pimelate	57	∿51	20.6	56.7	∿26.000	32.3-59.8°C
d. polybisphenol-A-50 iso : 50 tere-phthalate	∿200	-	19.2	-	30.000	29.5-53.5°C

The activation parameters for the final linear decolouration step are :

		i.Butylmethacrylate cop.	n.Propylmethacrylate cop.
below Tg	E_a kcal/mol	15.1	17.2
	$\Delta S^{\neq}$ e.u.	-25.5	-19
above Tg	E_a kcal/mol	32.3	36.9
	$\Delta S^{\neq}$ e.u.	+25.8	+41

Remarkable are the activation entropies passing from strongly negative values below Tg to highly positives above it.

iii) Several new photochromic polycondensates containing the photochrome within the polymer back-bone were prepared by a two step method as described by Smets and Evens (8); it prevents the presence of photochromes at the chain ends.
Using this method a photochrome polyester has been prepared containing 3 molar percent of component A i.e. N-(p-hydroxymethyl-benzyl)-6-nitro-8-hydroxymethyl-DIPS within poly(bisphenol-A-pimelate) sequencies.

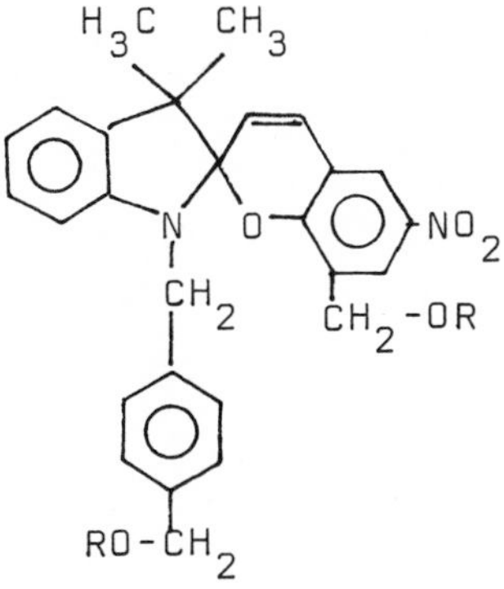

H_3C CH_3 ... N ... O ... NO_2 ... CH_2 ... CH_2-OR ... RO-CH_2

A (R=H)

$$R = [CO-(CH_2)_5-CO-O-C_6H_4-C(CH_3)_2-C_6H_4-O]_x \quad (A')$$

i.e. $[\text{polyester}]_x$

Tg of this polyester is situated between 50 and 55°C; it makes it possible to follow easily the photoresponse of the photochrome in a temperature domain where the segment mobility is assumed to be extremely sensitive to any temperature variation.
In figure 2 the optical density after five seconds irradiation of a film of 65 micron thickness is reported as a function of the temperature.

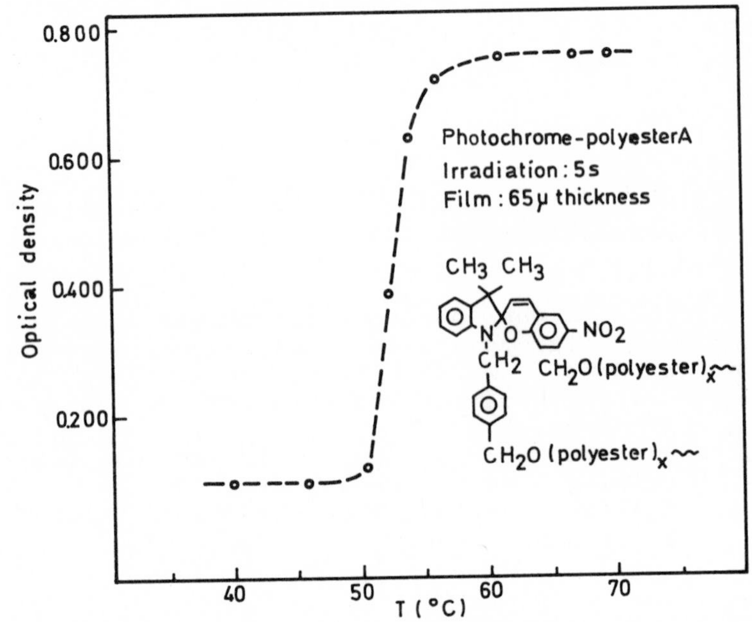

Fig.2: Photoresponse of photochrome polyester A'. Dependence on temperature.

As can be seen, a sharp increase of photoresponse occurs between 50-55°C in the neighbourhood of Tg.
At 33°C irradiation of 45 seconds are neaded for obtaining an optical density of 0.2 while above Tg 2 seconds are sufficient to induce an optical density of 1. Around 20°C, the activation energy of decolouration is equal to 15.5 kcal; around Tg, it amounts to about 76 kcal.
The importance of the incorporation of the photochrome into the polymeric back-bone as compared to the corresponding model dissolved in a similar matrix, is clearly illustrated with the bisphotochrome C, condensed with pimelic acid and bisphenol-A (4 molar %) and its bispropionate embedded (2%) in a polybisphenol-A pimelate matrix.
The lnk versus 1/T plots are given in figure 3.

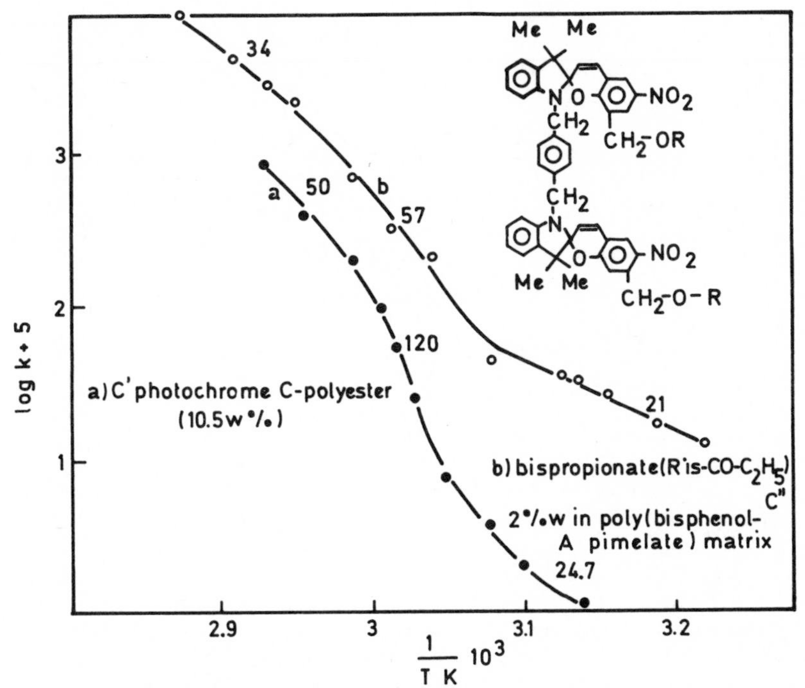

Fig.3: Dependence of lnk versus 1/T for photochrome C'-polyester (curve a) and photochrome dipropionate dissolved in poly(bisphenol-A-pimelate) matrix.

The figures along the curves indicate the variation of the apparent activation energies in function of the temperature.
The most striking difference between the curves lies undoubtedly in the much higher rate constant for the matrix C" than the polyester itself; moreover the mixture colours easily on irradiation, while the copolyester C' is practically light-insensitive at room temperature. Furthermore, the temperature dependence in the neighbourhood of Tg is much more strongly pronounced for C'. This comparison demonstrates clearly that the photoresponse of a system depends very strongly on the incorporation or not of the photochrome residue in the polymeric backbone.

iv) The strong temperature dependence of lnk, resulting in S-shape Arrhenius diagrams, has been interpreted by assuming a pre-exponential term proportional to the jump frequency of a molecular segment from one position to another, i.e. proportional to the reciprocal of the internal viscosity.

On the basis of the WLF equation (10) relating the viscosity-temperature coefficient of a polymeric system to its Tg, one can derive a final equation

$$\log(k_{Tg}/k_T) + E_a(T-Tg)/RTTg = -17.44(T-Tg)/(51.6+T-Tg)$$

where k_{Tg} and k_T are the rate constants at Tg and T respectively, while E_a represents the energy of activation of photochromic isomerization.

v) It may be worthwhile to point out that photochromic decolouration rate may be sensitive to :

a) secondary transition phenomena of the polymeric matrix, where the segment mobility increases e.g. for PMMA a rate increase of 40% is found at 56°C (7);
b) non-equilibrium states of the glassy polymer, due to an unequal distribution of the free volume, which are responsible for lower value of the activation energy below Tg (11); this anomaly can be reduced by annealing the polymeric film, and is similar to that observed with azo-group containing polymers (see above);
c) the degree of crystallinity (D.C.) of the matrix. Though this crystallinity leads to light scattering and may decrease the significance of the spectral data, S. Dumitriu has shown that an increase of D.C. causes an increase of rate constant in the cases of polyethylene and polypropylene matrices (12). For polyethylene a variation of D.C. from 0.62 to 0.81 causes a rate increase of 40%; for polypropylene going from a D.C. of 0.32 to 0.42 causes a rate increase of 60%. The higher rigidity of the structure resulting from a smaller intermicellar thickness of the amorphous phase favours the rate of decolouration on account of some internal pressure within the system, the dimensions of the ring closed spirobenzopyran being smaller than those of the open merocyanine form.

vi) A peculiar phenomenon related to photochromism and chain segment mobility, which should be mentioned here and upon which will be reported elsewhere, is the photochemical and thermal contractile behavior of crosslinked photochromic polyethylacrylates. On irradiation samples contract appreciably (2 to 5%) and reverse in the dark to their original length; the cycle photocontraction/dark recovery can be repeated several times and is reproducible.
The light-induced photoisomerization of the incorpora-

ted spirobenzopyran photochrome is responsible for these reversible changes of dimensions of the samples, i. e. light energy is converted into work.

C. RELATED ISOMERIZATION PHENOMENA

A few other isomerization reactions carried out in the bulk should be mentioned here; although they have not been described in as much details than the previous ones, nevertheless their importance should not be underestimated.

1. Kamogawa (13) applied to polymeric systems the orange-blue photochromism of mercurydithizonate, which is based on the following isomerization reaction (14) :

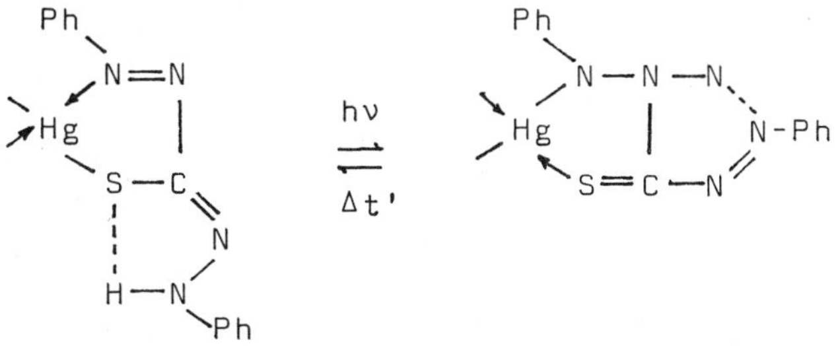

$$\text{Ph-N=N-C(=N-NH-Ph)-S-Hg} \underset{\Delta t'}{\overset{h\nu}{\rightleftharpoons}} \text{Ph-N-N=C(S\to Hg)-N=N-Ph}$$

Therefore he synthesized copolymers containing p-aminophenyl-mercuric-diphenylthiocarbazonate side groups, with styrene or methylacrylate as comonomers :

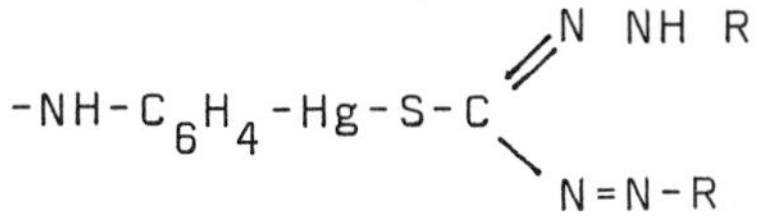

$$-NH-C_6H_4-Hg-S-C\begin{matrix}\nearrow N\ NH\ R \\ \searrow N=N-R\end{matrix}$$

The spectral recovery occurs thermally.
On the basis of decolouration halflifetimes, the reaction was found about one hundred times slower in a film than in solution; moreover, a higher glass transition of the copolymer (styrene comonomer) causes a longer decolouration time than that for copolymer with methylacrylate.

2. The photochromic behavior of thioindigo derivatives is based on the cis-trans photoisomerization around the C=C double bond.
This system is characterized by an extremely slow thermal cis-trans isomerization and thus practically the photostationary state is independent of the intensity of irradiation. Its properties for storage and read-out

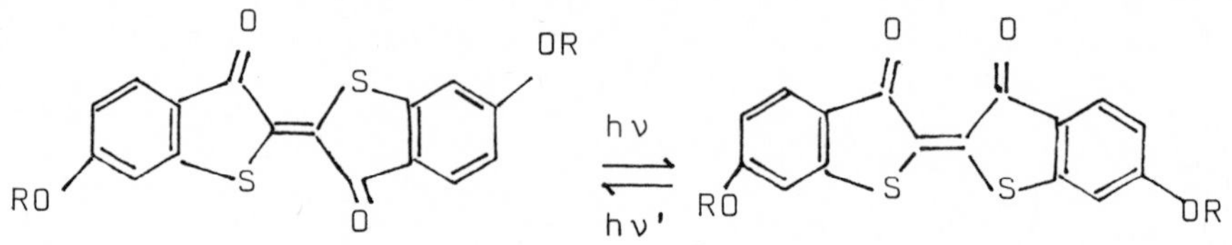

$$\text{trans} \underset{h\nu'}{\overset{h\nu}{\rightleftharpoons}} \text{cis}$$

of optical information using argon-ion laser have been examined by Ross (15,16).
Diisopropoxythioindigo is sufficiently soluble in organic solvents and matrices to make possible to measure the quantum yields for trans-cis Φ_t and cis-trans Φ_c isomerizations. It was embedded in a heat-cured epoxy-resin at 10^{-3} to 10^{-5} molar concentration. Though Φ_t and Φ_c are markedly lower than those measured in solution, they show a considerable increase above 35°C; these quantum yield values decrease with time, and this decrease is the faster the lower the temperature at which the sample is kept. Upon rewarming around 35°C, the quantum yields again rise to their previous values. The author assumes the existence of a glass transition around 30-35°C for accounting these hysterisis phenomena, which indeed do not occur in a polystyrene matrix. It is interesting to point out that in the cross-linked epoxy-resin the cycles trans-cis and cis-trans can be repeated 25.000 times using adequate irradiation wavelengths without noticeable loss of colour; in a polystyrene matrix, after an hundredfold repeated process 30% of the colour is already lost.

3. The thermal racemization of optically active derivatives, although this phenomenon is not reversible, is also directly related with internal chain segment mobility, and should therefore be considered here. Schulz and Jung examined the racemization kinetics in solution of the polyvinylester of the (+)2.methyl-6-nitro-biphenyl-2' carboxylic acid and found a biphasic process, of which the second step was ten times slower than the first one (17,18). No data on the solid state were reported.
P. Hermans (19) used the thermal interconversion (T° > 120°C) of asymmetric cyclooctatretaëne conformers (20), and prepared esters and polyesters of optically active (+) 2-bromo-dibenzo(a.e.)cyclooctatetraene 6.11 dicarboxylic acid (DBCOT). (Figure 4).

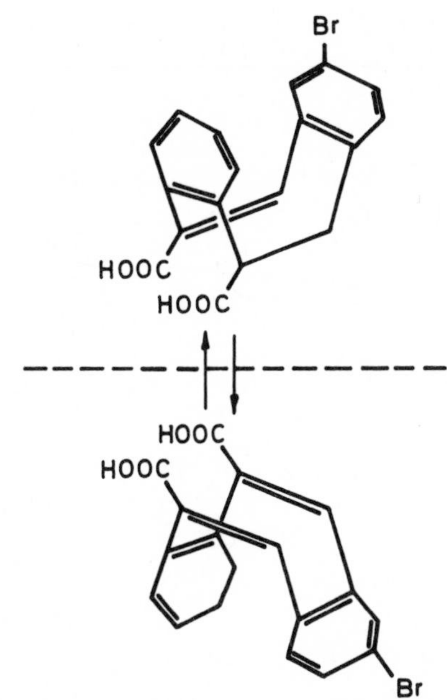

Fig.4: Racemization of 2-bromo-dibenzo (a.e.) cyclo-octatetraene-6,11-dicarboxylic acid.

In the bulk a much higher temperature (180-210°C) is needed for the racemization of these derivatives than in solution (120-150°C). While in solution the rates of racemization follow first-order kinetics and are very similar for the diphenylester and the bisphenol-polyester of DBCOT, again in the solid state two successive reaction steps are found for the polymers, the second slow phase resulting in an asymptotic limit value, depending of the temperature of reaction, e.g. 58% racemization at 180°C for a copolyester of bisphenol-A-carbonate containing 14.3 molar % DBCOT (mol.w. 8160). The behavior is illustrated in figure 5 and compared with the homopolyester of bisphenol-A with DBCOT (mol.w. 8.300).

DBCOT-diphenylester was embedded (15 weight %) in amorphous polystyrene and in partially crystalline polycarbonate; both polymers have a glass-transition temperature below the racemization temperature. The rates of racemization are sensibly equivalent in both

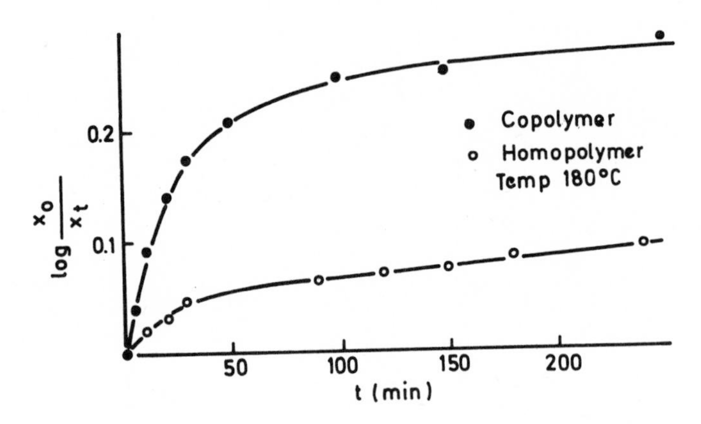

Fig.5: Rate of racemization of BCT-polymers.

matrices, and therefore independent of the nature of the matrix, as well as of its crystallinity.
An oligomer of following composition (molecular weight: 3.500); racemizes in bulk about 30 times slower than

$$\left[-\overset{O}{\overset{\|}{C}}-\bigcirc-\overset{O}{\overset{\|}{C}}-O-\bigcirc-\underset{C_2H_5}{\overset{CH_3}{C}}-\bigcirc-O-\right]_4-\left[-\overset{O}{\overset{\|}{C}}-O-\bigcirc-\underset{C_2H_5}{\overset{CH_3}{C}}-\bigcirc-O-\right]_4$$

the above mentioned mixture. The data summarized in table IV, stress again the importance of the chemical incorporation of the reactive group inside the polymeric back-bone.

Table IV : Thermal racemization of DBCOT-derivatives in bulk.

	in solid state (film)		
	Pstyr matrix	Polycarbonate matrix	oligomer amorphous
$k_{150} \times 10^4 sec^{-1}$	5.6	3.5	0.15
$E_{a_{act}}$ kcal/mole	33.3	33.6	52.1
log A	13.9	13.9	22
$\Delta S^{\neq}$	+2.5	2.5	+39.9
$\Delta F^{\neq}$	31.4	31.7	33.8

Besides the isomerization phenomena we have choosen for the present discussion,some bond-dissociation reactions which have been described recently should now be briefly considered, as far as they are also influenced by the physical state of the matrix.

a) Chandross, Lamola and al. (21) considered the reversible photodimerization of polycyclic aromatic hydrocarbons, e.g. anthracene and its derivatives.
The structures of the anthracene monomer and dimer are represented as follows.

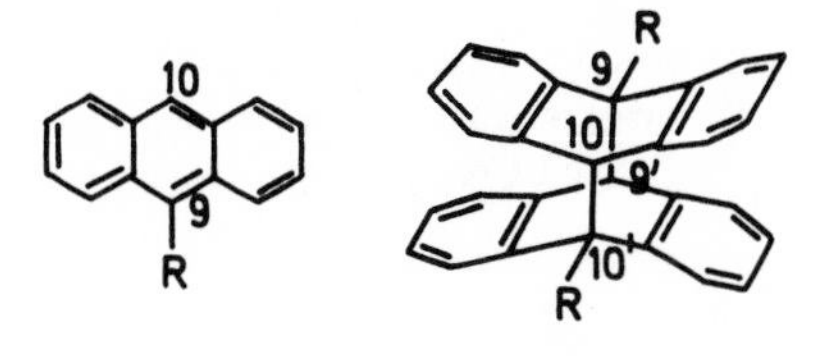

The dimer can be cleaved by irradiation with wavelength <310 nm, while the photodimerization occurs with light at 365 nm, but must strongly depend on the relative positions of the two monomers.

It is to be expected that in a rigid matrix, the pairs of monomers obtained by photolysis will remain near each other, in the solid cage, and therefore will remain favorably oriented for redimerization. As a consequence, the efficiency of redimerization in a solid matrix should be independent of the concentration, and be much higher than in solution.
In fact in PMMA matrix the quantum efficiency of redimerization was much to low (∿ 0.003) while the quantum efficiency of breaking the dimer was 0.3, i.e. hundred times larger. This means that PMMA matrix does not keep the broken dimer in adequate alignment for redimerization. Hardening of the matrix by annealing or eventually by including cross-linking agents causes only a slight increase of efficiency; on the contrary with crystalline dimers, the quantum efficiencies of dimer scission and redimerization are practically equal.
Using similar techniques Somers (22) examined the photodissociation of the n.hexyl anthracene-2-carboxylate in poly-isobutyl- and polymethylmethacrylate matrices. The dissociation/dimerization cycles can be repeated, but the redimerization is only partially reversible, some broken dimers being not enclosed in a matrix cage, and having diffused apart from each other.
Some data are given in table V. All films contained 0.48 weight % n.hexyl-dimeric ester, and were irradiated under vacuum during 25 minutes at 310 nm.
The data show that the dissociation of the dimer is similar in both matrices and that annealing of the films in the dark after first irradiation improves the diffusion of the broken dimers from each other; the same effect was obtained by keeping the samples for long time before photochemical redimerization.
These preliminary results confirm at least partially the data of Chandross and Lamola and have to be further worked out on a more quantitative basis.

b) Dan and Guillet (23) examined very recently the chain scission reaction of polymeric ketones in the solid state. They found that the main chain scission reactions (Norrish type II reaction) have a quantum yield

$$-\text{C}-\text{C}-\underset{\underset{\text{R}}{|}}{\underset{\text{C=O}}{|}}{\text{C}}-\text{C}-\text{C}- \xrightarrow{h\nu} -\text{C}-\text{C}-\underset{\underset{\text{R}}{|}}{\underset{\text{C=O}}{|}}{\text{C}}- \quad + \quad \text{C=C}-$$

Table V : Dimer-Dissociation and redimerization of anthracene derivatives.

annealing	% dissoc.	irrad.time/ min. 360 nm	% residual monomer	time for 20% dimeriz.	matrix
-	78	150	37.5	12	PtBuMA Tg 58°C
2 hrs.- 66-67°C	78.1	150	48	39	
-	66.8	180	38	47	PMMA Tg 105°C
1 hr. - 120°C	64.2	180	48.2	126	

Φ_{cs} that increases greatly in the glass transition region, and attributed this to the great increase in polymer segment mobility which occurs at this temperature. For example for copolymers of styrene and phenylvinylketon Φ_{cs} increases from 0.068 at 100° to 0.293 at 104°. Analogous results were obtained with methylmethacrylate methylvinylketon copolymers and with polyphenylvinylketon. Above Tg-region, the type II quantum yields remain constant, and equal to those obtained in solution for the same polymers and low molecular weight homologues. Below Tg, some much less important temperature dependence was also found (2-3 kcal/mole); it was attributed to changes in segment mobility which is required for the formation of six (or seven) membered transition state, and could be related with secundary transition domains (side group mobility).

c) In analogy with these experiments, mention should be made of the interesting work of Takahashi, Hasegawa (24) and coworkers on photodegradable polyamides, obtained from diphenyl α-truxillate (or δ-truxinate) with ω,ω' alkylenediamines. The polymers depolymerize photochemically (224 nm) and generate cinnamyl groups besides some stilbene and fumaramide; such facing cinnamyl groups remain in pairs close to each other, and can redimerize quite favorably in cyclobutane ring, besides some cross linking reaction.
The photodegradation of such polymers must therefore be considered, at least partially, as a reversible reaction, and to be limited by diffusion of the scission products out of the cage, where they were formed.
The reaction scheme for the dibutyl model compound must be written.

n-Bu-NHCO, CONH-n-Bu, C_6H_5, C_6H_5 (cyclobutane)

10× symmetrical scission → n-Bu-NHCO–CH=CH–C_6H_5 + n-Bu-CONH–CH=CH–C_6H_5

1× asymmetrical scission → n-Bu-NHCO—CH=CH—CONH-n-Bu + C_6H_5—CH=CH—C_6H_5 (3)

d) Finally, special mention should be made on a spectacular effect of matrix rigidity on the reactions of aromatic nitrenes in polymers (25). Reiser, Leyshorn and Johnston found that the matrix rigidity affects strongly the distribution of primary and secondary amines which are the main products of the photolysis of aromatic azides. The triplet nitrenes, which are produced by the photolysis of the azides, abstract an hydrogen atom and are converted into an amino radical

$$R\overset{\uparrow}{\dot{N}}{\cdot}^{\uparrow} + H-\overset{|}{\underset{|}{C}}- \longrightarrow R-\overset{\uparrow}{\dot{N}}H + {}^{\uparrow}\cdot\overset{|}{\underset{|}{C}}-$$

In fluid solution, the two radicals having parallel spins cannot couple and move apart before spin inversion. The amino radical abstracts a second hydrogen from another site and gives primary amine

$$R\overset{\uparrow}{\dot{N}}H + H-\overset{|}{\underset{|}{C}}- \longrightarrow RNH_2 + -\overset{|}{\underset{|}{C}}-$$

In the rigid polymer, the amino radical and the substrate radical remains together, long enough for electron spin inversion, and radical coupling occur giving a secondary amine. The yield of secondary amine depends thus on the competition between spin conversion, and segment mobility of the polymeric substrate.
The authors found a yield of secondary amine equal to 0.93 in polystyrene, 0.81 in polyvinylacetate, 0.67 in polyisoprene, 0.33 in hard parafinn wax, and only 0.04 in paraffin oil.

CONCLUSION

Chemical reactions carried out in polymeric matrices are strongly influenced by the physical state of the polymers. Their efficiencies and rates are very often different in the glassy and in the rubbery states on account of large differences in chain segment mobility. The chemical incorporation of the reactive groups inside the polymer molecule increases strongly these effects, when compared to mixture in polymeric matrices.

REFERENCES

(1) H. Kamogawa, M. Kato, H. Sugiyama; J.Polym.Sci.A-1, 6 2967 (1968)
(2) W.J. Priest, M.M. Sifain; J.Polym.Sci.A-1, 9 3161 (1971)
(3) C.S. Paik, H. Morawetz; Macromolecules 5 171 (1972)
(4) G.H. Brown; Techniques of Chemistry, Vol.III Photochromism, John Wiley & Sons N.Y. 1971, pp.49-288
(5) Z.G. Gardlund; J.Polym.Sci.B, 6 57 (1968)
(6) Z.G. Gardlund, J.J. Laverty; J.Polym.Sci.B, 7 719 (1969)
(7) G. Smets; Pure & Applied Chemistry 30 1 (1972)
(8) G. Smets, G. Evens; Pure & Applied Chemistry, 1973 in press
(9) J.B. Flannery Jr.; J.Amer.Chem.Soc. 90 5660 (1968)
(10) M.L. Williams, R.L. Landel, J.D. Ferry; J.Amer. Chem.Soc. 77 3701 (1955)
(11) M. Kryszewski, D. Lapienis, B. Nadolski, in press
(12) S. Dumitriu, G. Smets; J.Polym.Sci., in press
(13) H. Kamogawa; J.Polym.Sci.A-1, 9 335 (1971)
(14) L.S. Meriwether, E.C. Breitner, N.B. Colthup; J. Amer.Chem.Soc. 87 4448 (1965)
(15) D.L. Ross; Applied Optics 10 571 (1971)
(16) R.C.A. Corp, D.L. Ross; DT-OS 2160149, 3.12.71; US : 7.12.70 (95.578)
(17) R.C. Schulz, R.H. Jung; Makromolekulare Chemie, 96 295 (1966)
(18) R.C. Schulz, R.H. Jung; Tetrahedron Letters 44, 4333 (1967)
(19) P. Hermans; Ph.D. Thesis, Katholieke Universiteit Leuven 1971
(20) K. Mislow, H.D. Perlmutter; J.Amer.Chem.Soc. 84, 3591 (1962)
(21) W.J. Tomlinson, E.A. Chandross, R.L. Fork, C.A. Pryde, A.A. Lamola; Applied Optics 11 533 (1972)
(22) A. Somers; Lic.Thesis, Katholieke Universiteit Leuven 1973
(23) E. Dan, J.E. Guillet; Macromolecules 6 230 (1973)
(24) H. Takahashi, M. Sakuragi, M. Hasegawa, H. Takahashi; J.Polym.Sci.A-1, 10 1399 (1971)
(25) A. Reiser, L.J. Leyshon, L. Johnston; Trans.Faraday Soc. 2389 (1971)

DISCUSSION SESSION

Discussion Leader- J. A. Moore

Bamford: You mentioned that, in Guillet's work on the photo-degradation of poly(phenyl vinyl ketone) just above and just below Tg, the dramatic change in the quantum yield for type II scission was due to recombination. I would have thought this unlikely, because cage effects of that kind are not normally obtained for type II processes. I would imagine that the difficult in forming the six-membered transition state in the glassy state to be a more likely explanation.

Smets: Below the Tg, Guillet finds a quite small activation energy of Type II chain scission (3kcal). He postulates that this is due to local segment mobility problems in making the necessary six-membered. The value of Φ_{cs} (quantum yield for chain scission) for each case changes drastically around Tg. He follows Φ_{cs} by viscometry.

Bamford: It still puzzles me why there is no similar effect with small ketones in solution.

Cho: Are the photomechanical effects reversible? If you, e.g., hammer the film do you see the formation of color?

Smets: Yes, but it requires very large pressures to achieve this.

Bamford: How does the extent of bulk shrinkage compare with what is calculated from the changes in molecular dimensions?

Smets: There still remains some difficulty in exlaining how such minute molecular changes cause such large bulk effects.

Challa: I would expect that the thermal history of the films would affect the reproducibility of your measurments

Smets: You are right. This is the reason that we anneal the films before measurment.

Carraher: In the photomechanical effects, do you observe any creep?

Smets: It is completely reversible. Recovery is 100%, but it takes longer time than shrinking.

Rempp: You indicated that decoloration can be caused by irradiating the samples at a wavelength different from that which caused coloration. Can the decoloration be caused thermally?

Smets: Yes. This is the so-called dark reaction. The fact that the back reactin can also be photochemically induced is a disadvantage is practical applications.

Lewin: Do you think that this technique could be used for study of the fine structure of polymers?

Smets: The effects observed are small changes in rate and are not at all dramatic.

CHEMICAL REACTIONS INVOLVING THE BACKBONE CHAIN OF MACROMOLECULAR CRYSTALS

Bernhard Wunderlich

Department of Chemistry
Rensselaer Polytechnic Institute
Troy, New York 12180

INTRODUCTION

Ideal macromolecular crystals should, as all other crystals, show no chemical reactivity inside the closely packed crystals. Reactions may, however, occur at the surface, where free access of reactants is possible. The three steps of any chemical reaction -- approach of reactants, formation of the activated state configuration, and separation of the products -- must be influenced by the stringent geometric requirements of the crystal surface. Special effects arise on macromolecular crystal surfaces because of the long chain nature of the constituent molecules.

The macroconformations in the solid state are summarized in Fig. 1. The conformations on a 50 to 1000 Å scale are called macroconformations, in contrast to the microconformations which involve only rotational isomerism about one to several backbone bonds. It could be shown[1] that for all presently known homopolymers the extended chain conformation in form of a helix or a more or less planar zig-zag is energetically most stable. Figure 1 indicates, however, that a crystal with a fully extended chain macroconformation (area C) is only one of the limiting possibilities. The surface structure, and with it the chemical reactivity and accessibility, must depend critically on the different possible macroconformations.

Area A of Fig. 1 represents the glassy macroconformation in which no long range order is detectable. The chains are over longer distances close to parallel, a geometric requirement to account for the observed densities[2]. The added volume of the glass above that

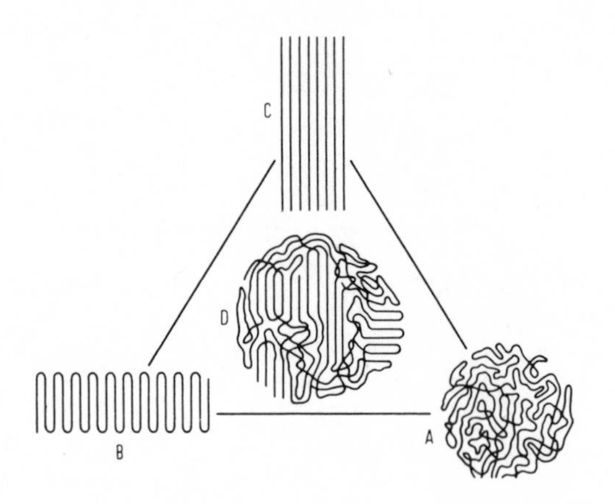

Fig. 1. Schematic drawing of the macroconformations of crystals of linear macromolecules. A. Glassy. B. Chain folded. C. Chain extended. D. Fringed micelle.

of the crystal (often 10-20 percent, for bulky helices occassionally less) can be thought of a holes added to the otherwise densely packed matter. The more open structure of the glass, its greater solubility (particularly in the molten state above the glass transition temperature, where the molecular chains become mobile) and the presence of strained molecules in the irregular structure account for the greater accessibility and reactivity of macromolecules in the amorphous state.

Area B of Fig. 1 represents the ideal, sharply folded macroconformation as it is approximated by slow crystal growth from dilute solution. As a matter of general principle,[1] flexible linear macromolecules must always fold in the molecular nucleation step on crystallization from the mobile state. Figure 2 illustrates solution grown folded chain crystals of polyoxymethylene. For such solution grown crystals typical lengths between folds are observed to be 50 to perhaps as high as 500 Å, with most fold-length observed at about 100 Å. A typical 100 Å fold-length lamellar crystal has now, when one neglects the small side surface areas, a specific surface area of about 10^6 cm^2/g. Such enormous surface area permits accessibility for chemical reaction. The detailed surface structure is, in addition, not as drawn in the schematic, idealized Fig. 1, but it may contain considerable roughness. Folds were shown to occur over a depth of as much as perhaps 30 Å on either side of a 160 Å thick polyethylene crystal.[3] In addition, it is known that this rough surface is covered with loose, amorphous material. Particularly on crystallization from the melt, one expects a fraction of the molecules to have uncrystallized portions in form of loops and chain ends as well as connections between different lamellae in form of tie-molecules. Reactivity of the material between the lamellae is enhanced because of easier accessibility and also because of strain on many of these surface chains. At the

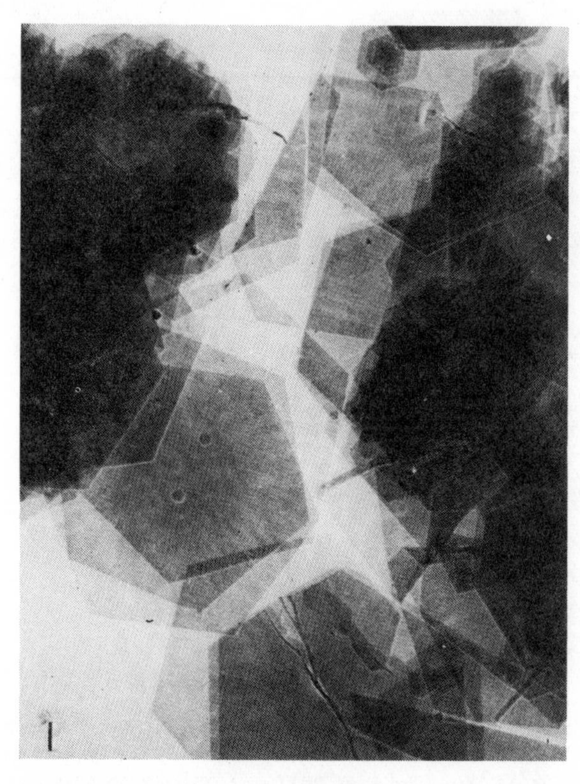

Fig. 2. Solution grown folded chain lamallae of polyoxymethylene. Electron micrograph. Scale, 1 µm. (Bassett, et al., 1964).

folds one can estimate that the potential energy of the chains is raised; for polyethylene, for example, by 3-5 kcal per fold, the energy being distributed over 3-5 bonds.

Area C of Fig. 1 represents the extended chain crystals. Figure 3 shows an electron micrograph of a fracture surface of polychlorotrifluoroethylene[4]. The molecular chains are aligned parallel to the striations, at right angles to the lamellar surfaces. The largest lamella is now more than 20,000 Å thick contrasting with the typical 100 Å of the folded chain lamellae. At such large thickness the surface has little influence on the crystal properties. Despite the relatively large chain extension, some molecules are, however, still larger and some folding and tie molecules may still exist on the surfaces. Chemical reaction is again most likely to occur on the lamellar (001)-surface, which, particularly after removal of folds, tie-molecules, and amorphous chain ends and loops, is suitable for depolymerization etching the crystal; or further polymerization to extend the crystal or alter its chemical nature. These reactions must naturally be studied in the light of the stringent geometric conditions imposed by the crystal structure.

Between the limiting areas A, B, and C lies a structure which is usually called the fringed micelle structure. One expects such

Fig. 3. Electron micrograph of a replica of a fracture surface of extended chain crystals of polychlorotrifluoroethylene. Scale bar, 1 μm. (Miyamoto, et al., 1972).

structure (which is characterized by small crystallites, tie molecules, irregular folds, and amorphous material) on relatively fast crystallization from a relatively immobile melt. Obviously the fringed micelle structure may assume all intermediate structures between the three limiting cases and represents a challenge for detailed characterization. Enhanced reactivity is expected as before on folds and tie molecules. Accessibility is helped by the usually relatively large amount of amorphous material. The crystalline dimensions are normally not larger than 10 to 100 Å.

The basic reactions to be discussed in more detail involve the chain breakage in the crystalline and amorphous areas and the chain lengthening. The first process will be called etching, in analogy to the well known treatment of crystals of low molecular weight. The second process will be called polymerization since one expects higher molecular weights as a result.

ETCHING

Etching techniques were first developed for cellulose[5]. Typically, cellulose can be etched by acid hydrolysis or alcoholysis at 80-100°C. Depending on reaction conditions, the molecular weight levels off after 15 minutes to several hours of treatment at a value of 2000 to 120,000 which corresponds to the reaction reaching the crystal surface. A good correlation exists between the length of the recovered crystals and the molecular weight measured[6]. The molecular weights given above correspond to a crystal length in the chain direction of 60 to 4000 Å. The end surface contains mainly -CHO- groups resulting from the hydrolytic cleavage of the (1→4)-β-glycosidic bonds. Table I shows some data indicating the changes occurring during the etching of cellulose[7]. Sample A represents native cellulose (cotton linter, 99.7% α-cellulose),

Table I

Changes in Cellulose on Etching
(C_2H_5OH in presence of H_2SO_4)

		Reaction time: 0h	1h	7h	12h
Weight loss (%)	A	0	-	9	8
	B	0	9	34	42
Deg. polymeriz.	A	1840	-	161	150
	B	396	51	30	29
X-ray crystallin.	A	0.71	-	0.82	-
	B	0.26	0.36	0.48	-

while sample B represents regenerated cellulose (super rayon cord). The more perfect native cellulose levels off at a higher molecular weight and the crystallinity observed after 7h of etching is close to the expected value for loss of amorphous polymer only (0.78). The picture of etching of the relatively perfect crystals of sample A is thus that chains are cut between the crystals and removed by alcoholysis down to the crystal surface. In case of the less perfect crystals of sample B, the crystallinity after 7h of etching is significantly higher than can be calculated on the assumption of weight loss of amorphous cellulose only (0.40). One must conclude thus that the crystals were strained before etching and reorganized to greater perfection during etching.

Other groups of crystalline macromolecules which have been etched by hydrolysis are polyamides and polyesters[1]. Poly(ethylene terephthalate) was analysed in some more detail[8]. In the presence of an excess of water, poly(ethylene terephthalate) (crystallized from the melt to a lamellar morphology) could be hydrolyzed in 2 to 5h at 210-180°C so that the initial mainly -OH terminated polyester changed to largely -COOH terminated oligomers. Figure 4 illustrates the fast decrease in molecular length which levels off as soon as the crystal surface is reached. The weight loss at the time etching levelled off approximated the weight fraction crystallinity. The final product is thus a quite uniform material of lamellar crystals of fully extended oligomers of polydispersity close to one. The crystals remained intact and could be studied by electron microscopy[8]. No other analysis method exists which permits separation of isolated crystal lamellae from

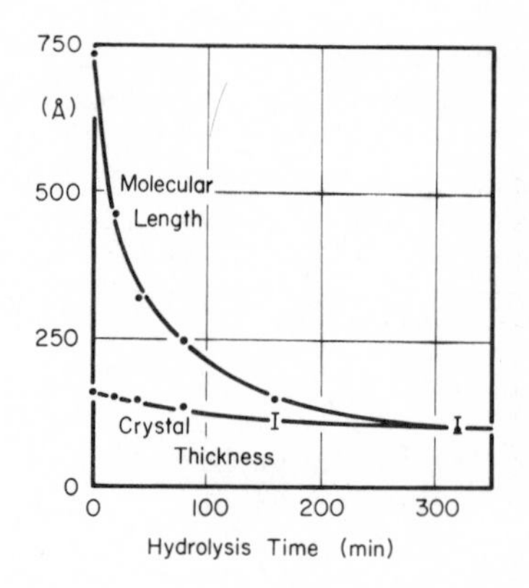

Fig. 4. Molecular length of poly(ethylene terephthalate) by viscosity, and crystal thickness as measured by low angle X-ray diffraction as a function of hydrolysis time at 180°C

the melt-grown crystal aggregates. Tie-molecules are frequent, so that the lamellae before etching are closely interconnected. Figure 5 illustrates the quick cutting of tie-molecules during the early stages of hydrolysis. The superheating on increasing heating rate, i.e. the increase in melting peak temperature due to the slowness of the melting process due to tie molecules[9], decreases to zero in approximately 50 minutes. At this time practically no weight loss has occurred, but the molecular weight decreased to about one-third the original value. At later times the molecular weight decrease is less and more material is lost in form of terephthalic acid and ethylene glycol. The melting behavior finally changes to a reorganization characteristic, i.e. on slow heating the oligomer lamellae repolymerize (see below) and indicate a higher melting point than the starting material. The true zero entropy production melting point at the temperature where the free enthalpy of

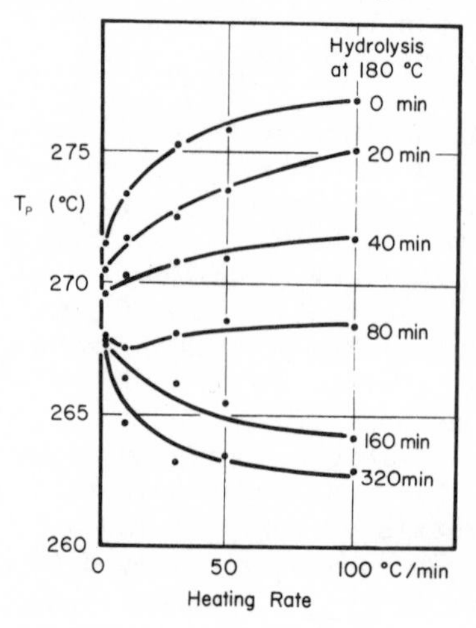

Fig. 5. Melting peak temperatures of poly(ethylene terephthalate) by differential thermal analysis as a function of heating rate after different times of hydrolysis.

the metastable crystal is equal to the free enthalpy of the super-cooled melt is indicated by the horizontal melting level at fast heating rate. It decreases slowly as the lamellae are etched to smaller thickness.

For the etching of polyolefins, a different approach is necessary because of the chemical stability of the backbone chain. Palmer and Cobbold[10] showed that oxidative attack with fuming nitric acid is selective enough to stop at the crystal surface. Table II shows the molecular weight, crystallinity, and weight loss as a function of etching temperature on etching of polyethylene at different

Table II

Changes in Polyethylene on Etching*
(100% HNO_3 after 24h, initial $M_n = 5300$, $w^c = 0.74$)

Temperature	Mol. Wt.	Wt. Loss	Crystallinity
45°C	2770	2.5%	0.76
50°C	2100	7.0%	0.85
55°C	1980	14.0%	0.91
60°C	1860	25.0%	0.91
70°C	1590	40.0%	0.92
80°C	1570	62.0%	0.87

*Data by Illers (1968)

temperatures. Similar results can be obtained with ozone and chromic acid, for example. The reaction leads to mainly chain scission with COOH end groups, which is followed by further attack of the α-carbon and loss of CO_2 and H_2O. Chain scission occurs already at low temperature as the fast reaction, the further oxidation is accelerated greatly at higher temperatures (see Table II). It could be shown that the COOH- content levels off at a concentration corresponding to two per remaining oligomer. Additional $-NO_2$ groups at a level of about one-eighth of the COOH were also observed, seemingly unrelated to the carboxyl end groups and of unknown location[11]. Following the weight loss as a function of time at constant temperature, again a definite break in the increase at a level corresponding to the starting crystallinity is observed. The slow-down of oxidation on approaching crystal surfaces was for the case of polyethylene attributed to the build-up of reaction product[11]. After washing and drying, it was always possible to restart the slowed-down reaction. Figure 6 illustrates the debris of a lamellar fragment of melt crystallized polyethylene. The etching product can again be used to analyze the morphology of the otherwise inseparable lamellae. Quantitative analysis of the molecular weight changes during etching permitted a chemical proof of chain folding and a more detailed

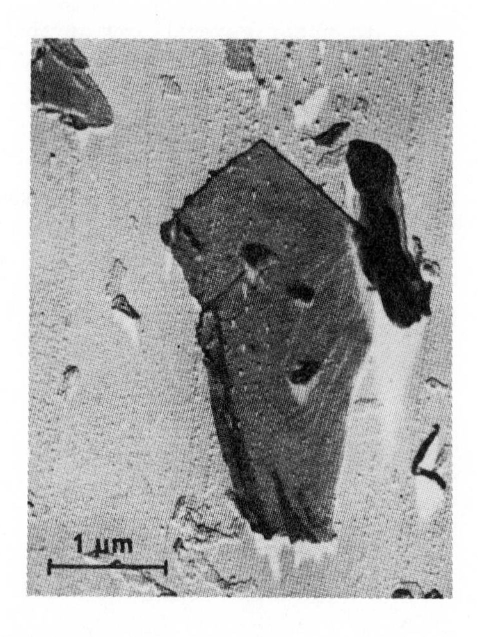

Fig. 6. Lamellar fragment of melt-crystallized polyethylene after etching with fuming nitric acid. The morphology is clearly visible[12]. (Electron micrograph).

analysis of fold surface configuration and roughness[1]. The uniform carboxylic acids thus obtainable have been converted via intermediate diiodides to paraffins of uniform length[13].

A number of more drastic degradation techniques were found to be unable to be stopped at the crystal surface. Two examples are the degradation of poly(ethylene terephthalate) with methylamine[14] and the degradation of polyoxymethylene by increased temperature[15]. Another approach to etching, which can at least reveal surface structure of crystalline macromolecules is etching through gas-discharge at pressures of 10^{-3} to 10^{-4} mm Hg pressure. Depending on the different perfection of crystal and amorphous areas, slightly different etching rates are established and electron microscopic observation of the surface topology permits conclusions about morphology. This technique[1,14] is naturally not able to lead to preparative amounts of the etched crystals.

The various etching techniques are not only useful for morphology and macroconformation analysis, but most of them also provide preparative routes to relatively sharp distributions of oligomers in the crystalline form suitable for further reaction (see below). Since the etching usually starts on tie molecules and folds, selective alteration of the intercrystalline structure is also possible, leading over many intermediate stages of increasing brittleness to the final microcrystalline powder of oligomers.

POLYMERIZATION

Polymerizations involving macromolecular crystals are based on the principles of crystallization during polymerization[14]. Before the reaction can occur, an active surface must be created. Two different starting materials are possible: fully extended chain crystals, usually produced by etching since direct growth to extended chain crystals from preformed macromolecules is rather difficult; and nuclei, created during the combined polymerization and crytallization process. First, the nucleation process followed by polymerization will be discussed, followed by the description of two examples of reactions on activated crystal surfaces.

The need of crystal nucleation seems to forbid direct transformation of liquid or gaseous monomer to the crystalline, macromolecular state. First, oligomers of the proper length for primary nucleation (20-100 Å) must be produced by normal solution or gas phase polymerization[15]. After a sufficient concentration of oligomers of the correct length is present, nucleation can occur. This is followed by monomer addition on the chain end surface of the oligomer nucleus. For further growth at right angles to the chain direction, new macromolecules must be nucleated (secondary nucleation). In certain cases this may be hindered, producing fibrous morphologies[1]. Figure 7 is an example of such fibrous morphology, produced by crystallization during polymerization of tetrafluoroethylene from the gaseous state[16]. Similar fibrillar crystals

Fig. 7. Electron micrograph of fibrous polytetrafluoroethylene crystallized during polymerization from the gas phase. Scale bar, 1.5 μm.

result from the biosynthesis of cellulose and poly(3-hydroxybutyric acid) and under proper conditions also durion trioxane and vinyl polymerization[1,14]. In such crystallizations during polymerization, chain folding is not an a priory condition for crystallization. It may occur, however, due to the specific reaction conditions. Often chain folding is irregular in this case.

One of the best analyzed reactions is the formation of polyphosphates on dehydration of dibasic alkali salts of the orthophosphoric acid[17]:

$$xLiH_2PO_4 \xrightarrow{\sim 200^\circ C} Li_xHO(PO_3)_xH + (x-1)H_2O$$

In this case, it could be shown by thin layer chromatography that in the early stages of the reaction increasing amounts of oligomers were produced. Since the reaction is a typical step reaction, appreciable amounts of the decamer, for example, are not expected before the reaction is half complete. In agreement with this statement no crystalline macromolecules were found until the first nonamer was identified. At later stages of the reaction, the higher oligomers disappeared again, being exhausted by nucleation. The reaction had thus shifted from a step mechanism to an addition mechanism as soon as nucleation was possible. The final molecular weight was independent of the degree of reaction, it depended only on the degree of perfection of the crystals grown[14].

A reaction under investigation in our laboratory is the crystallization during polymerization of poly-p-xylylene[18]. The amorphous polymer has a glass transition temperature of 80°C, polymerization from the p-xylylene monomer is possible to a large degree, however, only below 80°C. The produced material is semicrystalline, so that crystallization must occur before the molecules are completed, since bulk crystallization below the glass transition temperature is impossible. The polymerization occurs only on the surface and the reaction rate is controlled by the absorption equilibrium, i.e., the reaction rate increases at lower temperature. Between -17 and 26°C the produced polymer is of the regular folded macroconformation with the chain axis parallel to the substrate surface. Such morphology is only reasonable when one assumes that the reaction occurs by successive polymerization and crystallization[14,15]. Separate polymerization and crystallization is impossible because of the much higher glass transition temperature. Simultaneous polymerization and crystallization is unlikely since it does not provide a reasonable mechanism for regular folding. Increasing the reaction temperature, the rate of reaction slows down because of decreasing reactant concentration, but the crystallization rate increases due to approach to the glass transition temperature. One would expect thus that the polymerization site on the molecule would, on increasing temperature, be approached by the crystallization site. Indeed, at about 80°C, the produced crystal morphology changes abruptly to the β-form, signaling a change in mechanism. Also, in agreement with such a temperature determined variation of the polymerization and crystallization sites is the decreasing perfection of the crystals at lower temperatures. To complicate matters, at liquid nitrogen temperature a solid state reaction of the monomer seems possible[19]. Low temperature thermal analysis revealed that at such low temperature p-xylylene seems able to crystallize without reaction, to polymerize later to an irregular folded, fairly well crystallized polymer of the β-crystal form.

The polymorphism, changes in crystal morphology, varying crystallinity, and different macroconformations of the produced poly-p-xylylenes are clear indicators of the changes in reaction and crystallization mechanism. Materials of greatly varying properties can be produced in this way by coupling crystallization and polymerization. Morphologies otherwise not accessible can be produced.

Turning now to the creation of an active surface and further chemical reaction, selenium may serve as a first example. It was possible to show that on crystallization from the melt selenium crystallizes first in a chain folded manner with about 100 Å fold length, followed by immediate extension to perhaps 10 to 20 times the initial thickness[20]. The selenium melt is typically only about half its weight polymeric, the rest is in form of Se_8 ring-molecules. Following the disappearance of the rings quantitatively, it was possible to show that initially practically only chains crystallize. After completion of the reaction, however, only chains exist. Somewhere in the course of crystallization, ring-chain conversion must take place at an enhanced rate. Since in the melt this rate is slow, it should occur in the vicinity of the crystal interface[21]. That the fold surface is not an inactive fold surface, but can open up folds and continue polymerization can be deduced from the final selenium morphology shown in the electron micrograph of Fig. 8. The lamallar boundaries between the lamellae are much less distinct than, for example, in Fig. 3, and in many places one can see fibers and crystal fragments, split out from the crystals on replication, which run over several lamellar thicknesses. This should only be possible if actual covalent connections exist between the lamellae. One can conclude[20,21] that after initial crystallization, the selenium fold surfaces open up and continue polymerizing, adding either new rings of Se_8 or connecting touching lamellae. These conclusions which were suggested first by Crystal[22] led us to an effort to duplicate similar crystal perfections through polymerization on active crystal surfaces on synthetic macromolecules.

Polyesters, -amids, -urethans, and -ureas are most suitable for such study of polymerization on the crystal surface. Since melt and

Fig. 8. Electron micrograph of a replica of a fracture surface of melt crystallized selenium[20]. Scale bar, 1 µm. The black fibers and crystal fragments are actual Se portions embedded in the replica. (As in Fig. 3 the molecular chains are parallel to the striations, normal to the lamellar surfaces.)

solution crystallized macromolecules are always first folded, it is necessary to activate the surface by etching, as described above. Our example was poly(ethylene terephthalate)[23]. The oligomer lamellae were treated at various temperatures below their melting temperature under vacuum so that reaction through ester interchange or anlydride formation could occur. Excess of terephthalic acid was noted to sublime. Figure 9 illustrates the reversal of the behavior on hydrolytic etching. The increase in molecular weight is illustrated in Table 3. The initial molecular weight before etching was 11,600, so that the annealed material more than recovered the starting molecular weight. Since under the chosen conditions the oligomer lamellae did not melt, correction of the molecules must have occured in the interlamellar region. During these reactions there exists no reason for introducing strained folds, it is much more likely that tie molecules or longer loops are produced. The presence of these loops and ties are indicated by the superheating behavior of the annealed lamellae (see Fig. 9). The strictly folded chain solution grown lamellae (dotted in Fig. 9) show none of such superheating. In fact, studying the melting behavior of the starting material which was melt crystallized at 250°C for 48 h, one cannot account for its melting behavior on increased crystal perfection due to high temperature crystallization alone. One must assume that the initial chain folds open up due to ester interchange even during and after crystallization. Annealing sample S1 directly, without initial etching, tends to similar perfection as annealing after etching (dotted curve S3).

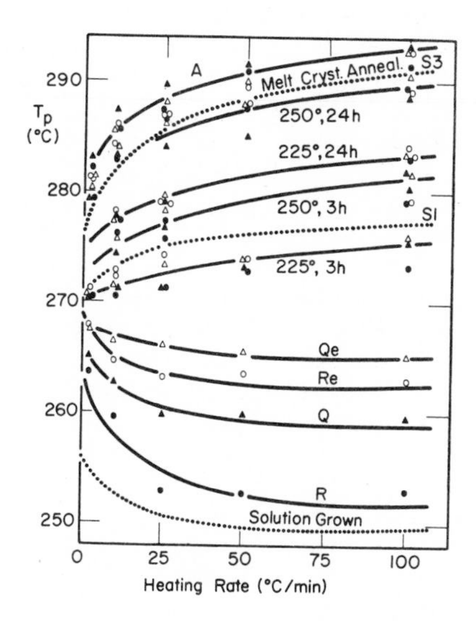

Fig. 9. Melting peak temperatures of poly(ethylene terephthalate) oligomer crystals after vacuum annealing. Qe, Re, Q, and R are different oligomers before annealing (compare to Fig. 5). Times and temperatures of the annealing under vacuum are given in the figures. All starting materials yield after sufficient times similar melting behavior. Dotted curves reference materials.

Table III

Increase in Molecular Weight of Poly(ethylene terephthalate) on Vacuum Annealing at 250°C.

Annealing time	Molecular Weight
0 h	2,300
3 h	9,700
6 h	14,200
12 h	19,600
24 h	25,200

These experiments have indicated that chemical reaction can occur during and after crystallization and alter the initial macroconformation of the polymer. It is, as a further extension of this work, even possible to carry out the interconnections of the lamellae with chemically different repeating units, leading to new block copolymers of otherwise unavailable block precision. The oxidatively etched vinyl polymer lamellae which also have a high COOH-content surface are similarly active for further chemical reaction and alteration. Since the initial oligomer molecules are held in place by the crystal structure, they represent ideal substrates for a (in this case truly) solid phase polymerization which may form the base for the production of more complicated, designed copolymers of exact sequence distribution and molecular weight[24]. The restrictions are those of the initial crystal geometry, presently a still largely unexplored topic.

SUMMARY

A large group of linear macromolecules can be affected chemically without destruction of the crystal by reaction on their surface. The new principles to be introduced to understand these reactions involve mainly geometry restrictions introduced by the crystal structure. The surface reactions provide a means for altering the chemical nature and the physical nature of the once produced crystals. It also provides a new path to make crystals of chain macroconformations not otherwise possible because of the chain folding principle. Finally, it seems to point a way to the production of advanced copolymer systems. The study of such reactions may also serve as an analog for many of the ill understood biological polymerizations leading to structural materials.

ACKNOWLEDGEMENTS

Support for the present work in our laboratory on etching and polymerization is derived from Grant GH33458 of the National Science Foundation.

REFERENCES

1. B. Wunderlich, "Macromolecular Physics, Vol. 1, Crystal Structure, Morphology, Defects." Academic Press, New York, London 1973.

2. R. E. Robertson, J. Phys. Chem., 69, 1575 (1965).

3. A. Keller, E. Martuscelli, D. J. Priest, and Y. Udagawa, J. Polymer Sci., Part A2, 9, 1807 (1971).

4. Y. Miyamoto, C. Nakafuku, and T. Takemura, Polymer J., 3, 120 (1972).

5. B. G. Rånby, and E. D. Ribi, Experimentia, 6, 12 (1950).

6. O. A. Battista, S. Coppick, J. A. Howsman, F. A. Morehead, W. A. Sisson, Ind. Eng. Chem., 48, 333 (1956).

7. J. Baudisch, J. Dechart, D. van Nghi, B. Philipp, and C. Ruscher, Faserforsch. Textiltech., 19, 62 (1968).

8. A. Miyagi and B. Wunderlich, J. Polymer Sci., Polymer Phys. Ed., 10, 2073 (1972).

9. A. Miyagi and B. Wunderlich, J. Polymer Sci., Polymer Phys. Ed., 10, 1401 (1972).

10. R. P. Palmer and A. J. Cobbold, Makromol. Chem., 74, 174 (1964).

11. A. Keller and Y. Udagawa, J. Polymer Sci., Part A2, 9, 1793 (1971).

12. A. Keller and S. Sawada, Makromol. Chemie, 74, 190 (1964).

13. A. Keller and Y. Udagawa, J. Polymer Sci., Part A2, 8, 19 (1970).

14. B. Wunderlich, Fortschr. Hochpolym. Forsch., 5, 568 (1968).

15. B. Wunderlich, Angew. Chemie, 80, 1009 (1968).

16. L. Melillo and B. Wunderlich, Kolloid Z. Z. Polymere, 250, 417 (1972).

17. E. Thilo and Grunze, Z. Anorg. Allg. Chem., 281, 262 (1955).

18. S. Kubo and B. Wunderlich, J. Polymer Sci., Polymer Phys. Ed., 10, 1949 (1972).

19. G. Treiber, K. Boehke, A. Weitz, and B. Wunderlich, J. Polymer Sci., to be published.

20. M. C. Coughlin and B. Wunderlich, Kolloid Z. Z. Polymere, 250, 482 (1972).

21. M. C. Coughlin and B. Wunderlich, J. Polymer Sci., Polymer Phys. Ed., 11, xxx (1973).

22. R. G. Crystal, J. Polymer Sci., Part A-2, 8, 1755, 2153 (1970).

23. A. Miyagi and B. Wunderlich, J. Polymer Sci., Polymer Phys. Ed., 10, 2085 (1972)

24. B. Wunderlich and C. Noll, Bull. Am. Phys. Soc., 15, 354 (1970).

DISCUSSION SESSION

Discussion Leader - J. A. Moore

Lewin: Prof. E. Baer has reported on the bromination of polyethylene single crystals. He has prepared single crystals of polyethylene of varying lamellar thicknesses and he found that bromine atoms could be introduced into the folds of the lamellae. This introduces reactive sites into polyethylene which may be used to introduce grafts as well as to perform a variety of chemical reactions. This may lead to the preparation of entirely new kinds of polymers.

Rempp: I would like to know if, when you recombine your etched lamellae of poly (ethylene terephthalate) you regain the original order and size?

Wunderlich: The crystallinity of the samples increases continuously during the etching to 100% and during the repolymerization it decreases again. The lamellar length found by low angle X-ray scattering increases a little but not very much. At the moment we are trying to perform the re-linking so carefully that we can make the crystal grow larger, as we have done for the polyphosphates. The difficulty is that there is an enormous heat of reaction which must be dissipated. On the other hand, the energy involved in creating or destroying a crystal is only about 1 kcal/repeating unit. If you therefore work too hard at trying to grow your crystal you merely destroy it.

Rempp: Can you really measure melting temperatures accurately using heating rates of 100° C/min?

Wunderlich: Yes. The smaller the sample the faster you can go. The measurements are made with standard commercial equipment on the mg. scale and we can measure with an accuracy of one-half to one degree. For Tg measurments we have pushed the heating rates to 600-700° C/min

Rivin: It is very likely that, because of the strong effects (ion-dipole interactions etc.) of the surfaces with which you are working on chemical reactivity, you make find processes occurring which do not take place in the absence of the surface.

Wunderlich: Yes, that is true.

Goethals: Would the partial hydrolysis of single crystals of polyesters constitute a preparative method for the formation of narrow distribution oligo-esters?

Wunderlich: Yes it is. And the distribution is very narrow,

particularly if you do a careful crystallization in the first place.

Bopp: In reference to the spherulites of selenium, what would you estimate to be the percent concentration of rings versus polymer chains? How do these rings incorporate themselves into the spherulite?

Wunderlich: If you allow enough time for the formation of the spherulites, there are no rings. The detailed mechanism for the incorporation of the rings into the spherulites is not yet known.

Challa: The crystallinity of isoactic polystyrene is rather low and therefore the chain-folds are probably rather long. We treated single crystals of isotactic polystyrene with benzoyl chloride in a Friedel-Crafts acylation. The reaction occurred in the large loops at the surface of the crystals. resulting in the formation of block copolymers. Reaction occurred only in the amorphous folds.

Wunderlich: In principle, it should be possible to accomplish similar results with almost any polymer. If the backbone contains enough CH_2 groups, the polymer will assume the poly ethylene single crystal structutre.

THE CHALLENGE TO POLYMER CHEMISTS IN SOLID PHASE PEPTIDE SYNTHESIS

John Morrow Stewart

University of Colorado School of Medicine
Denver, Colorado 80220

The use of an insoluble polymeric support as a "handle" to facilitate the operations of peptide synthesis (called "solid phase peptide synthesis" by Bruce Merrifield, its inventor) is the most important advance in peptide synthesis since the introduction of the carbobenzoxy group by Bergmann more than thirty years ago. In the decade since the initial announcement of the method by Merrifield (1), its application in several laboratories has led to the synthesis of a wide range of peptides, and even proteins. Several of the more outstanding of these syntheses are listed in Table 1.

Table 1. Progress of Solid Phase Synthesis

Date	Peptide	Amino Acids
1963	A Tetrapeptide	4
1964	Bradykinin	9
1965	Angiotensin II	8
1966	Insulin	21 + 30
1968	*Clostridium* Ferredoxin	55
1969	Ribonuclease A	124
1969	Phe^{59} Cytochrome *c*	104
1970	Pancreatic Trypsin Inhibitor	58
1971	"Human Growth Hormone"	188
1971	Parathormone 1-34	34
1971	Lysozyme	129
1971	Acyl Carrier Protein	74
1971	Salmon Calcitonin	$32\text{-}NH_2$

While some of these peptides duplicate work already done by the classical solution method of peptide synthesis and the chemistry in some of the syntheses is questionable, this new method is clearly a very powerful tool which has greatly increased the scope of peptide synthesis while simultaneously greatly reducing the labor involved in the operations. Although the achievments of the method have been impressive, there are many problems associated with its application, particularly to large peptides and proteins. Many of these problems are directly or indirectly associated with the nature of the polymeric support used during the synthesis. It is clear now that new types of polymeric supports are needed, and it is in the hope of encouraging the development of such supports that I wish to take this opportunity of explaining solid phase synthesis to you and outlining some of its present problems which can most likely best be solved by application of the expertise of polymer chemists.

The basic chemical operation in peptide synthesis is the formation of amide bonds. This very simple operation is complicated by several factors. Outstanding among these are the polyfunctional nature of amino acids, the problems associated with repetitive operations to synthesize macromolecules, the insolubility of the intermediates obtained, and the difficulty of purification of these intermediates.

All amino acids contain at least two functional groups, the amine and carboxyl functions, and for selective peptide synthesis at least one of these groups in each component of a reaction must be blocked by selectively removable groups. In addition, many amino acids contain reactive side chain functional groups which must also be blocked during the synthesis. The proper design of a system of blocking groups is one of the major considerations in peptide synthesis, since several categories of groups must be available with different orders of stability. This is true both of classical and solid phase methods of synthesis. Closely associated with this problem is that of strategy of synthesis. There are several ways in which a large peptide chain may be assembled. For example, amino acids may be added one at a time, from one end or the other of the chain. This is the stepwise approach. Alternatively small peptides may be synthesized and then these small peptide fragments may be combined into a large final product; this is the fragment condensation approach. Each of these methods has its advantages and disadvantages, which will not be discussed here, but have been adequately treated in reviews. Most large peptides synthesized by classical solution methods have been synthesized by the fragment condensation approach, while solid phase synthesis is a special case of the stepwise method, in which the polymeric resin support serves as blocking group for the carboxyl of the C-terminal acid, and amino acids are added one at a time to this residue bound to the polymer.

As amino acids with blocked side chain functional groups are assembled into a peptide, the molecular weight increases rapidly, and the solubility usually shows a concomitant decrease. Once the peptide has reached any significant size, only powerful solvents such as dimethylformamide can be used as reaction media. In many cases even this is not adequate, and insolubility of intermediate products frequently causes very serious problems.

Purification of intermediates in the synthesis is often very troublesome in classical solution methods. Coupling reactions frequently do not go to completion, the products (blocked peptides) usually do not crystallize, and the purification of these intermediates and separation from side products and unreacted starting materials can be very laborious, usually involving countercurrent distribution or chromatography.

SOLID PHASE PEPTIDE SYNTHESIS

It was in the hope of overcoming some of these serious problems of solution peptide synthesis that Bruce Merrifield developed the solid phase method. He reasoned that if the desired product of every reaction could be made insoluble (by supporting the product on an insoluble polymer), while all byproducts and unreacted started materials remain soluble, then washing would serve as the only and sufficient purification step. This approach is outlined in Figure 1.

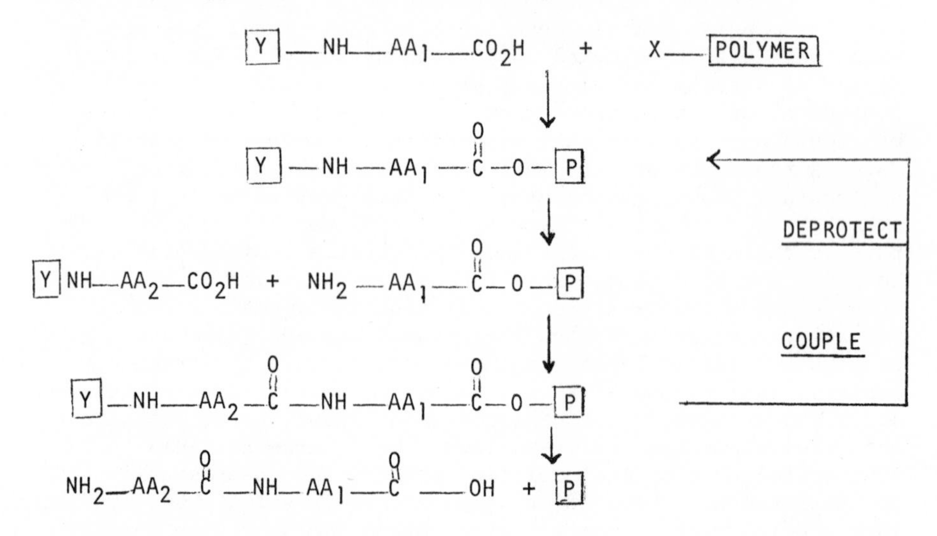

Figure 1. The Principle of Solid Phase Peptide Synthesis.

The amino acid which will form the carboxyl-terminal residue of the peptide is attached to a suitably functionalized insoluble polymer through its carboxyl group. During this process the amino group of the amino acid is blocked with the easily removable group Y. After the blocking group Y is removed, a new amino acid, with its amino group also blocked by a readily removable group, may be coupled to the residue on the polymer by any suitable coupling reaction. Repetition of this process of deprotection and coupling, using in each coupling reaction the desired amino acid, enables the chemist to assemble on the polymer support the desired sequence of amino acids. Finally, another reagent which will remove the peptide from the polymeric support is applied, and the peptide is dissolved and washed away from the polymer. The support most widely used in solid phase peptide synthesis so far has been a copolymer of styrene and divinylbenzene. Early work was done with a 2% DVB polymer, while recently a 1% crosslinked resin has been found to be more satisfactory. These resins swell very extensively in the solvents used for the synthesis. This is a crucially important point, since the synthesis takes place throughout the resin beads. There is hope of obtaining a homogenous product from a synthesis only if all the reactions go fully to completion. In order to achieve this goal reagents must be able to penetrate freely into the swollen resin matrix. This point will be discussed in greater detail later.

The chemistry used in the most common system of solid phase peptide synthesis (2,3)is shown in Figure 2.The polystyrene is functionalized by chloromethylation, and the chloromethyl polymer is caused to react with a salt of the N-protected amino acid. The most common protecting group for the amino function is the tertiarybutyloxycarbonyl (BOC) group. Reaction of the salt (usually with triethyl amine) of the BOC amino acid with the chloromethyl polystyrene attaches the amino acid to the resin as an ester. This resin ester, in effect a substituted benzyl ester, is stable to all of the reactions used in the synthesis. The BOC protecting group can be completely removed in most cases without loss of a significant amount of the first amino acid (and resulting peptide) from the resin. However, in the synthesis of large peptides or protines, loss of the peptide from the resin by acidolysis can be a significant problem. Anhydrous strong acid in a suitable medium (such as HCl in dioxane or trifluoroacetic acid in chloroform or dichloromethane) will remove the BOC group. The protonated form of the amino group on the aminoacyl polymer is converted to the free base by treatment with a tertiary amine, usually triethylamine, and the next BOC amino acid is coupled to the amino group. The most commonly used coupling reagent is dicyclohexylcarbodiimide. This is a satisfactory procedure for coupling all amino acids except glutamine and asparagine. In these cases active esters, such as p -nitrophenyl esters, are usually used to avoid

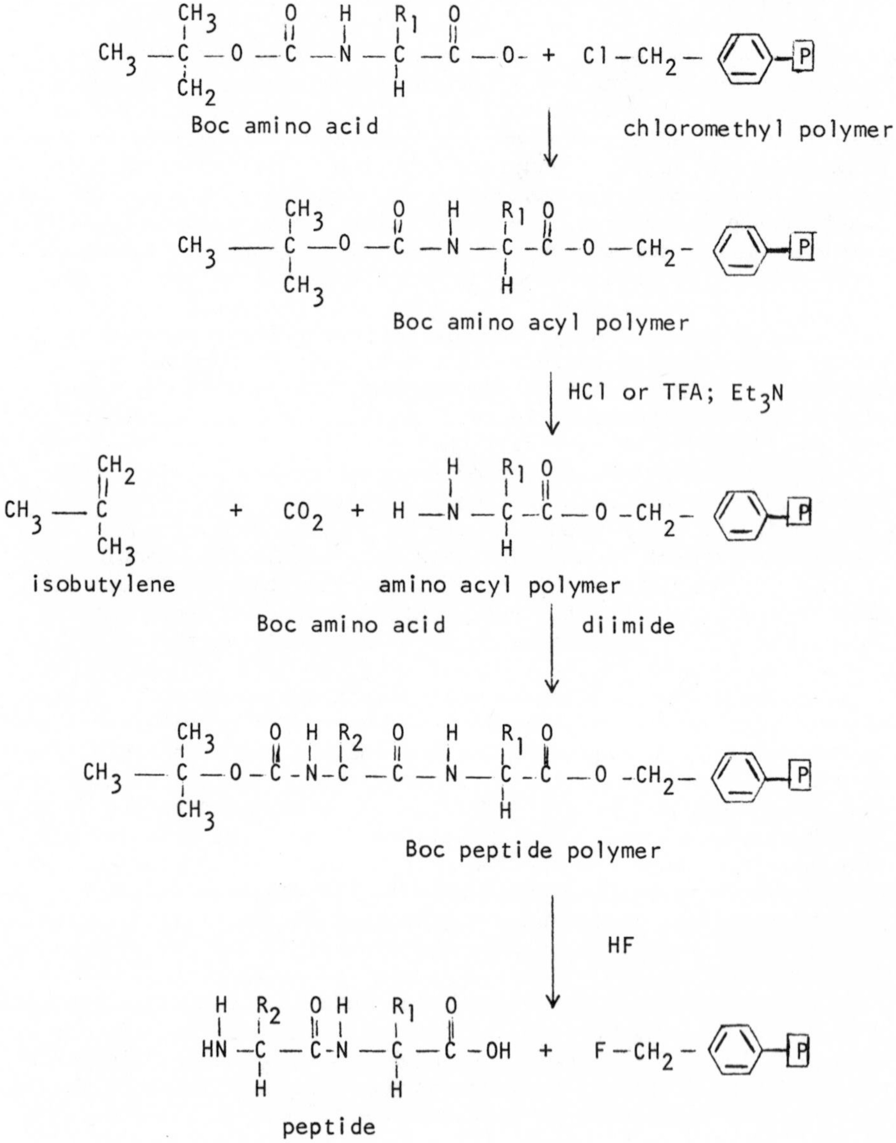

$$CH_3-\underset{CH_2}{\overset{CH_3}{C}}-O-\overset{O}{\overset{\|}{C}}-\overset{H}{N}-\underset{H}{\overset{R_1}{C}}-\overset{O}{\overset{\|}{C}}-O^- \;+\; Cl-CH_2-C_6H_4-\text{(P)}$$

Boc amino acid chloromethyl polymer

$$\downarrow$$

$$CH_3-\underset{CH_3}{\overset{CH_3}{C}}-O-\overset{O}{\overset{\|}{C}}-\overset{H}{N}-\underset{H}{\overset{R_1}{C}}-\overset{O}{\overset{\|}{C}}-O-CH_2-C_6H_4-\text{(P)}$$

Boc amino acyl polymer

$$\downarrow \text{HCl or TFA; } Et_3N$$

$$CH_3-\underset{CH_3}{\overset{CH_2}{\overset{\|}{C}}} \;+\; CO_2 \;+\; H-\overset{H}{N}-\underset{H}{\overset{R_1}{C}}-\overset{O}{\overset{\|}{C}}-O-CH_2-C_6H_4-\text{(P)}$$

isobutylene amino acyl polymer

Boc amino acid | diimide

$$\downarrow$$

$$CH_3-\underset{CH_3}{\overset{CH_3}{C}}-O-\overset{O}{\overset{\|}{C}}-\overset{H}{N}-\overset{R_2}{C}-\overset{O}{\overset{\|}{C}}-\overset{H}{N}-\underset{H}{\overset{R_1}{C}}-\overset{O}{\overset{\|}{C}}-O-CH_2-C_6H_4-\text{(P)}$$

Boc peptide polymer

$$\downarrow \text{HF}$$

$$H\overset{H}{N}-\underset{H}{\overset{R_2}{C}}-\overset{O}{\overset{\|}{C}}-\overset{H}{N}-\underset{H}{\overset{R_1}{C}}-\overset{O}{\overset{\|}{C}}-OH \;+\; F-CH_2-C_6H_4-\text{(P)}$$

peptide

Figure 2. The Chemistry of Solid Phase Peptide Synthesis.

side reactions. Following the coupling reaction excess BOC amino acid and byproducts can be washed out of the polymer suspension leaving the pure peptide attached to the polymer. An excess (usually 2.5 equivalents) of each new BOC amino acid and coupling agent must be used to help assure complete coupling. The cycle of deprotection, neutralization, and coupling of the new amino acid is repeated as necessary to assemble on the polymer the desire peptide chain. Finally, the peptide is removed from the polymer by treatment with a stronger acid,usually anhydrous HF, or HBr, in trifluoroacetic acid. The advantage of HF is that it also simultaneously removes all of the commonly used blocking groups from side chain functions of the peptide, yielding the free peptide in one step. When HBr is used for cleavage of the peptide from the resin, some further step must often be used to remove some blocking groups. It is noteworthy that all of the reactions of the synthesis and cleavage take place under anhydrous conditions, thus assuring that no hydrolysis of the peptide chain can occur. Further details of the chemistry of solid phase synthesis can be obtained from any of the reviews on the subject (3,4,5).

Several different kinds of peptide derivatives can be obtained from the synthetic peptide-resin. The different kinds of cleavage commonly used and the derivatives formed are shown in Figure 3. Other types of linkage of peptide to the resin support have been developed and used for certain special purposes. One area that has received special emphasis is the synthesis of peptide amides. This is important since so many peptide hormones are amides. This fact, coupled with the problems which arise in the synthesis of peptide amides containing glutamic and aspartic acids by ammonolysis of the peptides from the resin, has led to development of resins which yield peptide amides directly by treatment with HF. The most widely used of these is the benzhydrylamine resin (6), in which the additional activation of the added phenyl group on the carbon bearing the amine function is adequate to cause cleavage of the C-N bond in (Fig. 4A). Although this resin has been successfully used for the synthesis of a number of peptide amides, certain problems which have arisen make it evident that further work is needed in this direction. Other investigators have worked on the synthesis of small peptides by the solid phase method which would subsequently be coupled together in solution to synthesize larger peptides. This was the goal of the hydrazine cleavage of the peptide-resin indicated in Fig. 3. A better approach to this problem is probably that of Wang and Merrifield (7),who synthesized a resin carrying the equivalent of BOC hydrazide groups on the polymer (Fig. 4B) which after peptide synthesis would yield protected hydrazides directly by acidolysis. Large peptides synthesized in this way should be easier to purify than those synthesized by the standard solid phase method, since the differences between possible components

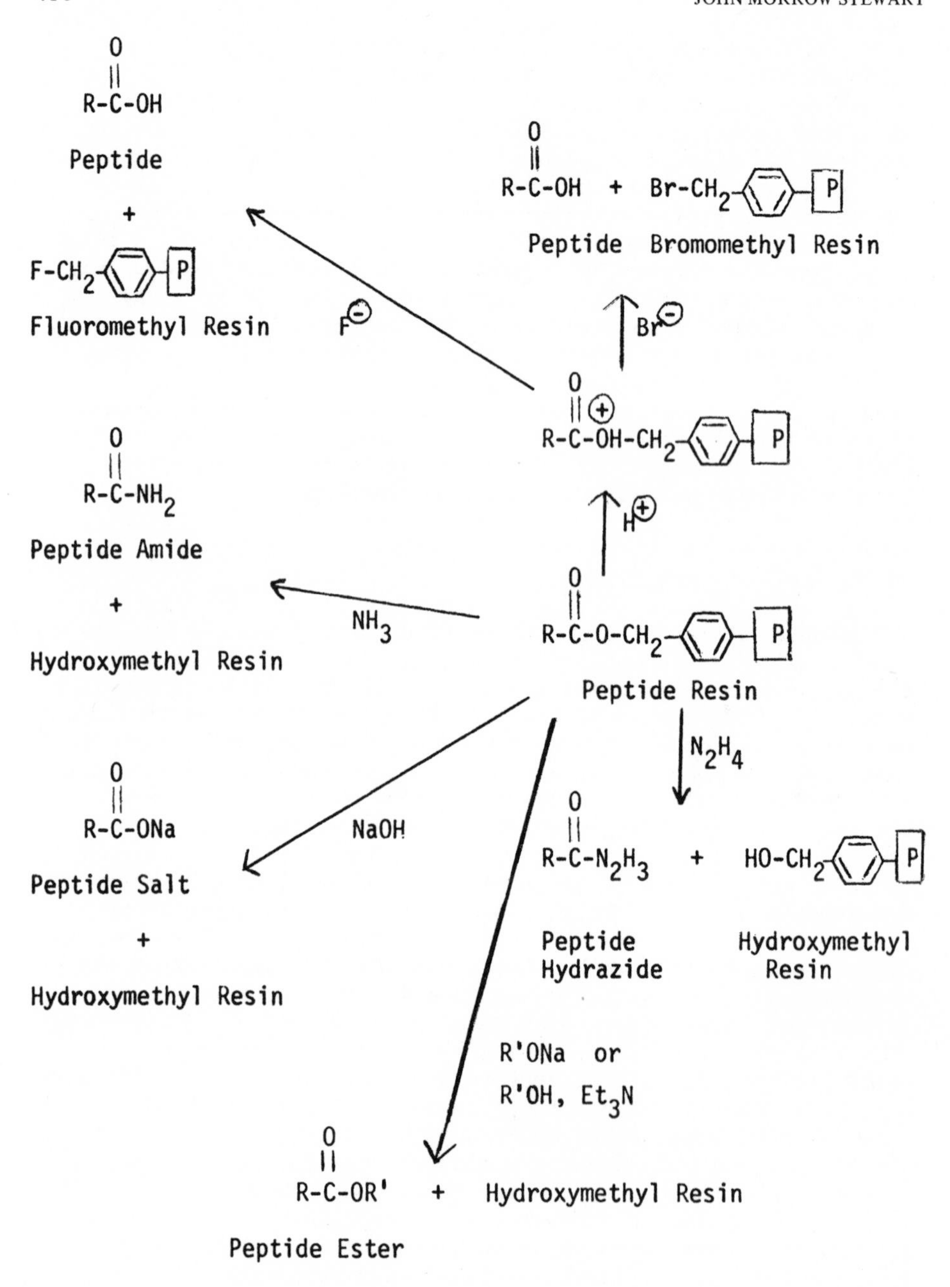

Figure 3. Cleavage of Peptides from the Resin.

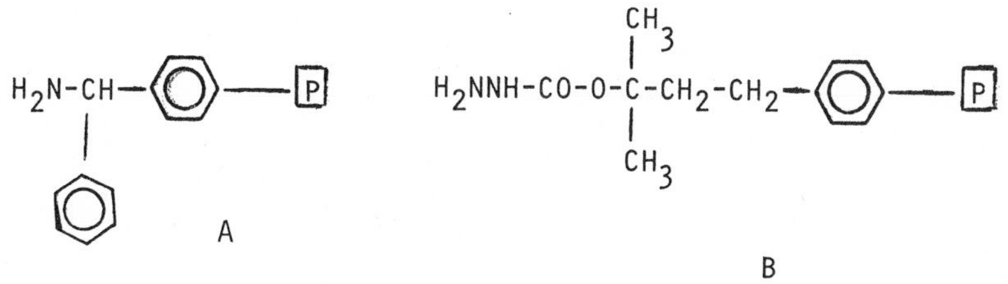

Figure 4. Special Purpose Resins. A, The benzyhydrylamine resin for synthesis of peptide amines. B, The resin for synthesis of protected peptide hydrazides.

of the crude product should be greater.

Perhaps the most outstanding achievement of the solid phase method was the synthesis by Gutte and Merrifield (8) of a protein (124 amino acid residues) with ribonuclease activity (Table 1). During the synthesis of this protein it became apparent that losses of the peptide from the resin at each step were significant, so that at the end of the synthesis only 16% of the original amino acid substituted on the resin still remained. Evidently one needs for this kind of synthesis a greater difference in the stability of the α-blocking group and the link attaching the peptide to the resin. A solution to this problem might be found either in the use of more labile α-blocking groups,or in the use a more stable bond linking the peptide to the resin. The latter approach is not practical since the currently used HF treatment is apparantly too harsh for some peptides, and will probably be undesirable for most proteins. Several α-blocking groups are available which are more labile than the BOC group (9), but there are also other problems associated with their use. Further research in this area is under way at the present time in several laboratories (10,11). Another approach to this problem is the use of what Sheppard has called the "Safety Catch" principle (12), which uses a modifiable bond linking the peptide to the polymer. In this approach, the first amino acid would be attached to the resin by a bond which is quite stable. At the end of the synthesis, the bond would be modified selectively by a chemical reaction to another form which could be very easily cleaved. One approach to such a method was described by Marshall and Liener (13), and Flanigan and Marshall (14), who used a thioether bond in the peptide-resin link. After the synthesis the thioether was oxidized to a sulfone which activated the ester bond linking the peptide to the support and made it quite labile. This example suffers from the need to use an oxidative reagent, and is not generally applicable, but it illustrates the principle. Other approaches have been described, but none appears to be widely applicable. Perhaps a successful application of this principle will be developed in the future.

PROBLEMS ASSOCIATED WITH THE RESIN

The desired goal of quantitative reaction at every step has been difficult to achieve reliably in solid phase peptide synthesis. Many small peptides have been synthesized very easily. The desired peptide is usually the major component by far of the crude reaction product, and the purification of such peptides has usually not been too difficult. On the other hand, the synthesis of large peptides and proteins is a very different matter. In work done so far, the products have been very heterogeneous, and isolation of the desired material has been difficult or impossible. The achievement of even 99.5% reaction at every step is wholly inadequate for the synthesis of large peptides. These failures to achieve complete reaction at each step appear to have been largely due to sequence-dependent steric interactions between the peptide chain and the polymer matrix. An early example of this type of problem was encountered by Merrifield in the synthesis of angiotensinyl bradykinin (15). At that time both angiotensin and bradykinin had been satisfactorily synthesized on 2% crosslinked resins using HCl in glacial acetic acid as the reagent for the deprotection. However, when the synthesis of the longer chain in which the sequences of angiotensin and bradykinin were added together was attempted, a difficulty was encountered during the second half of the synthesis. This difficulty was failure to remove the BOC group completely. The problem was overcome by making two changes in the synthesis. First, the crosslinking of the resin was reduced to 1%, giving a product which swells much more extensively in dichloromethane and which apparently allows better penetration of reagents into the resin, the reagent for deprotection was changed from HCl-HOAc to HCl in dioxane. Whereas the previous reagent does not swell the resin extensively, the dioxane reagent swells the resin to the maximum attainable with any solvent. Four years later, a claim was made in the literature (16) that a group of peptides could not be synthesized by solid phase because of failure to deprotect. It was later shown (17) that this difficulty could be overcome when reagents were used which swelled the resin maximally and when the resin crosslinkage was reduced to 1%. Very recently a 0.5% crosslinked resin has been described by Birr (18), who has said that it improves the synthesis of certain peptides. It is not yet possible to give a complete evaluation of this resin. This new resin swells twice as much as the 1% crosslinked resin in dichloromethane, but is very gelatinous and may require special equipment for handling.

The logical ultimate development in this direction would be the use of a non-crosslinked resin. However, linear polystyrene is soluble in the solvents used for synthesis, and serious problems arise in handling such material. Some work has been done in this area (19), the most recent being a proposal for use of polyvinyl alcohol as the polymeric suppot (20), with ultrafiltration as the means for separating the peptide-polymer from low molecular weight materials. The use of such soluble polymeric carriers introduces several new problems, and greatly complicates the mechanical operations of synthesis. These materials have not as yet received any extensive application.

It has been suggested by Sheppard (21) that a major difficulty in solid phase synthesis may be the different nature of the polymeric support and the peptide chain which has grown on that support. For successful synthesis presumably both the polymeric carrier and the peptide chain should be fully extended and solvated (Fig. 5A). The polystyrene matrix usually used is nonpolar, and swells in nonpolar solvents. On the contrary, the peptide chain is polar and should be expanded by polar solvents which may collapse the polystyrene matrix (Fig. 5B). The nonpolar solvents which expand the matrix may collapse the peptide chain (Fig.5C), thus causing many of the serious problems, apparently due to steric interaction, which have been observed. It has been observed that the nature of the resin beads changes greatly during a synthesis. After attachment of several amino acids the resin no longer swells as extensively in nonpolar solvents such as dichloromethane and chloroform. It has been found that with certain peptide resins coupling reactions which could not be forced to completion in dichloromethane as solvent did go to completion when dimethylformadide was added to the reaction mixture(22).

In order to evaluate difficulties such as those described, one must have methods for accurate monitoring of the deprotection and coupling reaction. At the present time, no fully satisfactory solution to these problems is available. Monitoring of deprotection is a very difficult problem, as the last percent of undeprotected amino group must be determined in the presence of a large amount of free amino groups. Neither method based on estimation of amount of blocking group removed or amount of free amino group produced on the resin appears to have the requisite sensitivity. The use of radioactive BOC groups has been suggested, but this appears to be a very expensive and impractical approach. Better success has been obtained in monitoring of coupling reactions, for in this case one goes to an end point of zero and small residual amounts of uncoupled amino groups are theoretically easier to determine. The widely used ninhydrin method (23) lacks sensitivity, the naphthaldehyde method (24) is slow, and the very sensitive fluorescamine method (25) does not detect proline amino groups and appears to suffer from spurious fluorescence. Improvements are needed in both these types of monitoring.

Since many of the difficulties encountered in solid phase

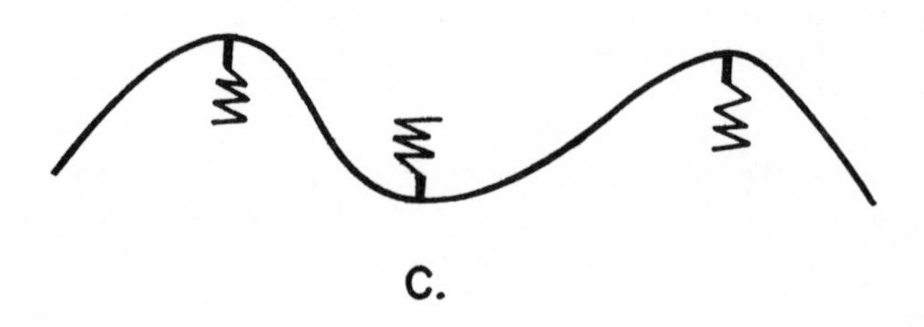

Fig. 5. Peptide-Resin Interactions in Solid Phase Peptide Synthesis. A, The ideal case, with both polymer matrix and peptide chain extended. B, In polar solvents, with the hydrophilic peptide chain extended but the matrix collapsed. C, In non-polar solvents , with the matrix expanded but the peptide collapsed.

synthesis appear to be related to penetration of reagents into the polymer matrix, several investigators have tried supports which are nonporous, and in which the reactive sites are limited to the surface. A pellicular layer of polystyrene coated on glass beads has been suggested (26), but appears to lack the requisite mechanical stability, and additionally did not appear to offer significant chemical advantages. Glass beads have also served as carriers for peptide synthesis in systems involving covalent linkage of the functional group to the silicon of the glass particles (27). At the present time there has been little evaluation of such systems. Another system which has received some attention consists of "linear" polystyrene grafted onto Kel-F particles (28). This type of carrier has been used for several syntheses but does not appear to offer the needed improvements in the system. Other investigators have reported on work with macroporous, macroreticular and "popcorn" types of polymers (5). Again, significant improvements did not appear to have been achieved. All of these surface-limited supports suffer from the disadvantage of having very low capacity.

SOME SUGGESTIONS FOR NEW APPROACHES

An ideal insoluble support for solid phase peptide synthesis would have several characteristics.

1. The resin should have good mechanical properties. It should be possible to conduct manipulations such as agitation, filtration, etc.,without mechanical degradation of the polymer or excessive filtration times.

2. The polymer should be compatible with a good chemical system of peptide synthesis. That is, a good selection of appropriate linkage of the peptide to the resin and compatability with suitable blocking groups for both the amine function as well as side chain functions is necessary. For example, a glass support is not compatible with the use of HF.

3. The polymer should not cause problems due to steric interactions with the peptide chain. This probably means that a very low degree of crosslinkage, or no crosslinkage at all, will be necessary.

4. A high level of substitution should be possible. This is desirable so that huge amounts of polymer need not be handled in order to synthesize a reasonable amount of peptide. This is difficult to attain in systems having the reaction restricted to the surface of the material, and division of the support into particles small enough to have a very large surface will probably cause serious problems in mechanical

handling.

5. The matrix and the peptide chain should probably be solvated to a similar degree by solvents suitable for reaction media. This requirement may mean that polystyrene-based resins are not ultimately desirable, and that new types of polymer need to be examined. A suitable reaction medium should probably be a aprotic, and should be helix-breaking for protected peptides.

It is difficult to visualize any conventional polymer which would meet all of these criteria, and new polymeric supports will probably need to be designed and synthesized. One novel suggestion was that solid phase synthesis might be carried out on continuous belts, films or fibers of support material, using automatic processing equipment. Perhaps a more realistic approach would to synthesize a new type of carrier such as that depicted schematically in Figure 6. This type of material, called a "Hippie" resin by Rudinger,would have a dense, impermeable core, with long hairs extending from it. These hairs should have reactive sites at suitable intervals for attachment of the first amino acid of the peptide chain. In this way the peptide chains would be able to grow freely outward from the center without encountering serious steric hinderence from a polymer matrix. The "whiskers" extending from the core should probably be relatively hydrophilic in nature, so that they would be solvated by the same types of solvents which would solvate the peptide chain. This might be achieved in a vinyl-type linear polymer by incorporating acrylamide groups,or it might be necessary to go to a polyamide type of chain. The reactive groups for attachment of the first amino acid should probably be incorporated into these whiskers during polymerization, rather than introduced later, as is the case with the present chloromethyl resin. One serious problem of the present system is that chloromethlation introduces additional crosslinking into the matrix and decreases the porosity of the polymer.

Perhaps this description of our problem and these suggestions may serve to inspire polymer chemists to design carrier materials specifically for peptide synthesis. I shall be happy to discuss these problems at greater length with any interested polymer chemist, and shall also be pleased to test any new types of carriers for their ability to function as support materials in solid phase peptides synthesis.

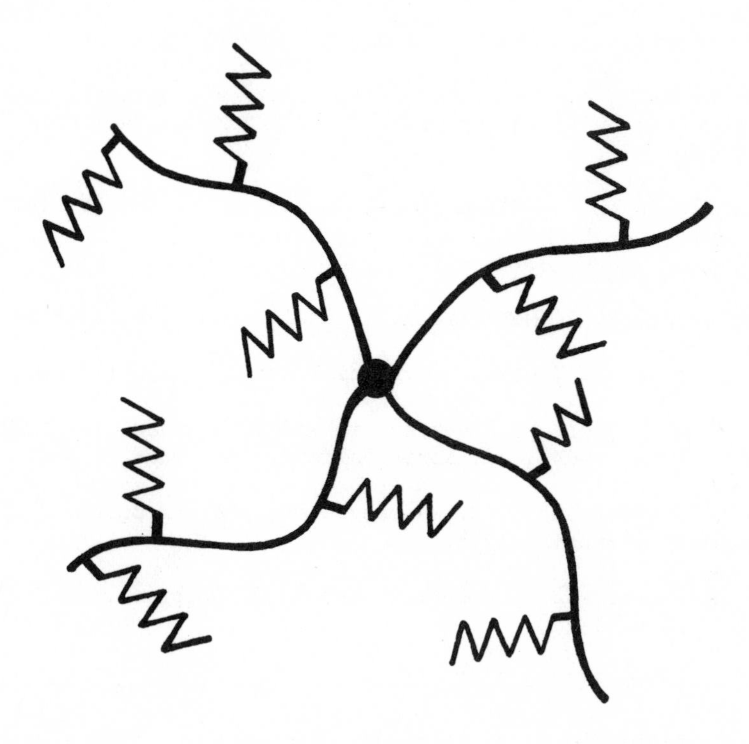

Fig. 6. A Proposed Ideal Support for Solid Phase Peptide Synthesis.

REFERENCES

1. R. B. Merrifield, J. Am. Chem. Soc. 85, 2149 (1963).

2. R. B. Merrifield, J. Am. Chem. Soc. 86, 304 (1964).

3. J. M. Stewart and J.D. Young, "Solid Phase Peptide Synthesis". W. H. Freeman, San Francisco, 1969.

4. R. B. Merrifield, Advan. Enzymol. 32, 221 (1969).

5. G. R. Marshall and R. B. Merrifield, in " Biochemical Aspects of Reactions on Solid Supports," G.R. Stark, ed, Academic Press, New York, 1971.

6. P. G. Pietta and G. R. Marshall, J. Chem. Soc. D, 650 (1970).

7. S. S. Wang and R. B. Merrifield, J. Am. Chem. Soc. 91, 6488(1969).

8. B. Gutte and R. B. Merrifield, J. Biol. Chem. 246, 1922 (1971).

9. P. Sieber and B. Iselin, Helv. Chim. Acta 51, 622 (1968).

10. C. Birr, F. Flor, P. Fleckenstein and T. Wieland, in "Peptides 1971," H. Nesvadba, ed. North Holland, Amsterdam, 1973, p.175.

11. J. M. Stewart and G. R. Matsueda, Madison Symposium on Peptide Synthesis, 1973.

12. G. W. Kenner, J. R. McDermott and R. C. Sheppard, J. Chem. Soc D, 636 (1971).

13. D. L. Marshall and I. E. Liener, J. Org. Chem. 35, 867 (1970).

14. E. Flanigan and G. R. Marshall, Tetrahedron Lett. 27, 2403 (1970).

15. R. B Merrifield, Recent Prog. Hormone Res. 23, 451(1967).

16. F. C. Chow, R. K. Chawla, R. F. Kibler and R. Shapira, J. Am. Chem. Soc. 93, 267 (1971).

17. J. M. Stewart and G. R. Matsueda, in"Chemistry and Biology of Peptides," J. Meienhofer, ed. Ann Arbor Science Publishers, Ann Arbor, Mich. 1972, p. 221.

18. C. Birr, Annalen, 1973 (in press).

19. Y. A. Ovchinnikov, A. A. Kiryushkin and I. V. Kozhevnikova, J. Gen. Chem. USSR 38, 2546(1969).

20. E. Bayer and M. Mutter; Nature 237, 512 (1972).

21. R. C. Sheppard, in Ref. 10 p. 111.

22. F. C. Westall and A. B. Robinson, J. Org. Chem. 35, 2842 (1970).

23. E. Kaiser, R. L. Colescott, C. D. Bossinger and P. I. Cook, Anal. Biochem. 34,595 (1970).

24. K. Esko, S. Karlsson and J. Porath, Acta Chem. Scand. 22, 3342 (1968).

25. S. Udenfriend, S. Stein, P. Bohlen, W. Dairman, W. Leimgruber and M. Weigle, Science 178, 871 (1972).

26. E. Bayer, H. Eckstein, K. Hagle, W. A. Konig, W. Bruning, H. Hagenmaier and W. Parr, J. Am. Chem. Soc. 92, 1735 (1970).

27. W. Parr and K. Grohmann, Angew. Chem. Internat. Ed. 11, 314 (1972).

28. G. W. Tregear, in Ref. 17, p. 175.

DISCUSSION SESSION

Discussion Leader- R. Williams

Daly: Are there any commercial applications in which solid phase peptide synthesis is used?

Stewart: There are several companies which are using the method to synthesize hormones and have been supplying researchers with these materials.

Ferruti: You mentioned a polyacrylamide matrix as a possible choice for peptide synthesis . Perhaps poly(acryloyl morpholine) would be a good choice. The monomer can be easily prepared from acryloyl chloride and morpholine and readily copolymerized with vinyl monomers. The Polymer dissolves in practically every solvent except aliphatic hydrocarbons. It may swell very well in all other solvents.

St.; What about halogenated solvents?

F.; It swells in these too.

Rempp I wonder if some of the problems connected with solid phase synthesis are due to something which might be called "heterocontact repulsions". When you start a synthesis the gel may swell easily. As the peptide chain grows the gel may not swell as before because of heterocontact interactions. Conformational changes in both elements of the network may serve to alter the quantitative nature of the synthetic reactions.

St.; It is clear that the system is changing a great deal as the peptide chains grow. The problems are obviously also sequence dependent since some very large peptides can be synthesized very easily whereas some small ones cause great problems, presumably because of steric hindrance. For example, a demonstration of this might be found in our "impossible peptide". This has a large number of hindered amino acids in it - valines,isoleucines, threonines, etc. Differing chain solubilities or aggregations play an important part during synthesis, as well.

Loucheux: If crosslinked copolypeptides were used as supports this previously-mentioned solvent -chain incompatibility would disappear.

St.; It would be very interesting to try. My personal preference would be to use a dense core and have supported polypeptide chains on it. All the normally used reactions would then have to be compatible with the system too.

R.; I wonder, in the case of successful synthesis of longer chains, if only the outer surface has played a role?

St.; The main reason for using gels, at least historically, has been to increase capacity. In the case of the RNAase synthesis, because the BOC deprotecting step was too vigorous, quite a bit of the peptide was lost. At the end of the synthesis only about 16% of the original amino acid still remained in the polymer. It is true that the final degree of substitution was compatible with the degree of substitution which one would find if the chains were located only on the surface. In this regard I think that it would be a desirable goal to make a very dense,impermeable core with freely accessible long chains.

Wunderlich: It has bothered me for many years that the term"solid-phase synthesis" is used. There is nothing solid about the process. It is done in a swollen gel and a gel is a liquid according to the mobility of the molecules and micro- Brownian motion. You could call it "fixed molecule" synthesis or better still "Merrifield synthesis". If you really want to use the term why not do the synthesis that way? Why not use the solid phase? You need a thin layer to provide accessiblity and to provide the necesarry reaction sites. This might be done with as ABA block copolymer where the center block crystallizes. The amorphous end units would then protrude from the surface and provide little "hairs" as carriers.

St.; These would be very interesting systems.

Wun.;These are available and some examples have been synthesized in Professor Remp's laboratory. They are ethylene oxide polymers.

Williams: Are these crystalline systems soluble or insoluble in common solvents? Merrifield synthesis would require them to be insoluble.

Wun,;You can make them either soluble or insoluble. DMF, for example, does not dissolve polyethylene. Polyethylene could be etched and then with the generated carboxyl groups pendent chains could be attached, maybe, through acid chloride groups. We have attached amines to such systems and found greater than 90% substitution.

Rivin: Would it be possible by increasing the surface area to achieve increased capacity? 0.3-0.4mM/gm would seem to be no problem on a 100-200 Å sphere with a surface area of about 100 m^2. You could consider an inorganic substrate with surface reactive groups. After attachment the remaining surface could be deactivated to prevent polypeptide absorption.

St.; Can these particles be filtered?

Rivin; Very easily.They tend to agglomerate but can be separated to expose all sites by a very small amount of shear.

Rempp: I would like to point out the work done by Hamann, Funke and Reichert in which the SiO group in silica were replaced by SiCl groups. They can then react phenyl lithium with them and afterwards nitrate them. They are using them for both radical and ionic polymerizations.

St.; Do you think silica gel offers any mechanical advantage over glass? Are the particles stable?

Rempp: I am not sure.The greatest advantage comes from the large surface area.

Carraher: Could you describe how the silyl ethers are made?

St.; The publications indicate that an alkoxyltrichlorosilane is used. Some of the problems that seemed to occur may be the result of localized build up and the spacer chains might need to be extended. The people at Corning are interested in functionalized glass and other inorganic materials of many kinds and have already marketed a functionalized amine-containing glass which they say can be used for immuno assays.

C.; Do they etch the glass?

St.; No etching was mentioned.

Fox: Silylation of glass is sometimes as adsorption phenomenon. This may be only a wetting phenomenon and not a functionaliation.

St.; I think in this case it was indeed a functionalizing process.

Parikh: That's true.

St.; At present, the most useful reagent for cleaving the peptide from the support is hydrogen fluoride. The common useage of glass would necessitate the use of an intermediate step such as HBr treatment to remove the peptide from the glass and then it could be treated with anhydrous HF to remove the other blocking groups. This would not be desirable.

Lewin: I was wondering if fibers had been used for these purposes? Actually you can make a fibrillation of the fibers. You are not limited in the chemical nature of the fibers. The surface of a fiber could be activated. Would polyester be suitable for this job?

St.; I know of no reason why it should not work. Merrifield first tried fibrous materials, cellulose and cellulose derivatives. They didn't seem to work very well. At a European Peptide Symposium, I believe in 1969, there was a presentation which described a hypothetical system for solid phase synthesis using a film and a routine was proposed in which the film would be processed more like a motion picture developing machine, dipping from one bath to the next.

L.; With a fiber, you could do it in a textile mill.

St.; I have also thought that some very fine beads of cross-linked polystyrene or divinylbenzene would serve as ideal nuclei on which to grow chains. If some of you have such material or know where I might get some I would be happy to know about it.

Ferr: I wonder if you have considered functionalizing carbon black particles. They can be nitrated and so on. I do not think you have any objection to the fact that they are black.   f

St.; I have not had any discussion along these lines.

Moore: Sakakibara has synthesized oligoprolines via a fragment coupling method. Do you think that this is a generally useful strategic manoeuvery?

St.; This fragment coupling type of system has been investigated to help separate " errors" at the end of a synthesis by maximizing polypeptide chain length differences. It would be more difficult to couple a peptide to the resins because of the problems associated with diffusion and steric hindrance of the peptide to the reactive sites. The fragments would have to also terminate in proline or glycine to avoid racemization.

Wil.; The cost of making the larger fragments and the necessity of using large amounts during coupling is also as important factor.

St.; Where you have repeating units you need to synthesize it only once but with the synthesis of a multiamino acid macromolecule this becomes more serious.

M.; Would you comment on current "error capping" or blocking techniques used in solid phase peptide synthesis to eradicate errors?

St.; Most serious solid phase synthesizers do this in syntheses routinely. We monitor at every coupling step and if the coupling does not go to completion as indicated by the ninhydrin reaction or by the newer Fluorescamine, (Fluram-TM) which seems to be more sensitive, one then recouples. If that is unsuccesful then one blocks the amines in some manner. There are many ways to do this. We generally find that acetylimidazole is best although there are some peptides that cannot be reacted completely with that reagent or any other acetylating reagent,just as they cannot be coupled completely with any other activated amino acid. Others have their favorite reagents.

M.; The Merck group also synthesized RNAase using NCA's (N-carboxyanhydrides) in solution. Would you care to compare the solid-phase and NCA synthesis of this enzyme?

St.; The NCA synthesis did not seem to be a satisfactory way. If you look at their syntheses they used it very, very little. They made a large point of using the synthesis and introduced it with a lot of advertising and fanfare but found that once you got past a few residues it just doesn't go well. People who have tried to reproduce it have found some very serious problems because it is a question of having your peptide in solution at high pH. If steric problems arise you've had it.

Wil.; I have a comment to make about the NCA syntheses. It is very difficult to get them to go to 100%. The best pH of coupling for each amino acid varies. For each particular system conditions have to be worked out to ensure maximal yields so that the synthesis is not as easy or universal as it might look. Some dipeptide syntheses we have done, using a modification of this method, indicated that amino acid $p^k{}_a$'s, α-amino group accessibility and carbamate stability play an important part in determining the yields (Can. J. Chem. 51, 1284 (1973).

St.; Several of the amino acid NCA's were not suitable for synthesis. They developed the thio derivatives for those. Since they did not use it extensively in the RNA'ase work I think that's the best evidence that it is not as useful as they would have us believe.

Wil.; Proline is one of the more difficult NCA's to make. Several others do require side chain blocking and are not readily synthesized.

St.; If the synthesis does not go the first time you've had it.

Wil.; Yes, once the unstable carbamate breaks down and carbon dioxide starts to come off the reaction must be terminated.

St.; You then cannot recouple as in solid phase syntheses.

Loucheux: What are the advantages of incorporating reagents into a solid, insoluble form?

Williams: The incorporation of the reagent into a insoluble form offers two advantages. One, large-scale industrial syntheses would be faciliated because very reactive or corrosive reagents are more easily handled in this form. Two, the isolation of products is facilitated because the by-products from the reactions, for example, with the reagent EEDQ, the quinoline can be difficult to remove using solvent extraction or simple evaporation.

Loucheux: It seems that recemization during the use of some of these reagents should be very high.

Williams: If urethane blocking groups are used when single amino acids are being coupled, the amounts of recemization are very much lower than we observed using the Izumiya procedure. Fragments could be coupled without carboxy terminal racemization if the C-terminal amino acids were proline or glycine.

Stewart: Another problem in solution peptide synthesis using DCC comes from the insolubility of the dicyclohexylurea. Longer peptide chains, which tend to be insoluble, are difficult to separate from the urea formed. In the synthesis of shorter chains the similarity in solubility between the urea and the small peptide causes problems of purification,as well.

Ferruti: You mentioned several polymers with pendent carbodi-imide groups. Could you tell me how they were obtained?

Williams: They are obtained through the urea by dehydration or by dehydrothiolation, that is removal of sulfur from the thiourea.

Moore: The in-chain cabodiimide most probably was obtained by polymerization of the diisocyanate with a phosphine oxide.

SUPPORT-AIDED PEPTIDE SYNTHESIS - REGENERABLE POLYMERIC REAGENTS, RECENT DEVELOPMENTS

R.E. Williams

Division of Biological Sciences
National Research Council of Canada
Ottawa, Canada K1A OR6

The support-aided synthesis of peptides can be classified under one of the three categories illustrated below:

TYPE A $\quad P + AA_1 \rightarrow P\text{-}AA_1 + AA_2 \rightarrow P\text{-}AA_1 - AA_2$

TYPE B $\quad P + AA_1 \rightarrow P\ AA_1 + AA_2 \quad P + AA_1 - AA_2$

TYPE C $\quad P \rightarrow P^* + AA_1 + AA_2 \rightarrow P + AA_1 - AA_2$

With type A resins product formation is carried out by linking successive units to the support P and when all units are linked in the correct sequence the product is removed from the support. With type B resins the product is formed by the nucleophilic attack of one of the amino acid or peptide units, AA_2, on the preformed polymer supported intermediate $P\text{-}AA_1$. The coupled product is liberated into solution and can be separated from the insoluble polymer. With type C resins the product is formed by utilization of an insoluble polymeric reagent P*. The polymer is acting in the normal sense as a reagent and is used to cause the joining of the two amino acid units. After product formation, the reagent is separated from the soluble product of the reaction.

The type A polymeric system in which the polymer is used as a support is best exemplified by the chloromethylated divinylbenzene crosslinked polystyrene first described by R.B. Merrifield (I) (1). The use of this resin in the synthesis of small, medium and large-sized polypeptides has increased rapidly since its inception.

Basically, the linkage to the support is via a benzyl ester bond which is stable to most synthetic operations employed. After linking all the amino acids in the required order this bond is cleaved and the product isolated.

Modification of the basic resin have been made to improve its use and to provide easy access to further intermediates in the peptide synthetic process (eg. C-terminal hydrazides (II, III) (2,3).

Most of the recent developments in the use of this resin and the problems associated with it have been reviewed (4-6) or will be discussed in other sections of the symposium. They will not be dealt with further here.

If we might turn our attention to the supports in which the polymeric system is acting as a chemical reagent we will see a basic similarity in their purpose and design to that of the Merrifield resin. The type B systems are all based upon active esters in which the benzyl alcohol group of the 'inactive' benzyl ester in the Merrifield resin is replaced by a much more easily displaced group, such as nitrophenol. The polymeric systems thus become good acylating agents. Many examples of this type of resin have appeared over the last few years and their structures are given in diagram

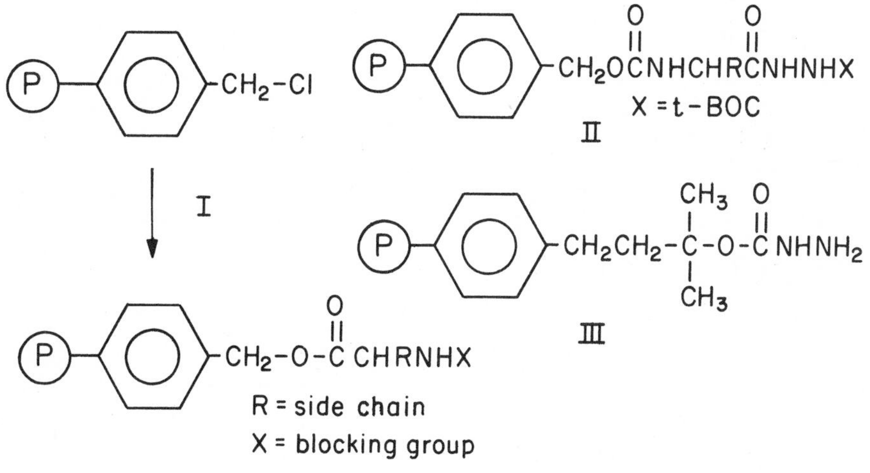

(P)–C₆H₄–CH₂–Cl

I

(P)–C₆H₄–CH₂–O–C(=O)CHRNHX

R = side chain

X = blocking group

(P)–C₆H₄–CH₂OC(=O)NHCHRC(=O)NHNHX

X = t-BOC

II

(P)–C₆H₄–CH₂CH₂–C(CH₃)₂–O–C(=O)NHNH₂

III

II (IV-IV) (7-15). All have the active ester base. Several of them use the Merrifield resin as a support upon which the alcohol portion of the active ester is suspended (eg. IV). Others have been custom-made via various synthetic routes, which either involve synthesis of a required monomer before being polymerized (IV, VI) (8,9,14,15) or by operating on a preformed polymeric system (IV, V) (7,10,11,12,13). In all cases, the resultant active ester, formed by reaction with the acid functionality of a suitably blocked amino acid, is used to acylate the free amino group of a growing peptide chain and, just as in the Merrifield procedure, the peptide chain is formed by the backing-off procedure, usually one unit at a time. This single step procedure is necessitated since the amino function of the acid must be protected by a urethane-type blocking group. Unusually large amounts of racemization of the carboxyl containing amino acid result during the coupling of larger fragments unless such a urethane group is used (16).

Carbodiimides (i.e. R-N=C=N-R) and EEDQ (N-ethoxycarbonyl-2-ethoxy-1,2-dihydroquinoline) are reagents which have been used in the synthesis of peptides (17). These two reagents have been incorporated into type C resins in which the support is acting as a true chemical reagent. If the reagents were regenerable they could be more simply classified as catalysts of amide bond

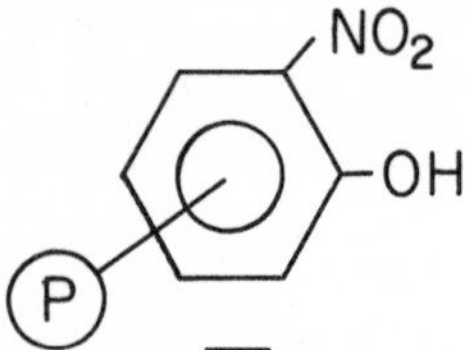

NO_2 OH (P) IV — N-OH (P) V — R-SO_2 OH (P) VI

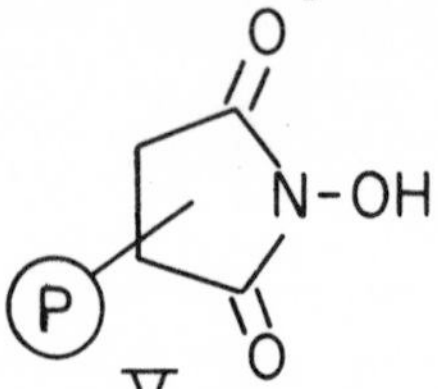

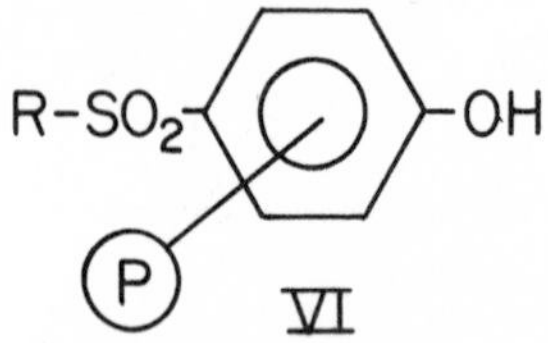

$-(CH_2N=C=N(CH_2)_5CH_2)-$

VII

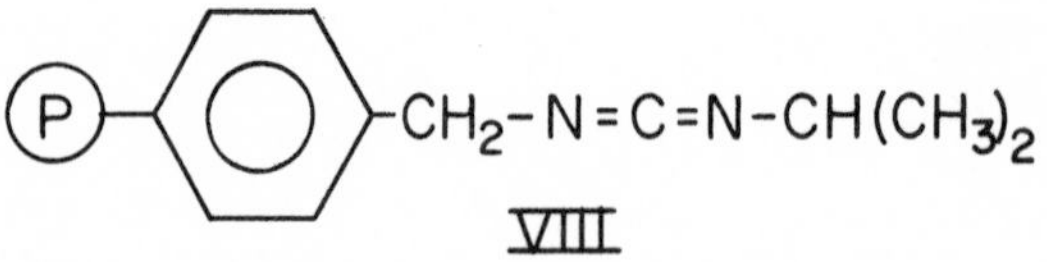

(P) $-CH_2-N=C=N-CH(CH_3)_2$

VIII

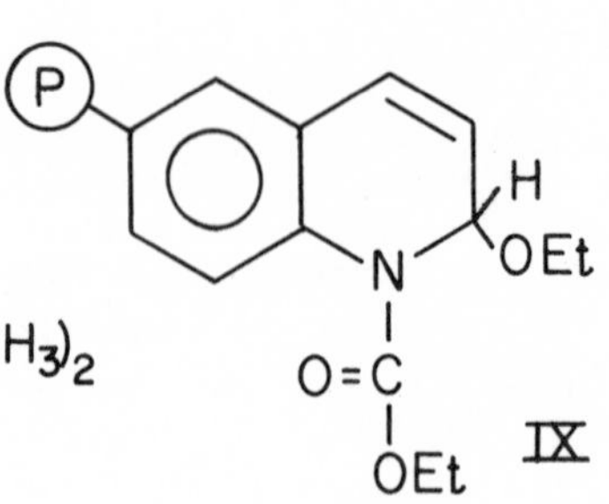

(P) H OEt N O=C OEt

IX

formation. The carbodiimide reagent has been either incorporated into the polymeric backbone (VII) (18) or attached to a carrier (VIII) (19). The EEDQ reagent was prepared from a preformed quinoline polymer (IX) (20,21).

Essentially, this type of reagent is added directly to the acid and amine in solution. After product formation the spent reagent is removed by filtration and the product isolated in the usual manner.

As can be seen from the above the chemical reagent type polymers, either active ester based or reagent based, do offer an alternative to the Merrifield type of resin. Their advantage lies in the fact that isolation and purification of the intermediate products in any synthetic sequence is possible. It is also possible to improve the yields in the normal chemical synthesis of peptides with these reagents since the reaction may be pushed to completion by using a large excess of the readily removed reagent. Their disadvantage lies in the fact that the final product of the reaction must be soluble in a solvent in which the resin remains insoluble. This product solubility requirement effectively limits the size of peptide which can be synthesized with them. The actual length of the peptide would depend, of course, on the type of side chains and on the blocking groups present in the peptide.

As well as these synthetic advantages and disadvantages the usual problems which plague dual phase reactions apply (5). Since all these processes are diffusion controlled, reaction site availability must rank highest amongst the most severe problems encountered. Inacessibility of sites can be caused by simple steric problems. The approach of one or both of the reactants can be hindered by the proximity of the large hydrophobic resin surface. Irreproducible swelling and contraction of the resin matrix can also influence strongly the site availability (22). Separation of pendant groups from the resin surface would go a long way towards solving the former problem. The latter problem could be more readily solved by use of preexpanded matrices in which swelling and contraction are prevented (i.e. macroreticular resins (23,24)). An alternate solution to the problem resides in the use of pellicular resins in which only the surface layer of beaded material is used. In this instance, the core materials could be organic (e.g. highly crosslinked polystyrene) or inorganic (e.g. glass (25,26)). The reduction in the distance through which the reactants have to migrate would improve quantitation of reactions and increase the speed with which they may be carried out. This approach has been shown to be of advantage in Merrifield type syntheses (27). This type of resin suffers, at present, from the fact that the number of reactive sites per gram of resin is very low when compared with the resins in which bead interiors are used.

Finally, future developments in this area should provide certain relief of the problems associated with syntheses using insoluble reagents and thus promote the industrial utilization of these reagents for the synthesis of biologically-active and thus pharmacologically-important peptides. Needless to say, experience gained in this area will also allow extension of the results into other areas of synthetic chemistry - a field which just seems to be opening up.

REFERENCES

1. R.B. Merrifield J. Amer. Chem. Soc. *85* 2149 (1963)
2. S.-S. Wang and R.B. Merrifield J. Amer. Chem. Soc. *91* 6488 (1969)
3. A.M. Felix and R.B. Merrifield J. Amer. Chem. Soc. *92* 1385 (1970)
4. A. Marglin and R.B. Merrifield Ann. Rev. Biochem. *39* 841 (1970)
5. G.R. Marshall and R.B. Merrifield in "Biochemical Aspects of Reactions on Solid Supports" G.R. Stark ed. Academic Press New York 1971
6. E. Wunsch Angew Chem. internat. Edit. *10* 786 (1971)
7. A. Patchornik, M. Fridkin and E. Kalchalski in "Peptides" H.D. Beyerman, A. Van de Linde and W. Maassen van den Brink eds. North-Holland, Amsterdam 1967
8. M. Fridkin, A. Patchornik and E. Katchalski J. Amer. Chem. Soc. *88* 3164 (1966); *ibid*, *90* 2953 (1968)
9. R.E. Williams J. Poly. Sci. Part A-1 *10* 2123 (1972)
10. M. Fridkin, A. Patchornik and E. Katchalski in "Peptides 1969" E. Scoffone ed. North-Holland, Amsterdam 1971
11. G.T. Panse and D.A. Laufer Tet. Lett. *1970* 4181

12. D.A. Laufer, T.M. Chapman, D.I. Marlborough, V.M. Vaidya and E.R. Blout J. Amer. Chem. Soc. *90* 2696 (1968)

13. M. Fridkin, A. Patchornik and E. Katchalski Biochem. *11* 466 (1972)

14. Th. Wieland and Ch. Birr Angew Chem. internat. Edit. *5* 310 (1966)

15. E. Flanigan and G.R. Marshall Tet. Lett. *1970* 2403

16. D.R. Lauren and R.E. Williams Tet. Lett. *1972* 2665

17. Y.S. Klausner and M. Bodansky Synthesis *1972* 453

18. Y. Wolman, S. Kivity and M. Frankel Chem. Comm. *1967* 629

19. N.M. Weinshenker and C.-M. Shen Tet. Lett. *1972* 3281

20. J. Brown and R.E. Williams Can. J. Chem. *49* 3765 (1971)

21. J. Brown, D.R. Lauren and R.E. Williams in "Chemistry and Biology of Peptides" J. Meinenhofer ed. Ann Arbor Science, Ann Arbor, 1972

22. W.S. Hancock, D.J. Prescott, P.R. Vagelos and G.R. Marshall J. Org. Chem. *38* 774 (1973)

23. S. Sano, R. Tokunaga and K.A. Kun Biochim. Biophys. Acta *244* 201 (1971)

24. M.A. Tilak and C.S. Hollinden Org. Prep. Proced. Int. *3* 183 (1971)

25. E. Bayer, G. Jung, I. Halász and I. Sebastian Tet. Lett. *1970* 4503

26. W. Parr and K. Grohmann Angew Chem. internat. Edit. *11* 314 (1972)

27. E. Bayer, H. Eckstein, K. Hägele, W.A. Konig, W. Brüning, H. Hagenmaier and W. Parr J. Amer. Chem. Soc. *92* 1735 (1970)

AFFINITY CHROMATOGRAPHY

I. Parikh and P. Cuatrecasas

Department of Pharmacology and Experimental Therapeutics
The Johns Hopkins University School of Medicine

In the last few years there has been much interest in the development and application of selective adsorbents based on biochemical specificity (1, 2). The use of such especially prepared adsorbents for the purification of enzymes (and other biological macromolecules) has been termed affinity chromatography (3). The isolation and purification of macromolecules by conventional means depend on differences in the physicochemical properties, like differential solubility, overall charge, and molecular size and shape of the various components in the mixture. These differences are frequently not sufficiently discriminative to permit facile separation of the macromolecules. Purification is, therefore, frequently laborious and often incomplete. Partly for these reasons certain important proteins have not yet been purified by the existing conventional procedures.

Basically, the methodology of affinity chromatography makes use of the specific and selective affinity of the protein to be purified for a second substance. In the case of an enzyme, a reversible competitive inhibitor (or substrate) is generally used, whereas for hormone receptors the corresponding hormone is used. The specific ligand (e.g., inhibitor) is insolubilized by attaching it by stable covalent bonds to a suitable polymer, usually beaded agarose. In practice, an alkylaminoagarose is first prepared by the reaction of cynogen bromide (CNBr)-activated agarose with a suitable diamine (3), such as diaminodipropylamine. The amino group of the alkylaminoagarose can be converted by relatively mild methods to various other functional groups, or it can be coupled directly to ligands containing a carboxylic group by known methods (2, 3).

The ligand must be coupled to the matrix in such a way that does not chemically or sterically interfere with its ability to be recognized by the enzyme. The early experience with α-chymotrypsin (4) clearly indicated the efficacy of having a sufficiently long spacer arm interposed between the ligand and the agarose matrix. The recently introduced macromolecular spacer arms (5, 6) have various advantages. Besides providing long spacers (up to 150 Å) between the ligand and the matrix, these spacers create multipoint attachment on the agarose matrix. This results in an increase in the overall chemical stability of the bonds. Macromolecules like poly-L-lysine, the branched chain copolymer, poly(L-lysyl-DL-alanine), native albumin and denatured albumin have been used successfully in the purification of estrogen receptor from calf uterus (5) and cholera toxin from the culture filtrates of *Vibrio Cholerae* (7).

Once a selective adsorbent is prepared, it is packed in a column. After washing and equilibrating the adsorbent with an appropriate buffer, the crude enzyme preparation is passed through the affinity column. If the system is properly designed, only the desired protein will be removed from the crude mixture. After washing away the contaminating proteins the specifically adsorbed protein is eluted from the column. Elution is generally achieved by an appropriate change in the buffer conditions (such as, pH, ionic strength, etc.) so as to perturb the binding of the adsorbed protein with the insolubilized ligand. Since purification is based on the unique functional features of the biological protein, the specific conditions used in a given affinity chromatographic purification must be highly individualized to the specific requirements of the system under study.

Some recent applications of affinity chromatography in this laboratory include the isolation and purification of insulin receptors (8), neuraminidases (9), estrogen receptors (5), and cholera toxin (7).

REFERENCES

1. P. Cuatrecasas, *in Advances in Enzymology 36*, 29-89 (1972)
2. P. Cuatrecasas and C. B. Anfinsen, *in Methods in Enzymology XXII*, 345-378 (1971)
3. P. Cuatrecasas, *J. Biol. Chem. 245*, 3059-3065 (1970)
4. P. Cuatrecasas, M. Wilchek and C. B. Anfinsen, *Proc. Nat. Acad. Sci. U.S.A. 61*, 636-643 (1968)
5. V. Sica, I. Parikh, E. Nola, G. A. Puca, and P. Cuatrecasas, *J. Biol. Chem.* (1973) in press
6. I. Parikh, S. March and P. Cuatrecasas, *in Methods in Enzymology, XXIIB* (1974) in press

7. P. Cuatrecasas, I. Parikh, and M. D. Hollenberg, Biochemistry 12 (1973) in press
8. P. Cuatrecasas, Proc. Nat. Acad. Sci. U.S.A. 69, 1277-1281 (1972)
9. P. Cuatrecasas and G. Illiano, Biochem. Biophys. Res. Comm. 44, 178-184 (1971)

DISCUSSION SESSION

Discussion Leader - R. Williams

Rivin: Dr. Parikh, if I understood it correctly you said that your adsorbed material had 1/1000 the activity in free solution. What is the cause of the loss of activity?

Parikh: I cannot give you a definite answer but I have a feeling that it is just steric factors playing a role.

Moore: What about the possibility of using other materials for affinithy chromatography such as silica gel,silica or activated charcoal?

Parikh: We have tried glass and other people have tried polystyrene. With glass we find non-specific absorption of proteins we are not interested in. Most polymers one could suggest would have some properties like non-specific adsorption unless they are coated with agarose or dextran. If one could get a support having none of these properties the chances are it would find broad use in systems with high affinities because the surface interactions might be just enough to catch the macromolecule from solution and you would not need to use bead interiors. You would also not need a large amount of the ligand which one would need in the case of low affinity systems.

Williams: Do these same comments refer to the polyacrylamides used as supports, that is, in so far as non-specific absorption is concerned?

Parikh: Yes. I think with polyacrylamides specifically the Biogels, there is a susceptibility to organic solvents which one quite often needs to attach small organic molecules. It seems to be unstable in organic solvents and after coupling in aqueous solvents, which is possible in the case of some molecules and proteins, the porosity of the bead is tremendously decreased. The flow rate in a column is also badly influenced in the case of these polyacrylamide gels.

Williams: What seems to present the most problem when one uses this method?

Parikh: The removal of the specifically adsorbed material in a fully biologically active form.